Undergraduate Lecture Notes in Physics

Series Editors

Neil Ashby, University of Colorado, Boulder, CO, USA

William Brantley, Department of Physics, Furman University, Greenville, SC, USA

Matthew Deady, Physics Program, Bard College, Annandale-on-Hudson, NY, USA

Michael Fowler, Department of Physics, University of Virginia, Charlottesville, VA, USA

Morten Hjorth-Jensen, Department of Physics, University of Oslo, Oslo, Norway

Michael Inglis, Department of Physical Sciences, SUNY Suffolk County Community College, Selden, NY, USA

Barry Luokkala , Department of Physics, Carnegie Mellon University, Pittsburgh, PA, USA

Undergraduate Lecture Notes in Physics (ULNP) publishes authoritative texts covering topics throughout pure and applied physics. Each title in the series is suitable as a basis for undergraduate instruction, typically containing practice problems, worked examples, chapter summaries, and suggestions for further reading.

ULNP titles must provide at least one of the following:

- An exceptionally clear and concise treatment of a standard undergraduate subject.
- A solid undergraduate-level introduction to a graduate, advanced, or non-standard subject.
- A novel perspective or an unusual approach to teaching a subject.

ULNP especially encourages new, original, and idiosyncratic approaches to physics teaching at the undergraduate level.

The purpose of ULNP is to provide intriguing, absorbing books that will continue to be the reader's preferred reference throughout their academic career.

Ravinder R. Puri

Modern Thermodynamics and Statistical Mechanics

A Comprehensive Foundation

 Springer

Ravinder R. Puri
Indian Institute of Technology
Gandhinagar, Gujarat, India

ISSN 2192-4791   ISSN 2192-4805 (electronic)
Undergraduate Lecture Notes in Physics
ISBN 978-3-031-54312-8   ISBN 978-3-031-54310-4 (eBook)
https://doi.org/10.1007/978-3-031-54310-4

Dedicated to
my students
who make me learn constantly

Preface

There are a number of books on thermodynamics and statistical mechanics at different levels. Then why this one? This book intends to put together in one place the fundamentals of thermodynamics and statistical mechanics for the students of physics by presenting the subjects in a way different from the existing texts.

In the books on statistical mechanics, thermodynamics is generally given at most a cursory glance. It is covered comprehensively in this book. Indeed, since statistical mechanics is the microscopic theory of thermodynamics, placing it alongside thermodynamics helps in providing a clearer and complete view of the subject.

Furthermore, fundamentals of thermodynamics are generally presented either in the traditional way in terms of the laws of thermodynamics or in Callen's postulatory form. This book presents both the approaches. For, the postulatory approach is built on the concept of entropy which emerges from the traditional approach. The traditional approach in that sense prepares one to understand and appreciate the postulatory approach better. The postulatory approach in turn provides a natural connection between thermodynamics and statistical mechanics.

After introducing the fundamentals of thermodynamics in first two chapters we turn to developing the concepts and methods of equilibrium statistical mechanics. We begin with the kinetic theory in Chap. 3. Though it is discussed routinely in the texts on statistical mechanics, this book provides a glimpse in to its historical developments too.

The kinetic theory leads to coupled time-dependent integro-differential equations finding whose asymptotic solution to describe equilibrium thermodynamic phenomena is generally a formidable task. Boltzmann's insight provided a way of linking thermodynamics with microscopic description without the need of solving those equations. It laid the foundation of modern statistical mechanics which links the two descriptions by means of the concept of statistical entropy. The book traces in Chap. 4 the development of Boltzmann's thought process as it emerges from his paper wherein he introduced the concept of what he called permutability measure as the mechanical analog of thermodynamic entropy.

Boltzmann's formalism applies to systems of non-interacting molecules. The general statistical mechanical formalism is developed in Chap. 5 by introducing the concept of statistical entropy. It is based on Shannon's information theoretic entropy. We construct the statistical entropy in terms of the phase space distribution function for classical systems and in terms of the probability of occupation of energy levels for quantum system. Since it depends on the probability of occupation of energy levels and not on the transition probabilities between them, the use of density matrix formalism is unnecessary. Based on the principle of maximum entropy, the classical and statistical entropies for systems in thermodynamic equilibrium are derived in Chap. 6 and their relation with thermodynamics established.

The phase space equilibrium statistical entropy is used in Chap. 7 to study system of non-interacting molecules in free space, including their internal degrees of freedom, as well as when the gas is in the gravitational field near the surface of earth. The equations of state for the freely evolving non-interacting gas obtained in this way are same as the empirical equations of state for the ideal classical gas.

The theory of non-interacting quantum gases in free space, called ideal quantum gases, is developed in Chap. 8. The concept of distinguishable and indistinguishable particles is introduced. The quantum theory of distinguishable particles is shown to be equivalent with the classical gas. The indistinguishable particles are categorized as Bosons and Fermions.

The theory of ideal Fermi gas at low temperature is outlined in Chap. 9. The thermodynamics of ideal Bose gas is studied in Chap. 10 where the phenomenon of Bose-Einstein condensation is studied in detail.

Chapter 11 is devoted to the thermodynamic theory of the phase transitions and critical phenomena.

Having studied ideal gases, the interaction between the molecules in the gas is included in Chap. 12. The equation for the simplest model of an interacting classical gas namely the Van der Waals gas is derived and analyzed as a microscopic model of phase transition and critical phenomenon.

Except for reference to time-dependence in classical formalism in the kinetic theory chapter, the statistical mechanical formalism till Chap. 12 is to understand the equilibrium properties, constructed based on the principle of maximum entropy. It is evidently desirable to ascertain that the equilibrium state so obtained is approached asymptotically as the solution of appropriate equation describing the evolution of the state. That issue is addressed for quantum systems in the Chaps. 13 and 14. Since the quantum mechanical state of a system interacting with others is described by the density matrix, Chap. 13 develops the density matrix formalism. The evolution equation for the density matrix of a system interacting weakly with a reservoir, called the master equation, is presented in Chap. 14 and shown to have same asymptotic solution as is predicted by equilibrium statistical mechanics.

Some mathematical topics necessary for the purpose of the book are summarized in the appendices.

I would like to take this opportunity to thank Mr. Akshat Khanna for helping me with drawing figures. I am thankful to the faculty at the Indian Institute of Technology Gandhinagar, in particular Prof. Rishi Narain Singh, for helpful discussions and encouragement. Special thanks to Prof. Rajat Moona, director Indian Institute of Technology Gandhinagar for providing invaluable support.

Gandhinagar, India   Ravinder R. Puri

Contents

Chapter 1
Fundamentals of Thermodynamics-I

A theory is the more impressive the greater the simplicity of its premises is, the more different kinds of things it relates and the more extended is its area of applicability. Therefore the deep impression which the classical thermodynamics made upon me. It is the only physical theory of universal content concerning which I am convinced that, within the framework of the applicability of its basic concepts, it will never be overthrown.
—Albert Einstein

If someone points out to you that your pet theory of the universe is in disagreement with Maxwells equations then so much the worse for Maxwell's equations. If it is found to be contradicted by observation well, these experimentalists do bungle things sometimes. But if your theory is found to be against the second law of thermodynamics I can give you no hope; there is nothing for it but to collapse in deepest humiliation.
—Arthur Eddington

Thermodynamics is a funny subject. The first time you go through it, you don't understand it at all. The second time you go through it, you think you understand it, except for one or two points. The third time you go through it, you know you don't understand it, but by that time you are so used to the subject, it doesn't bother you anymore.
—Arnold Sommerfeld

The quotations above underline the character of thermodynamics: robust and tricky. As expressed in first two quotations, why this unwavering faith in the robustness of thermodynamics? And, as alluded to in the third quotation, why is it tricky?

The edifice of thermodynamics draws its robustness from the strength of the pillars of the postulates it stands upon: conservation of energy and the impossibility of heat flowing on its own from a cold to a hot body even though it is allowed by energy conservation. The edifice can crumble only if either of these pillars weakens. Can that happen? Our firm belief is: it can not.

© The Author(s), under exclusive license to Springer Nature Switzerland AG 2024
R. R. Puri, *Modern Thermodynamics and Statistical Mechanics*, Undergraduate Lecture Notes in Physics, https://doi.org/10.1007/978-3-031-54310-4_1

The origin of the challenges faced in applying thermodynamics, according to Feynman ([1], pp. 44–49), lies in the non-unique choice of independent variables from among several possible to describe a thermodynamic transformation. Indeed it is tricky to negotiate the maze of equations to describe a thermodynamic phenomenon.

Thermodynamics was developed to understand the thermal properties of macroscopic systems at a time when the atomic theory was not in vogue. It describes systems in contact with its surroundings with which it may exchange heat and work. It identifies the state of a macroscopic system in terms of amazingly small number of variables when in equilibrium with its environment: pressure, density, and temperature. It builds relationships between the change in variables between different states of equilibria under thermodynamic transformations. Since the principles on which they are based, as asserted in the beginning of this chapter, are kind of "immortal", so are those relations. Thermodynamics as such describes systems in thermodynamic equilibrium and not how it is achieved when the conditions are altered. In that sense thermodynamics is said to be thermostatics. Study of time-dependent behavior of macrosystems is the subject of non-equilibrium thermodynamics.

Even after the atomic theory started growing its roots raising the possibility of describing macroscopic systems in terms of the motion of their constituent molecules governed by Newtonian laws, the problem of providing mechanical theory of thermodynamics was challenging. For, since the number of molecules in a macroscopic body may be as large as $\sim 10^{20}$, there is no way to construct and solve that many Newton's equations. Even if, by some magic, we construct and solve those equations, what do we do with the solution? How do we identify temperature and heat to relate the solution with thermodynamics? The mechanical description did not seem to have place for heat and hotness. Clearly, different concepts were required to relate mechanical and thermodynamic descriptions leading to the emergence of the kinetic theory followed by the theory of statistical mechanics.

We begin by outlining the fundamentals of thermodynamics, presented in two ways: one is the traditional approach and the other the axiomatic one. The traditional approach presented in this chapter is meant to describe the basic concepts of thermodynamics along with their origin. The axiomatic approach, presented in the next chapter, builds upon those concepts. It provides natural connection between thermodynamics and statistical mechanics.

Apart from using number of molecules as a fixed parameter in this chapter and a variable in Chap. 2, the theory of thermodynamics is described without any reference to the molecular motion. Starting with the kinetic theory in Chap. 3, the mechanical theory is developed in the rest of the chapters.

1.1 Brief History

We describe briefly the main historical developments which shaped thermodynamics in its present form. For detailed history see, for example, [2, 3].

What is heat was a question which engaged people since ancient times. It was regarded by philosophers as an element. Later it came to be regarded as a fluid, a belief which continued for a long time. That fluid was named *Caloric* by Guyton de Morveau in 1787. According to this view, two bodies, one hotter than the other, attain same temperature on contact with each other due to the flow of Caloric fluid from the hotter to the colder body. However, various experiments showed that heat is not a fluid but is a form of energy. Other forms of energy, like mechanical and electrical, can be converted to heat and vice versa. For example, generation of heat during the motion of an object on a rough surface is a common experience. Churning of a fluid also results in an increase in its temperature. These are the examples of conversion of mechanical energy to heat. The traditional unit of heat is Calorie, defined as the amount of heat required to raise the temperature of one gram of water by $1\,°C$ at the pressure of one atmosphere. In a number of experiments, Joule in 1843 determined the amount of mechanical work needed to raise the temperature of water by $1°$ C at the atmospheric pressure and thus found the ratio $J = W/Q$ between the amount of work W required to produce the amount of heat Q. The standardized modern value of J, called the mechanical equivalent of heat, is $J = 4.186\mathrm{J/calorie}$.

Besides the concept of heat, interest was also in quantitative understanding of how the density of a gas changes under variation of pressure and temperature. The experiments in that direction showed that every gas when dilute, called an *ideal gas*, obeys following laws: (1) Boyle's law (1662). According to it, the pressure P of a fixed mass of an ideal gas at constant temperature is inversely proportional to its volume V: $PV = $ constant. (2) Charles's law (1787). According to it, when the pressure of a fixed mass of an ideal gas is kept constant then its volume is directly proportional to its temperature T on the Kelvin scale: $V/T = $ constant where T is the Kelvin temperature related with the temperature T_{celsius} in Celsius by the relation $T = T_{\mathrm{celsius}} + 273.15$. (3) Gay-Lussac's law (1802). According to it, when the volume of a fixed mass of an ideal gas is kept constant then its pressure is directly proportional to its temperature, i.e. $P/T = $ constant. (4) Combined gas law (Émile Clapeyron, 1834) obtained by combining three afore-stated laws: $PV = kT$, where k is a constant whose value depends on the mass of the gas. Invoking the notion of the molecular composition of gases which came much later, the combined gas law for a sufficiently dilute gas, called the *ideal gas law*, for a gas consisting of one kind of molecules is now written as

$$PV = nRT \quad \text{or} \quad PV = Nk_\mathrm{B}T \quad \Longrightarrow \quad R = Nk_\mathrm{B}/n \equiv N_\mathrm{A}k_\mathrm{B}, \qquad (1.1)$$

where in the first form n is the number of moles of the gas, R is the gas constant and in the second form N is the number of molecules in the gas, k_B is the Boltzmann constant. The number N_A, called Avogadro's number, is the number of molecules per mole. The values of the constants in (1.1) are

$$R = 8.315\,\mathrm{joule/mole\ K} = 1.986\,\mathrm{cal/mole\ K}$$
$$N_\mathrm{A} = 6.022 \times 10^{23}\,\mathrm{molecules/mol},$$
$$k_\mathrm{B} = 1.38 \times 10^{-16}\,\mathrm{erg/K} = 8.617 \times 10^{-5}\,\mathrm{eV/K}. \qquad (1.2)$$

The symbol R for the gas constant was used first by Clapeyron. The thermodynamic state of the gas of given mass is thus specified in terms of its pressure, volume, and temperature. Equation (1.1), depicting relation between pressure, volume, and temperature, is called the equation of state of an ideal gas. It, however, does not determine relation of those quantities with heat and work put into the gas. In that sense the equation of state does not provide complete description of thermodynamic properties of a gas.

The questions raised above regarding relationship between the thermodynamic state variables, work and heat, were of crucial importance in designing efficient steam engines. The interest in steam engines has its origin in the discovery in 1698 by Thomas Savery of a steam-powered pump. Since then the design of steam engine underwent several changes culminating in a design by James Watt in 1763 which was later used for making railroad locomotives. Finding ways of improving efficiency of engines was a challenging problem. There was, however, very little systematic work in that direction till the work of Sadi Carnot in 1824. His work, published as a book entitled (translation of the original French title) *Reflections on the Motive Power of Fire* [4], laid the foundation for building modern thermodynamics. In Sect. 1.2, we outline Carnot's visualization of the working of an engine, its efficiency, and the way it leads to the concept of entropy using the terminology summarized below:

1. Heat is considered a form of energy. It can be converted to work, and in turn can be produced by consuming work.
2. Work and heat together obey law of conservation of energy.
3. The state of a thermodynamic system having a fixed number of moles of different kinds of molecules is characterized in terms of the set of *thermodynamic parameters* or thermodynamic *state variables*: pressure P, volume V, and temperature T. The formalism can be extended to systems in which number of moles of one or the other kind is not fixed.
4. The energy of a system has a unique value for a given thermodynamic state of the system.
5. The state of a thermodynamic system which does not change with time is called the state of *thermodynamic equilibrium*. It is assumed that, when disturbed, the system will attain an equilibrium state in time called its relaxation time. The systems of interest here are the ones whose relaxation times are much smaller than the time scale of observation.
6. The state variables in the state of equilibrium are not independent. The relationship between them is known as the *equation of state*. The laws of thermodynamics do not specify the equation of state; it has to be determined empirically or by other theoretical means. Recall that the equation of state of an ideal gas, determined empirically, is given by (1.1). A widely employed equation of state of a non-ideal gas is the van der Waals equation of state (see Sect. 1.6).
7. The state of a system changes when the external conditions are changed. When its state changes, the system is said to undergo *thermodynamic transformation*. The laws of thermodynamics, four in number, specify what state of equilibrium does the system attain when the external conditions are changed. They do not

specify the rate at which the system transits from one to the other equilibrium state. In this sense, what we call thermodynamics is truly speaking thermostatics.

8. If the state variables change so slowly that the system can be assumed to be in thermodynamic equilibrium at any instant then the transformation of the system is said to be *quasi-static*.

9. If the transformation is carried such that it is possible to retrace the steps from its final to the initial state by reversing the external conditions then it is called *reversible*. Else the transformation is called *irreversible*. A reversible transformation is quasi-static but not all quasi-static transformations are reversible.

10. If the transformation is such that the system does not exchange heat with its surroundings then the transformation is said to be *adiabatic*: from the Greek *a* (not)+dia (through)+bainein (to go) (see [1], pp. 39–45). It may be mentioned that the meaning of adiabatic transformation in mechanics is different.

11. A transformation in which temperature of the system remains unchanged is called *isothermal*.

12. A transformation in which pressure of the system remains unchanged is called *isobaric*.

13. A *thermal or heat reservoir* is a system so large that addition or removal of a finite amount of heat from it does not change its temperature. The thermodynamic processes in a heat reservoir are reversible.

14. If a transformation is such that it restores the system to its initial state then it is called *cyclic*.

1.2 Carnot Engine

The Fig. 1.1 depicts schematic operation of a Carnot engine. In it, the gas in a specified state (that is having specified values of P, V, T) receives an amount Q_h of heat from a thermal reservoir at temperature T_h, converts a part of it to work W, rejects the remaining into a heat reservoir at temperature T_c ($T_c < T_h$) and returns to its initial

Fig. 1.1 Carnot engine

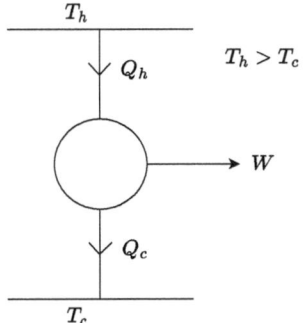

state. The working fluid thus undergoes a cyclic process. It is assumed that the process is reversible. An example of realization of Carnot's engine is discussed in Sect. 1.5.

To find efficiency of the engine note that, according to the principle of conservation of energy, the work produced in one cycle of the process is the difference in the amount of heat absorbed and the heat given out:

$$W = Q_h - Q_c. \tag{1.3}$$

Hence efficiency η, defined as the ratio of work W produced by the engine to the amount of heat Q_h received by it, is given by

$$\eta = \frac{W}{Q_h} = 1 - \frac{Q_c}{Q_h}. \tag{1.4}$$

The expression for efficiency given above is based only on the definition of efficiency and the principle of conservation of energy. It holds not only when the processes in the engine depicted in the Fig. 1.1 are reversible but also when they are irreversible. Whereas the value of η depends on the details of the processes in case they are irreversible, we will see that the expression (1.4) for efficiency of a Carnot engine in case the processes are reversible reduces to the form (1.16) which depends only on the temperatures T_h and T_c and not on the details of the processes involved.

Since the Carnot engine depicted in Fig. 1.1 is assumed reversible, by definition of a reversible process, the reversal of the processes in it would lead to the reversed Carnot engine in which the directions of the flow of heat and work are reversed (see Fig. 1.2) which means, operating in the reverse direction, work W is put into the engine while it draws amount Q_c of heat from the reservoir at lower temperature T_c and deposits the amount Q_h in the reservoir at temperature T_h higher than T_c. The engine in reverse operation thus acts as a refrigerator. The reverse Carnot engine is therefore also called Carnot refrigerator.

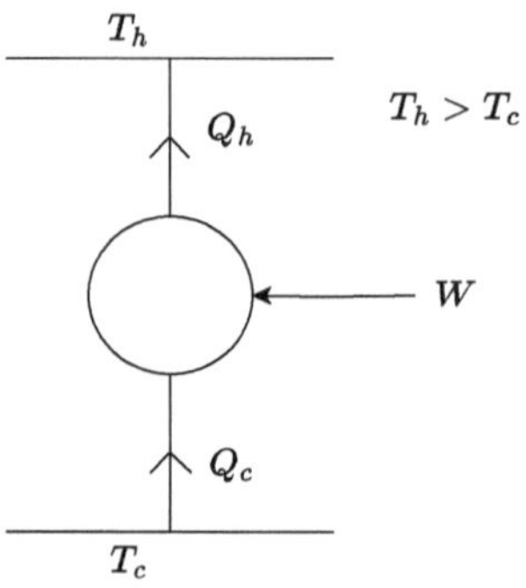

Fig. 1.2 Reverse Carnot engine or Carnot refrigerator

Next we describe how the concepts of absolute temperature and entropy arise from the analysis of combinations of Carnot engines operating in forward and reverse directions.

1.2.1 Absolute Temperature

We prove the following proposition and show how the concepts of absolute temperature and entropy emerge therefrom:

If it is assumed that heat cannot be transported from a cold to a hot reservoir without doing any work then

1. *No engine is more efficient than reversible Carnot engine.*
2. *All reversible Carnot engines operating between same reservoirs have same efficiency.*

This implies that the efficiency of a reversible Carnot engine is (a) independent of what the operating fluid is and (b) independent of the details of the thermodynamic processes inside the engine.

It may be mentioned that Carnot assumed heat to be a massless fluid, a concept which is since abandoned. The analysis given below is essentially due to Clausius (in 1850s) based on Carnot's work.

To prove the propositions above, consider two engines operating as shown in Fig. 1.3. The engine on left (numbered 1) need not be reversible whereas the one on the right (numbered 2) is reversible Carnot engine operating in reverse direction. The engine 1 absorbs the amount Q_h of heat from the reservoir at temperature Θ_h, performs work W, and deposits the amount Q_c into the reservoir at temperature Θ_c ($\Theta_h > \Theta_c$). The work produced by first engine is used as input to the engine 2 which uses it to absorb heat Q_c' from the low-temperature reservoir and deposit heat Q_h' in the high-temperature reservoir. Assume that the efficiency of engine 1 is η_1 and that of engine 2 is η_2. Clearly $\eta_1 = W/Q_h$. Engine 2 is operating in the reverse direction.

Fig. 1.3 $\Theta_h > \Theta_c$. The engine on left (numbered 1) need not be reversible whereas the one on the right (numbered 2) is reversible Carnot engine operating in reverse direction

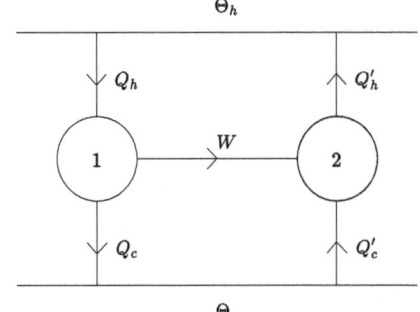

Since it is reversible its efficiency is obtained simply by reversing the directions of flow of work and heat. It then follows that $\eta_2 = W/Q'_h$.

Now, contrary to the proposition to be proved, assume that the efficiency of possibly irreversible engine is more than that of the reversible one, i.e. $\eta_1 > \eta_2$ so that

$$\frac{W}{Q_h} > \frac{W}{Q'_h} \quad \Longrightarrow \quad Q'_h - Q_h > 0. \tag{1.5}$$

Clearly, net work done in the process is zero and the operating fluids in the two engines are left in their initial corresponding states. Hence, due to conservation of energy, net amount of heat absorbed must also be zero so that

$$Q_h + Q'_c = Q'_h + Q_c. \tag{1.6}$$

On invoking (1.5) this yields

$$Q'_c - Q_c > 0. \tag{1.7}$$

The condition $Q'_h - Q_h > 0$ in (1.5) means that a net amount of heat is deposited in the high-temperature reservoir whereas the condition $Q'_c - Q_c > 0$ in (1.7) means that a net amount of heat is drawn from the low-temperature reservoir. Since no work is done externally and, since the operating fluids are left in their initial states at the end of the cycle, the internal energy of the operating fluid is also unchanged. The conclusion drawn above then means that a net amount of heat has been drawn from the colder reservoir and delivered to the hotter one without performing any work. This is contrary to our experience. Hence, the assumption on which the said conclusion is based, namely, that the efficiency of possibly irreversible engine is greater than that of the reversible one is erroneous. We have thus proved that $\eta_1 \not> \eta_2$.

Next we show that it is possible to have $\eta_1 \leq \eta_2$. Assuming that to be the case, it is readily seen that the conditions (1.5) and (1.7) would then reverse and read $Q'_h - Q_h \leq 0$, $Q'_c - Q_c \leq 0$, respectively. These conditions state that a net amount of heat is drawn from the hotter reservoir and delivered to the colder one without performing any work. This is an acceptable situation. We thus conclude that $\eta_1 \leq \eta_2$ where η_2 is the efficiency of the reversible Carnot engine and η_1 that of the engine which may or may not be reversible. This proves the first of the two propositions listed above.

To prove the second of the said two propositions, assume that the engine 1 is also reversible. Since the inequality $\eta_1 \leq \eta_2$ proved above applies when engine number 1 is any engine, it is applicable also when that engine is reversible. Now, if engine number 1 reversible, it can be operated in the reverse direction. We therefore reverse the operations in Fig. 1.3 to get Fig. 1.4. The role of η_1 and η_2 is now interchanged. The arguments similar to those for arriving at the result $\eta_1 \leq \eta_2$ in case of the operations in Fig. 1.3 would lead to the conclusion $\eta_2 \leq \eta_1$. Thus when both the engines are reversible, one having efficiency η_1 and the other η_2 then we find that both the inequalities, viz., $\eta_1 \leq \eta_2$ and $\eta_2 \leq \eta_1$ should hold simultaneously, which

Fig. 1.4 $\Theta_h > \Theta_c$. Recall that engine number 2 in Fig. 1.3 is reversible but that numbered 1 need not be. When engine number 1 in Fig. 1.3 is also reversible then the operations in that figure can be reversed to get this one

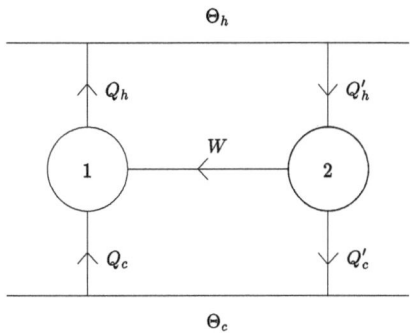

is possible only if $\eta_1 = \eta_2$. We have thus proved the second proposition, namely, all reversible Carnot engines operating between same reservoirs have same efficiency.

What the result derived above means is: Consider a reversible cyclic process operating between reservoirs at temperatures Θ_h and Θ_c ($\Theta_h > \Theta_c$), drawing an amount Q_h of heat from the reservoir at temperature Θ_h, performing work W and depositing the amount Q_c of heat in the reservoir at temperature Θ_c. Consider another reversible cyclic process operating between same reservoirs drawing a different amount Q'_h of heat from the reservoir at temperature Θ_h, performing different amount W' of work and depositing different amount Q'_c of heat into the reservoir at temperature Θ_c. The result derived above, namely, the efficiency of all reversible Carnot engines operating between same reservoirs is same, on invoking (1.4), implies

$$\frac{Q_c}{Q_h} = \frac{Q'_c}{Q'_h}. \tag{1.8}$$

Referring, for example, to the process depicted by the closed cycle $ABCDA$ in Fig. 1.10, we may consider a different closed cycle $A'B'C'D'A'$ in which the values of pressure and volume at the primed points are different from the corresponding unprimed ones but the primed cycle having the values of the temperatures T_h and T_c along the two isotherms same as the corresponding values on the unprimed cycle. The values Q'_h and Q'_c of the heat drawn and the heat deposited and the work output W' in the primed cycle will be different from the corresponding quantities Q_h, Q_c and W in the unprimed cycle but they will be such that (1.8) holds.

In arriving at the conclusions above, no reference has been made to the equation of state of the operating fluid or to the processes employed. Hence the conclusion that the efficiency of all reversible Carnot engines is same depends neither on the working fluid, nor on its initial state, and nor on the details of the process. Hence, it follows that the efficiency of a reversible Carnot engine can be a function only of the remaining parameters in the process, namely, the temperatures Θ_h and Θ_c of the two reservoirs. We therefore let

$$1 - \eta \equiv \frac{Q_c}{Q_h} = f(\Theta_h, \Theta_c), \tag{1.9}$$

Fig. 1.5 The engine 1 is equivalent with the combined operation of the engines 2 and 3 ($\Theta_1 > \Theta_2 > \Theta_3$). All engines are reversible

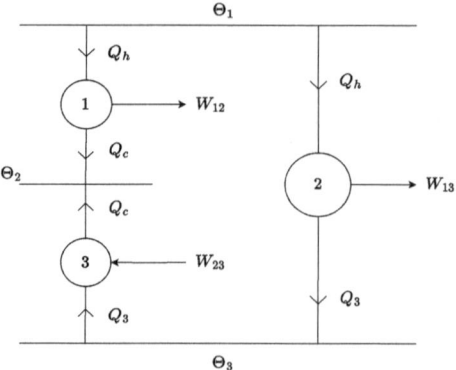

where $f(\Theta_h, \Theta_c)$ at this stage is an arbitrary function of its arguments. Since $\eta \leq 1$, $f(\Theta_h, \Theta_c) \geq 0$. Thus if a Carnot engine operates between reservoirs at temperature Θ_a and Θ_b ($\Theta_a > \Theta_b$) drawing heat Q_a from the reservoirs at Θ_a and depositing heat Q_b into the reservoirs at Θ_b then

$$\frac{Q_b}{Q_a} = f(\Theta_a, \Theta_b), \quad \Theta_a > \Theta_b. \tag{1.10}$$

The form of the function may be determined as follows.

Since the function $f(\Theta_a, \Theta_b)$ is universal we can consider any process convenient for its determination. To that end we consider the combination of Carnot cycles depicted in Fig. 1.5 which is self-explanatory. All the engines in the figure are reversible with $\Theta_1 > \Theta_2 > \Theta_3$. The figure depicts three engines: (1) engine 1 which absorbs heat Q_h from the reservoir at temperature Θ_1, deposits heat Q_c into the reservoir at temperature Θ_2, and generates work W_{12}; (2) engine 2 which absorbs heat Q_h from the reservoir at temperature Θ_1, deposits heat Q_3 into the reservoir at temperature Θ_3, and generates work W_{13}; (3) engine 3 which absorbs heat Q_3 from the reservoir at temperature Θ_3, deposits heat Q_c into the reservoir at temperature Θ_2 by absorbing work W_{23}. Since engine 3 draws as much heat from the reservoir at temperature Θ_3 as the engine 2 deposits in it, the combination of the engines 2 and 3 may be considered as operating effectively between the reservoirs at the temperatures Θ_1 and Θ_2 which draws heat Q_h from the reservoir at Θ_1 and deposit heat Q_c into the reservoir at Θ_2. The combined operation of the engines 2 and 3 is thus same as that of engine 1.

Now, the work generated by the combination of the engines 2 and 3 is evidently $W_{13} - W_{23}$. Using the principle of conservation of energy we have $W_{13} = Q_h - Q_3$, $W_{23} = Q_c - Q_3$ and hence the work W_{12} generated by the engine 1 may be written as

$$W_{12} = Q_h - Q_c = (Q_h - Q_3) + (Q_3 - Q_c) = W_{13} - W_{23}. \tag{1.11}$$

Thus the engine 1 generates as much work as the combination of the engines 2 and 3. Since we have already shown that the combination of the engines 2 and 3 operates effectively between the same reservoirs as does the engine 1 and draws/deposits the same amount of heat from/into each of the said reservoirs as does the engine 1, it follows that the engine 1 is equivalent with the combined operation of the engines 2 and 3. Now,

$$\frac{Q_c}{Q_h} = \frac{Q_3}{Q_h} \Big/ \left(\frac{Q_3}{Q_c} \right). \tag{1.12}$$

Each of the ratio of the quantities of heat in the equation above is between the quantities of heat drawn/deposited in the reservoirs in one or the other Carnot cycle. We can therefore use the relation (1.10) in (1.12) to obtain

$$f(\Theta_1, \Theta_2) = \frac{f(\Theta_1, \Theta_3)}{f(\Theta_2, \Theta_3)}. \tag{1.13}$$

There is no Θ_3 on the left side. Hence Θ_3 must cancel in the right side. This can be achieved if $f(\Theta_a, \Theta_b) = g(\Theta_b)/g(\Theta_a)$. With $\theta \equiv g(\Theta)$, (1.9) reads

$$\frac{Q_c}{Q_h} = \frac{\theta_2}{\theta_1}. \tag{1.14}$$

The quantity θ is called the *absolute temperature*. Since the relation above is universal, one way of fixing the value of θ in terms of measured temperature is to consider a Carnot cycle with such operating fluid and transformations corresponding to which we can evaluate Q_c/Q_h analytically exactly to determine thereby θ in terms of measured temperatures. In Sect. 1.5, we have considered one such process taking ideal gas as the operating fluid. We have shown that $Q_c/Q_h = T_c/T_h$ where T is temperature on the Kelvin scale. Hence, for a reversible Carnot engine operating with any fluid and process,

$$\frac{Q_h}{T_h} = \frac{Q_c}{T_c}. \tag{1.15}$$

We refer to the equation above as *Carnot's formula*. Feynman (see [1], pp. 44–49) calls (1.15) the center of universe of thermodynamics just as Force = mass × acceleration is the center of universe of mechanics. According to him, the relation (1.15) is all to thermodynamics as Force = mass × acceleration is all to mechanics!
On using (1.15), the expression (1.9) for the efficiency of the Carnot engine operating between heat reservoirs at temperatures T_h and T_c reads

$$\eta = 1 - \frac{T_c}{T_h}, \qquad T_h > T_c. \tag{1.16}$$

The efficiency can be unity if $T_c = 0$ or $T_h = \infty$. Since $0 \le \eta \le 1$, $T \ge 0$.

To underline the significance of (1.15), note that if Q ($Q > 0$) is the amount of heat delivered by the system to the reservoir then we may equivalently say that $-Q$ is the amount of heat received by the system. The relation (1.15) for the reversible cyclic process may then be rewritten as

$$Q_h/T_h + (-Q_c/T_c) = 0. \tag{1.17}$$

This states that, if Q is the amount of heat absorbed reversibly by the system at temperature T then in the reversible cyclic process involving two heat reservoirs, the sum of Q/T is zero.

1.2.2 Entropy

Equation (1.17) shows that the quantity Q/T is a constant in the reversible Carnot cycle. It is called *entropy* of the system. Entropy is generally denoted by the letter S:

$$S = \frac{Q}{T}, \qquad Q \text{ absorbed reversibly at temperature } T. \tag{1.18}$$

The name entropy was coined by Clausius in 1868. It is derived from the combination of English "en" meaning inside and Greek "tropḗ" meaning transformation to indicate that it describes some transformation taking place inside the system. Explaining his reason for choosing to name Q/T as entropy, Clausius states that he wanted to choose such a name which is as close as possible to energy. It is believed that he chose S as the symbol for entropy to honor Sadi Carnot as it is the first letter of his first name.

On using (1.18) in (1.17) we see that total change in entropy of the system in a reversible Carnot cycle is zero:

$$\Delta S_{sys} \equiv Q_h/T_h + (-Q_c/T_c) = 0. \tag{1.19}$$

The equation above gives change in entropy of the working fluid in a reversible Carnot cycle exchanging heat with reservoirs at temperatures T_h and T_c. Let us find change in entropy of the reservoirs. To that end note that, by definition, a heat reservoir absorbs or gives away heat reversibly. Hence, since it gives away the amount Q_h at temperature T_h, change in entropy of the hotter heat reservoir is $-Q_h/T_h$. Similarly change in entropy of the colder heat reservoir is Q_c/T_c. We call the two reservoirs together as constituting the *environment*. Hence, change in entropy of the environment during the process in question is

$$\Delta S_{env} \equiv -Q_h/T_h + Q_c/T_c = 0. \tag{1.20}$$

Fig. 1.6 Engine operating between the reservoirs at temperatures T_h and T_c by irreversible cyclic process. It draws heat Q_h from the reservoir at temperature T_h, performs work W', and releases heat Q'_c into the reservoir at temperature T_c

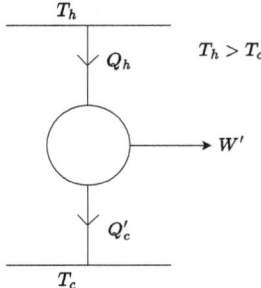

Total change in entropy of the combined system of the environment and the fluid in the reversible Carnot engine is

$$\Delta S_{sys} + \Delta S_{env} = 0, \qquad \text{reversible process.} \tag{1.21}$$

The system under consideration and its environment together are said to constitute the *universe*. The equation above shows that there is no change in entropy of the universe in a reversible Carnot cycle.

Next we find change in the value of Q/T in an irreversible Carnot-type cycle. To that end, consider an engine operating cyclically by some irreversible process, drawing heat Q_h from the reservoir at temperature T_h and depositing heat Q'_c into the reservoir at temperature T_c by generating work W' (see Fig. 1.6). The efficiency of the process is η_{irr}. We have shown that η_{irr} is less than the efficiency η of a reversible Carnot engine operating between same reservoirs drawing same amount of heat from hot reservoir. With $\eta_{irr} = W'/Q_h = (Q_h - Q'_c)/Q_h$, and η given by (1.16), the inequality $\eta_{irr} < \eta$ yields

$$\frac{Q_h}{T_h} - \frac{Q'_c}{T_c} < 0, \qquad \text{irreversible cyclic process.} \tag{1.22}$$

This shows that there is net loss of the quantity Q/T in an irreversible cyclic process. It should be emphasized that if the amount Q of heat absorbed by the system at temperature T is by an irreversible process then Q/T is *not* the change in entropy of the system.

However, even when the system absorbs and gives away heat irreversibly from the reservoirs, the heat absorbed or given away by a heat reservoir is a reversible process for the reservoir. Hence $-Q_i/T_i$ is the change in entropy of the reservoirs if it gives away the amount of heat Q_i while remaining at the temperature T_i all through the process. In the case of the process under discussion in which the reservoir at temperature T_h gives away the amount Q_h and the one at T_c receives the amount Q'_c of heat, total change in entropy of the two reservoirs which constitute the environment is given by

$$\Delta S_{env} \equiv -Q_h/T_h + Q'_c/T_c > 0, \tag{1.23}$$

where last inequality is due to (1.22). This shows that entropy of environment increases when the system coupled with it performs irreversible processes.

The significance of Q/T in an irreversible process lies, as proved next, in the fact that it is a measure of the reduction in the amount of work generated by an irreversible process compared with that generated in a reversible process operating between same heat reservoirs and drawing the same amount of heat. It is called the *lost work*. To that end, note that the work generated by irreversible engine is $W' = Q_h - Q'_c$ whereas the work generated by the reversible engine is $W = Q_h - Q_c = Q_h(1 - Q_c/Q_h) = Q_h(1 - T_c/T_h)$. It is straightforward to see that

$$W' - W = T_c \left(\frac{Q_h}{T_h} - \frac{Q'_c}{T_c} \right) < 0, \tag{1.24}$$

where the last inequality is by virtue of (1.22). This shows that the reduction in the value of Q/T in an irreversible cyclic process is a measure of the loss of work that could have been available were the process reversible. For another way of describing the lost work, recall (1.23) to rewrite (1.24) as

$$W' = W - T_c \Delta S_{\text{env}}. \tag{1.25}$$

Thus increase in entropy of the environment may be considered as the measure of the lost work.

We will derive the equivalent of Q/T for a general transformation as part of the second law of thermodynamics.

1.3 Laws of Thermodynamics

In this section we list the laws of thermodynamics.

1.3.1 Zeroth Law

After the formulation of three laws of thermodynamics to be listed below, the need for stating now named zeroth law was realized. It was apparent that logically the said law should be stated before the other three already formulated and numbered laws. Hence the name. The *zeroth law* states that

If a system A is in thermal equilibrium with B and also separately with C then B and C are in thermal equilibrium with each other.

To see the consequences of this law, assume that P_A, V_A are the pressure and volume that specify the state of the system A and P_B, V_B are those for the system B when it is in thermal equilibrium with A. The equilibrium between the two systems implies that the two sets of variables are related. Let that relationship be expressed in the form $P_A = f(V_A, P_B, V_B)$. Similarly, when A is also in thermal equilibrium with C then we will have the relationship $P_A = f(V_A, P_C, V_C)$ between the state variables of A and C. Consequently we arrive at the equation

$$f(V_A, P_B, V_B) = f(V_A, P_C, V_C). \tag{1.26}$$

Note that the relationship above is arrived at by assuming A to be in thermal equilibrium with B and with C. Now, according to the law stated above, B and C must also be in thermal equilibrium with each other. Hence, there must be a relationship between the state variables of B and C. Equation (1.26) will be the desired relation provided it is independent of V_A because the equilibrium condition between the systems B and C cannot depend on the variables of any other system. That will be the case if $f(V_A, P, V)$ is of the form

$$f(V_A, P, V) = g(P, V)h_1(V_A) + h_2(V_A). \tag{1.27}$$

On using this, (1.26) reduces to

$$g(P_B, V_B) = g(P_C, V_C). \tag{1.28}$$

This shows that there exists a function $g(P, V)$ which has the same value for the systems in thermal equilibrium. We call that function temperature $\tilde{T}$: $\tilde{T} = g(P, V)$. The last equation defines the equation of state. However neither this law nor the other laws of thermodynamics determine the form of the function $g(P, V)$. We know the equation of state under some special conditions on the fluids. For example, for an ideal gas, we know that the equation of state is $T = PV/Nk_B$ where T is the temperature on the Kelvin scale. In this case $\tilde{T} = T$ and $g(P, V) = PV/Nk_B$. A different form of $g(P, V)$ would result if the gas is not ideal.

1.3.2 First Law

The first law is the statement of conservation of energy. It states that the internal energy U of a thermodynamic system is a state function such that

If δQ is the amount of heat absorbed and δW the amount of work done by the system in an arbitrary transformation then change δU in its internal energy is given by

$$\delta U = \delta Q - \delta W. \tag{1.29}$$

The amount of heat absorbed and the work done by the system to transform from one state to another depends not only on the initial and final states, but also on the path that the transformation follows in the space of state variables. To indicate the said path dependence, the amount of heat absorbed and the work done by the system in an infinitesimal transformation are denoted generally by $đQ$ and $đW$, respectively. The differentials $đQ$ and $đW$ are imperfect (see also Sect. 2.8).

However, U is a state function in the sense that it has a unique value for the system in a given state. Hence, the change in it during a transformation is simply the difference in its values in the final and the initial states, independent of the path that the transformation follows in the space of state variables. Infinitesimal change in internal energy is therefore denoted by dU, a perfect differential. The infinitesimal form of the first law therefore reads

$$dU = đQ - đW. \tag{1.30}$$

Though change in the internal energy is due to transfer of heat and/or work, the internal energy cannot be partitioned as heat and work energies.

The work may consist of several components each caused by change in some macroscopic control parameter, say, ξ_i $(i = 1, 2, \ldots m)$ so that

$$đW = -\sum_{i=1}^{m} F_i d\xi_i. \tag{1.31}$$

The F_i is the "force" associated with the change in ξ_i. The negative sign in (1.31) is because $đW$ in (1.30) is work performed by the system whereas $F_i d\xi_i$ is the work performed on it by the force F_i.

In particular, if the volume V is the only macroscopic control parameter then $đW = PdV$ where P is the pressure and (1.30) reads

$$dU = đQ - PdV. \tag{1.32}$$

The gas gains energy when its volume decreases and loses energy when its volume increases.

1.3.3 Second Law

The first law is the statement of conservation of energy. It is, however, not sufficient to determine the permissible thermodynamic processes. For example, if two bodies at different temperatures are placed in contact, the first law does not rule out the possibility of transfer of heat from cold to hot body on its own, a process we never observe. The second law, to be stated below, determines permissible processes in terms of the concept of entropy.

There are two equivalent ways of stating the second law:

1. **Kelvin statement**: There does not exist any thermodynamic transformation whose *sole* effect is to extract heat from a heat reservoir and convert it entirely into work.

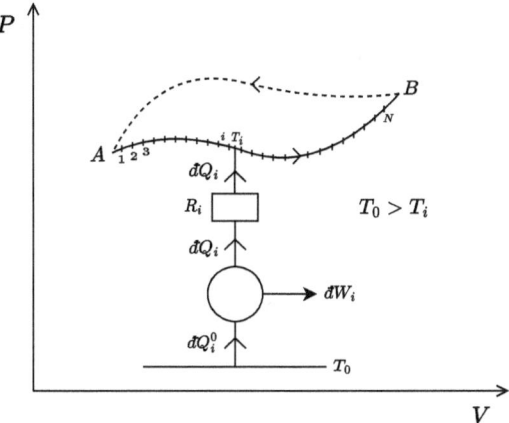

Fig. 1.7 The figure shows the solid line path from A to B divided into segments. Though not shown the dotted path is divided similarly. The ith segment is connected with the reservoir R_i at temperature T_i which in turn is connected to a reversible Carnot engine operating between R_i and another reservoir at temperature T_0 generating work dW_i. Though not shown in the figure, similar construction is assumed for all points on the solid as well as dotted return path in the figure

2. **Clausius statement**: There does not exist any thermodynamic transformation whose *sole* effect is to extract heat from a body at lower temperature and deliver it to the one at higher temperature.

We do not address the question of showing equivalence of the two statements (see, for example, [Huang]).

In Sect. 1.2, we used the Clausius statement to arrive at the concept of entropy and its properties in a reversible cyclic process in which flow of heat into the system or out of it involved only two reservoirs at different temperatures. We generalize those considerations to an arbitrary quasi-static cyclic transformation in which the system may be at different temperatures during the transformation.

To that end, referring to Fig. 1.7, consider a quasi-static cyclic process that takes the system from the state represented by A to that represented by B along the solid line path and back to A along the dotted line path. Divide the paths from A to B and also from B to A, in infinitesimal elements. Let T_i be the temperature of the system at the point i on the path, as if in equilibrium with a reservoir R_i at that temperature. From i it evolves to the neighboring state as if by drawing the amount of heat dQ_i from the reservoir R_i while performing work $dW_i^{(s)}$. The change in the internal energy dU_i in the infinitesimal transformation at the point i on the path is then, according to the first law, given by

$$dU_i = dQ_i - dW_i^{(s)}. \tag{1.33}$$

Now, imagine a reversible Carnot engine operating between the reservoir R_i and the reservoir R_0 which is at some fixed temperature T_0 ($T_0 > T_i$ for all i) such that it draws an amount of heat dQ_i^0 from R_0 and delivers the amount of heat dQ_i to R_i while producing work dW_i for all $i = A \rightarrow B \rightarrow A$ so that

$$dW_i = dQ_i^0 - dQ_i. \tag{1.34}$$

Since the Carnot engine between R_i and R_0 is reversible, Carnot's formula (1.15) holds so that

$$\frac{dQ_i}{T_i} = \frac{dQ_i^0}{T_0}. \tag{1.35}$$

Summing over all i we get

$$\sum_{i=A \rightarrow B \rightarrow A} \frac{dQ_i}{T_i} = \sum_{i=A \rightarrow B \rightarrow A} \frac{dQ_i^0}{T_0}. \tag{1.36}$$

In the limit of the number N of points becoming infinitely large, we can convert the summations to integrals to get

$$\oint \frac{dQ}{T} = \frac{1}{T_0} \oint dQ_i^0 \equiv \frac{Q_0}{T_0}, \tag{1.37}$$

where Q_0 is total amount of heat drawn from the reservoir at temperature T_0 when the process completes a cycle. The closed integral is from A to B along the solid line path and then back to A along the dotted line path. In the following we show that Q_0 must be negative.

To that end, add (1.33) and (1.34) to get

$$dU_i = dQ_i^0 - dW_i^{(s)} - dW_i. \tag{1.38}$$

Sum this over $i = A \rightarrow B \rightarrow A$ and note that

1. Since, while summing over $i = A \rightarrow B \rightarrow A$, the system starts from A and returns to it, the change in its internal energy is zero:

$$\sum_{i=A \rightarrow B \rightarrow A} dU_i = 0. \tag{1.39}$$

2. Total work done by the cyclic process in going from A to B and returning back to A, together with the work done by all the Carnot engines, is

$$W_T = \sum_{i=A \rightarrow B \rightarrow A} \left[dW_i + dW_i^{(s)} \right]. \tag{1.40}$$

Hence summation of (1.38) over i along with the use of Equations (1.39) and (1.40) and conversion of summation over i to integral yield

$$W_{\mathrm{T}} = \oint dQ_i^0 \equiv Q_0. \tag{1.41}$$

We can understand this result by describing the process depicted in Fig. 1.7 as follows:

While evolving from the point i to the neighboring point, the system receives the amount dQ_i of heat from the reservoir R_i which is the same as the heat that R_i receives from the Carnot engine, imagined to be connected to it. We can therefore say that all the heat given away by the Carnot engines is received by the system under consideration. The combination, consisting of the system under considerations, the reservoirs $R_i's$, and the Carnot engines constitute the entire system exchanging amount Q_0 of heat with the reservoir at temperature T_0 through the Carnot engines connected to it.

As regards work, total work done by the entire system is evidently W_T, the sum of the work done by the Carnot engines and that by the system under consideration while evolving cyclically from A to B and back to A.

At the end of the cyclic process, each part of the entire system returns to its initial state. Hence, depending on the sign of Q_0, and due to the relation $W_T = Q_0$ derived in (1.41), the overall process in the entire system interacting with the reservoir at temperature T_0 is equivalent with one or the other process in Fig. 1.8. The process depicted on the left in that figure represents overall process in Fig. 1.7 in case $Q_0 > 0$. It describes a cyclic process in which the amount Q_0 of heat is extracted from the reservoir at temperature T_0 and converted entirely to work without any other change. Such a process is ruled out by the second law.

On the other hand, the process depicted on the right in Fig. 1.8 represents overall process in Fig. 1.7 in case $Q_0 < 0$. It describes a cyclic process in which the amount of work W_T is converted entirely into heat which is deposited in the reservoir at temperature T_0. Such a process is not ruled out by the second law. Consequently due to (1.37) we must have

$$\oint \frac{dQ}{T} \leq 0. \tag{1.42}$$

This is a form of the second law of thermodynamics.

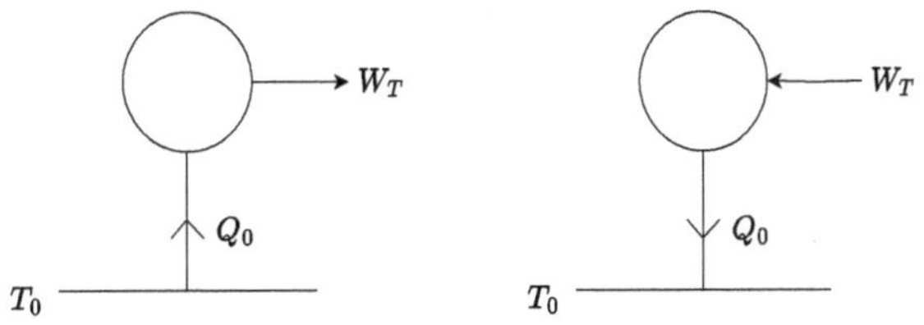

Fig. 1.8 Figure on the left depicts the overall process of Fig. 1.7 when $Q_0 > 0$ and that on the right when $Q_0 < 0$

We now show that the equality in the equation above holds if the process is reversible. To that end, note that if the process is reversible then its reverse is described by the reversal of signs of heat and work flows. Hence the condition equivalent with (1.42) for the reversed process will be obtained by the transformation $dQ \rightarrow -dQ$ leading to the equation

$$-\oint \frac{dQ}{T} \le 0, \quad \text{if the process is reversible and is reversed.} \quad (1.43)$$

Equations (1.42) and (1.43) can be satisfied simultaneously if

$$\oint \frac{dQ}{T} = 0, \quad \text{reversible closed path.} \quad (1.44)$$

It is straightforward to prove that (1.44) implies

$$\int_A^B \frac{dQ}{T} \quad \text{is independent of path between A and } B \text{ if reversible.} \quad (1.45)$$

To prove this, refer to Fig. 1.7 to write (1.44) as

$$\left[\int_{\text{A to B along solid path}} + \int_{\text{B to A along dotted path}} \right] \frac{dQ}{T} = 0. \quad (1.46)$$

If the process is reversible then B to A can be reversed in which case

$$\left[\int_{\text{B to A along dotted path}} \right] \frac{dQ}{T} = - \left[\int_{\text{A to B along dotted path}} \right] \frac{dQ}{T}.$$

$$(1.47)$$

On substituting this in (1.46) we see that

$$\left[\int_{\text{A to B along solid path}} \right] \frac{dQ}{T} = \left[\int_{\text{A to B along dotted path}} \right] \frac{dQ}{T}.$$

$$(1.48)$$

Since, for a given reversible solid path the choice of the dotted path is arbitrary, it follows that the value of the integral of dQ/T on a reversible path does not depend on the path. It depends only on the end points. This proves (1.45). This in turn means that dQ/T on a reversible path is a perfect differential which we denote by dS:

$$dS = \frac{dQ}{T}, \quad \text{along reversible path.} \quad (1.49)$$

Fig. 1.9 The process from B
to A along the path I is
reversible whereas that from
A to B along the path II is
irreversible

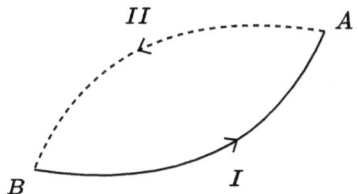

Hence, with O as any reference point, the quantity

$$S(A) = \int_O^A \frac{dQ}{T}, \qquad \text{along reversible path,} \qquad (1.50)$$

depends only on the state A. The $S(A)$ is called entropy of the state A. Since it
depends only on the state variables characterizing a state, entropy is a state variable.
The change in entropy in going from A to B is given by

$$S(B) - S(A) = \int_A^B \frac{dQ}{T}, \qquad \text{along reversible path.} \qquad (1.51)$$

Now, assume that the path from B to A is reversible but that from A to B is irreversible
(see Fig. 1.9). Equation (1.42) then reads

$$\left[\int_{\text{along reversible path I}} + \int_{\text{along irreversible path II}} \right] \frac{dQ}{T} < 0. \qquad (1.52)$$

Since path I is reversible, the first integral is nothing but $S(A) - S(B)$ leading
thereby to the inequality

$$\int_{A \to B} \frac{dQ}{T} < S(B) - S(A), \qquad \text{irreversible path from } A \text{ to } B. \qquad (1.53)$$

This may also be rewritten as

$$\int_{A \to B \text{ irr}} \frac{dQ}{T} < \int_{A \to B \text{ rev}} \frac{dQ}{T}. \qquad (1.54)$$

This shows that the change in dQ/T over an irreversible path between two states is
less than that on any reversible path between same states.

Some Consequences of Second Law

1. Consider a system interacting with a reservoir, drawing the amount dQ of heat
   from it at temperature T by reversible or irreversible process. The quantity dQ/T
   for the system therefore may or may not stand for change in its entropy. However,
   the processes inside a reservoir are reversible. Hence, since dQ is the amount of

heat given out by it at temperature T, $-dQ/T$ stands for the change in entropy of the reservoir. Since all the reservoirs from which the system draws heat in going from state A to state B constitute its environment, it follows that the change in entropy of the environment is given by

$$\Delta S_{\text{env}} = -\int_A^B \frac{dQ}{T}. \tag{1.55}$$

On substituting this in (1.53) it follows that, as the state of the system changes from A to B,

$$\Delta S_{\text{universe}} \equiv \Delta S_{\text{env}} + \Delta S_{\text{sys}} \geq 0, \qquad \Delta S_{\text{sys}} \equiv S(B) - S(A). \tag{1.56}$$

The equation above shows that entropy of the universe never decreases.

2. Assume that the system is thermally isolated. In that case it does not exchange heat with its environment, i.e. $dQ = 0$. Hence $\Delta S_{\text{env}} = 0$ as a consequence of which it follows from (1.56) that, as the system transforms from state A to state B, its entropy cannot decrease:

$$S(B) - S(A) \geq 0, \qquad \text{thermally isolated system.} \tag{1.57}$$

The system in question may be the universe itself which is evidently thermally isolated. Like the conclusion drawn from (1.56), the equation above when applied to the universe shows that the entropy of universe never decreases.

3. Consider a thermally isolated system in some state A. Relax some of its constraints. Equation (1.57) shows that it will evolve to the state B whose entropy is more than that in the state A. However, subject to the relaxed constraints, if there is another state C available whose entropy is more than that in the state A then it will evolve to C. Continuing this argument we see that a thermally isolated system would evolve to the state of maximum entropy consistent with the constraints on it. This is called the principle of *maximum entropy*. This principle will constitute one of the postulates of the postulatory approach to thermodynamics outlined in Chap. 2.

4. Consider a reversible process which takes a system from state A to state B. From the first law we know that the change in internal energy of the system is given by

$$dU = dQ - dW_{\text{rev}} = T d S_{\text{sys}} - dW_{\text{rev}}, \tag{1.58}$$

where the second equation is due to the fact that the process is reversible so that $dQ = T d S_{\text{sys}}$. The change in the internal energy in going from A to B is given by

$$U_B - U_A = \int_A^B \left[T d S_{\text{sys}} - dW_{\text{rev}} \right], \qquad \text{reversible path.} \tag{1.59}$$

Consider another process which connects the same two states as in the said reversible process but now by an irreversible path so that

$$dU = đQ - đW_{irrev}.$$ (1.60)

Since the process is irreversible, $đQ \neq T dS$. The change in internal energy in going from the state B to the state A in this case would be

$$U_B - U_A = \int_{A \to B \text{ irrev}} [đQ - đW_{irrev}].$$ (1.61)

Since internal energy is a state variable, the value of $U_B - U_A$ is same whether the process is reversible or not. Hence, on equating (1.59) and (1.61) it follows that

$$W_{lost} = \int_A^B T dS_{sys} - \int_{A \to B, \text{ irrev}} đQ,$$ (1.62)

where

$$W_{lost} = W_{rev} - W_{irrev},$$ (1.63)

with W_{rev} denoting the work done along the reversible path,

$$W_{rev} = \int_A^B đW_{rev},$$ (1.64)

and

$$W_{irrev} = \int_{A \to B \text{ irrev}} đW_{irrev}$$ (1.65)

is the work done when transformation is irreversible. The difference between the said two works is the lost work denoted by W_{lost}. In particular, if the process is isothermal then T can be taken into or out of the integral so that

$$W_{lost} = T \left[\int_A^B dS_{sys} - \int_{A \to B \text{ irrev}} \frac{đQ}{T} \right]$$
$$= T \left(\Delta S_{sys} + \Delta S_{env} \right) \equiv T \Delta S_{universe},$$ (1.66)

where the second equation is due to (1.55). By virtue of (1.56), $\Delta S_{universe} > 0$. The equation above relates lost work in an isothermal irreversible process with change in entropy of the universe.

5. If the work done is only by change of volume, then (1.58) for a reversible process reduces to

$$dU = TdS - PdV, \qquad (1.67)$$

   where the quantities in the equation above are for the system under observation.
6. Let a system undergo thermodynamic transformation from some state A to B by a process which may be irreversible. In that case we cannot evaluate change in its entropy by using the formula in (1.51) as that is applicable when the process is reversible. However, since entropy is a state function, we can imagine some reversible process connecting the states in question and evaluate entropy using (1.51).
7. We know that an adiabatic process is defined as the one which does not involve exchange of heat with the environment. We also know that $\bar{d}q = TdS$ if the process is reversible. Hence entropy is unchanged in a reversible adiabatic process. Such a process is called *isentropic*. However, $\bar{d}Q \neq TdS$ if the process is irreversible. Hence an irreversible adiabatic process is not isentropic. An example of irreversible adiabatic process is free expansion of a gas. The change in entropy in this case for an ideal gas is evaluated in Ex. 1.1.

The second law determines the change in entropy during a transformation and hence determines the value of entropy in a given state up to an arbitrary constant. That arbitrary constant is fixed by the third law.

1.3.4   Third Law

The second law determines change in entropy but not its absolute value. Entropy can be assigned absolute value in any state by assigning it a value in a particular state. Third law achieves that end by asserting that
The entropy of any system at $T = 0$ is zero.
This is Planck's formulation of the Third law. It is stronger version of Nernst's *heat theorem*, formulated in 1905, according to which the change in entropy is zero as $T \to 0$:

$$Lt_{T \to 0} \Delta S = 0. \qquad (1.68)$$

It is equivalent with the statement that entropy is independent of external parameters, i.e. $S(T, x_1) = S(T, x_2)$ as $T \to 0$ where x_1, x_2 are different sets of values of external parameters. Nernst theorem, on its own or combined with the assumption of boundedness of the derivative of entropy with respect to temperature at $T = 0$, predicts vanishing of several physical quantities (see the exercises below which will need use of Maxwell relations compiled in (2.98)). An interesting consequence of it, not elaborated here, is the question of unattainability of $T = 0$ (see [Callen]).

The applicability and limitations of the third law are, however, understood best in the framework of quantum statistical mechanics (see Sect. 6.4.4).

Exercises

Ex. 1.1. Show that in adiabatic free expansion of a gas

$$\Delta U = 0, \quad \text{adiabatic free expansion.} \tag{1.69}$$

Hint: An adiabatically expanding gas does not exchange heat with its surroundings so that $\Delta Q = 0$. Since expansion is free, it does not perform any work even. Consequently (1.69) follows using first law.

Ex. 1.2. Show that, as a consequence of (1.68), as $T \rightarrow 0$, (i) $\alpha_P \rightarrow 0$ where α_P is the coefficient of isobaric thermal expansion defined in (2.111), (ii) $(\partial P/\partial T)_V \rightarrow 0$. Hint: Use Maxwell relations $(\partial V/\partial T)_P = -(\partial S/\partial P)_T$ and $(\partial P/\partial T)_V = (\partial S/\partial V)_T$ given in (2.98).

Ex. 1.3. Consider a reversible process. The heat capacities at constant volume and that at constant pressure are given, in terms of entropy, by $C_V = T(\partial S/\partial T)_V$ (see (2.101)) and $C_P = T(\partial S/\partial T)_P$ (see (2.107)). Show that $C_V, C_P \rightarrow 0$ as $T \rightarrow 0$ if, in addition to (1.68) it is assumed that $(\partial S/\partial T)_V$ and $(\partial S/\partial T)_P$ are bounded as $T \rightarrow 0$.

1.4 Ideal Gas Equations of State

The zeroth law tells us that, at thermodynamic equilibrium, there exists a relationship between the thermodynamic variables P, V, T. It is called the equation of state. However, that law, or even the other laws put together, cannot determine the equation of state for any gas. That equation must be found either phenomenologically or by some theory like statistical mechanics. As mentioned circa (1.2), the said equation of state does not provide complete description of a system in thermodynamic equilibrium. We will formally define an equation of state in Chap. 2 and show that complete thermodynamic description of a one component system (i.e. a system composed of one kind of molecules) is provided by two independent equations of state. We therefore need to construct second equation of state for the ideal gas. The second equation of state we construct is the relation between U, V, T.

Like the equation of state relating P, V, T, the second equation of state must also be determined either phenomenologically or theoretically. We address the question of constructing the second equation of state in detail in Sect. 2.14 and find that, though the two equations of state are independent, the thermodynamic relations impose consistency requirement on them.

In the following subsection, we outline the phenomenological approach to derive the additional equation of state $U = U(V, T)$ obeyed by an ideal gas.

1.4.1 Internal Energy of Ideal Gas

The first law of thermodynamics determines the change in the internal energy U of a substance when it is supplied heat and work. The question is: can that law or its combination with other laws determine U as a function of the thermodynamic state variables? To answer that question note that, since P, V, T are related by some equation of state, U may be expressed as a function of two of the three said variables. Let $U = U(V, T)$. Invoking the first law (1.32) we have

$$\left(\frac{\partial U}{\partial T}\right)_V = \left(\frac{\partial Q}{\partial T}\right)_V \equiv C_V, \tag{1.70}$$

where, by definition,

$$C_V = \left(\frac{\partial Q}{\partial T}\right)_V \tag{1.71}$$

is the heat capacity at constant volume.

To determine dependence of internal energy on volume, Joule conducted following experiment. He took a vessel, having conducting walls, divided into two parts, one part was kept in the state of vacuum and the other had air in it at high pressure. The two parts were separated by an immovable partition which had an opening with a stopcock. The vessel was immersed in water at the same temperature as air and stopcock opened to allow air to expand. At equilibrium air fills the whole volume of the vessel. Since gas does no external work while expanding, any change in its internal energy will result in its exchanging heat with water thereby changing its temperature. No change in temperature of water was observed which showed that the internal energy of air when it occupied a part of the volume of the box is same as when it occupied its whole volume. This established independence of internal energy of ideal gas on volume. However, doubts arose as to what if change in temperature of the gas is too small to have been observed due to much lower heat capacity of the gas compared with that of the vessel and water. More accurate experiments were conceptualized and performed by Thomson and Joule (in 1845) which involved direct measurement of temperature of the gas (see Sect. 2.15). Those experiments confirmed the finding of Joule's experiment.

As a result of the experiments, we conclude that internal energy of an ideal gas is independent of volume. It is also known experimentally that the heat capacity of ideal gas is independent of temperature. Consequently C_V is a constant so that (1.70) leads to the following expression for internal energy:

$$U(V, T) = C_V T, \tag{1.72}$$

where integration constant has been taken as zero.

Of interest is also the heat capacity C_P at constant pressure:

$$C_P = \left(\frac{\partial Q}{\partial T}\right)_P. \tag{1.73}$$

Invoking the first law (1.32) we see that

$$\left(\frac{\partial U}{\partial T}\right)_P = C_P - P\left(\frac{\partial V}{\partial T}\right)_P. \tag{1.74}$$

On using (1.1) and (1.72) the equation above may be rewritten as

$$C_P - C_V = Nk_B. \tag{1.75}$$

In terms of the parameter γ, called *adiabatic constant*, defined by

$$\gamma = \frac{C_P}{C_V}, \tag{1.76}$$

Equation (1.75) yields

$$C_V = cNk_B, \quad c = \frac{1}{\gamma - 1}. \tag{1.77}$$

Using this, the expression (1.72) for U assumes the form

$$U = cNk_B T. \tag{1.78}$$

The physics meaning of the constant c is best appreciated in the framework of statistical mechanics (see Sect. 7.4). We will see that $c = 3/2$ for a monatomic gas

$$\gamma = \frac{5}{3}, \quad \text{ideal monatomic gases.} \tag{1.79}$$

The multi-atom molecules have internal degrees of freedom too, like rotation and vibration which contribute to the value of c, evaluated in Sect. 7.4.

Summary

An ideal gas is described by following two equations of state:

$$PV = Nk_B T, \quad U = cNk_B T, \quad c = \frac{1}{\gamma - 1}, \quad \gamma = \frac{C_P}{C_V}. \tag{1.80}$$

We apply the results derived above to a cyclic process to realize the Carnot engine
with ideal gas as the working fluid. We find the efficiency of the process and show
that it agrees with Carnot's formula. In the exercises at the end of the chapter, we
evaluate efficiency of the said cyclic process with non-ideal gases as the working
fluid to confirm Carnot's formula.

1.5 A Cyclic Process to Realize Carnot Engine

Consider the cyclic process depicted in Fig. 1.10. The working fluid in it can be any
gas—ideal or non-ideal. For now we assume the gas to be ideal. Cases of non-ideal
gases obeying particular equations of state as also the case of a gas obeying arbitrary
equation of state as a working fluid are discussed in the exercises at the end of the
chapter. The gas is initially in equilibrium at temperature T_h and its pressure and
volume P_a, V_a are represented by the point A on the $P - V$ diagram. Let the gas
expand isothermally to volume V_b with P_b as its pressure. This state is represented by
the point B on the $P - V$ diagram. The temperature is kept constant by keeping the
gas in contact with a heat reservoir at the desired temperature. As the gas expands, it
does work against pressure and loses internal energy. The temperature of the gas is
maintained at constant value by the heat it receives from the heat reservoir in contact
with it. We denote by Q_h the amount of heat absorbed.

At B the gas is isolated from the heat reservoir. It expands further but without
exchanging heat with the environment. This is adiabatic expansion. It continues till
its volume and pressure become V_c and P_c (point C on the $P - V$ diagram). Let its
temperature at C be T_c.

At C it is brought in contact with the reservoir at temperature T_c and compressed
isothermally till its volume and pressure become P_d, V_d (point D on the $P - V$
diagram). It gains energy as its volume decreases but it gives away heat energy to

Fig. 1.10 $P - V$ diagram of
a cyclic process to realize
Carnot engine with $T_h > T_c$.
The processes are assumed
to be reversible

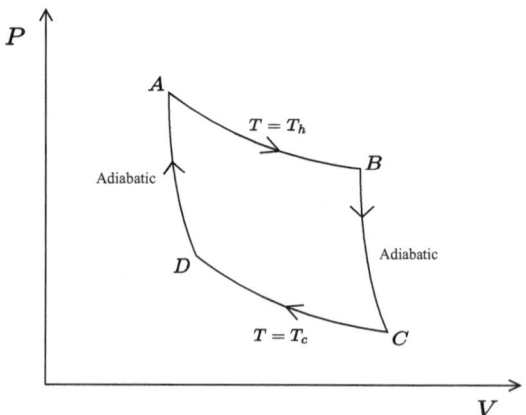

the reservoir so that its temperature is maintained at constant value T_c. We denote by Q_c the amount of heat it loses to the reservoir.

At D it is again isolated and compressed adiabatically till its pressure, volume, and temperature attain their initial values P_a, V_a, and T_h. Since the process from D to A is adiabatic, no heat is exchanged with the surroundings.

All the processes in the said cycle are carried reversibly. The gas absorbs the amount Q_h of heat along the isothermal path AB and gives out the amount Q_c along the isothermal path CD. Since the transformations along BC and DA are adiabatic, no heat is absorbed or given out along these segments. The cyclic process depicted in Fig. 1.10 is therefore equivalent with the Carnot engine operating between two heat reservoirs, absorbing heat Q_h from the reservoir at temperature T_h and depositing heat Q_c into the reservoir at temperature T_c. Invoking the laws of thermodynamics, we show that Carnot's engine formula (1.15) holds.

The amount of heat absorbed along AB may be evaluated using the first law in (1.32) along with the equations of state (1.80). Since temperature has the constant value T_h along AB, (1.80) reduces to $P = Nk_B T_h/V$, and $dU = 0$, which on substitution in (1.32) yields

$$dQ = Nk_B T_h \frac{dV}{V}. \tag{1.81}$$

On integrating between initial and final volumes V_a and V_b, the heat absorbed turns out to be given by

$$Q_h = Nk_B T_h \ln(V_b/V_a). \tag{1.82}$$

The same procedure applied along the segment CD on which the temperature has the constant value T_c and volume varies from V_c to V_d leads to the following expression for the heat received by the gas:

$$Q'_c = Nk_B T_c \ln(V_d/V_c). \tag{1.83}$$

Since $V_d < V_c$, the heat absorbed by the gas is negative, which means that a positive amount of heat is delivered to the reservoir. Hence the amount of heat delivered to the reservoir at temperature T_c is

$$Q_c = -Q'_c = Nk_B T_c \ln(V_c/V_d). \tag{1.84}$$

We have thus evaluated heat absorbed along AB and that lost along CD.

Consider now the transformation along BC. Since it is adiabatic, it does not involve exchange of heat, i.e. $dQ = 0$. The first law (1.32) then reduces to $dU + PdV = 0$ which, on using the equations of state (1.80), reads

$$\frac{dT}{T} + (\gamma - 1)\frac{dV}{V} = 0. \tag{1.85}$$

On integrating the equation above we obtain

$$TV^{\gamma-1} = \text{constant}. \tag{1.86}$$

The relation between P and V, derived by using first equation in (1.80) to express T in terms of P and V, reads

$$PV^{\gamma} = \text{constant}. \tag{1.87}$$

Similarly, on expressing V in terms of P and T, (1.86) leads to the relation

$$TP^{-\nu} = \text{constant} \qquad \nu = \frac{\gamma - 1}{\gamma}. \tag{1.88}$$

Equations (1.86)–(1.88) are equivalent descriptions of a reversible adiabatic process, called *adiabatic equations of state*.

On applying (1.86) to the path BC we get

$$\left(\frac{V_b}{V_c}\right)^{\gamma-1} = \frac{T_c}{T_h}. \tag{1.89}$$

In similar manner, the adiabatic transformation along DA gives

$$\left(\frac{V_d}{V_a}\right)^{\gamma-1} = \frac{T_h}{T_c}. \tag{1.90}$$

On comparing (1.89) and (1.90) it is seen that

$$\frac{V_c}{V_d} = \frac{V_b}{V_a}. \tag{1.91}$$

Use of the relation above in (1.82) and (1.84) leads to

$$\frac{Q_h}{T_h} = \frac{Q_c}{T_c}. \tag{1.92}$$

This is same as the Carnot engine formula (1.15).

To find efficiency of the process under consideration from first principles note that total work done by the gas in the cycle, due to first law, is

$$W = \oint (dQ - dU) = \oint dQ, \tag{1.93}$$

where second equation is due to the fact that change in internal energy in a cyclic process is zero. Now

$$\oint dQ = \left[\int_A^B + \int_B^C + \int_C^D + \int_D^A \right] dQ. \tag{1.94}$$

Since no heat is exchanged on adiabatic paths BC and DA, (1.94) yields

$$\oint dQ = \left[\int_A^B + \int_C^D \right] dQ = Q_h - Q_c. \tag{1.95}$$

Substitution of this in (1.93) gives

$$W = Q_h - Q_c. \tag{1.96}$$

The relation $W = Q_h - Q_c$ derived above is independent of the equation of state obeyed by the working fluid. It is a consequence of the law of conservation of energy and is same as the formula of work output in the Carnot engine. Using (1.92), efficiency is given by

$$\eta = \frac{W}{Q_h} = 1 - \frac{Q_c}{Q_h} = 1 - \frac{T_c}{T_h}. \tag{1.97}$$

We have thus demonstrated realization of Carnot engine with an ideal gas as the working fluid. We will confirm the Carnot engine formula (1.92) for arbitrary equation of state in the exercises.

Summary

The adiabatic equations of state of the ideal gas are

$$TV^{\gamma-1} = \text{constant}, \quad PV^\gamma = \text{constant},$$
$$TP^{-\nu} = \text{constant}, \quad \nu = \frac{\gamma - 1}{\gamma} \tag{1.98}$$

Exercises

Ex. 1.4. Find change in entropy of an ideal gas when it expands isothermally at temperature T from volume V_1 to V_2. What is the lost work? Hint: Change in internal energy of the ideal gas in isothermal expansion is zero. Hence, due to first law, when gas expands reversibly from volume V_1 to V_2 at temperature T, change in its entropy is

$$\Delta S_{\text{gas}} = \frac{1}{T} \int_{V_1}^{V_2} P dV = N k_B \int_{V_1}^{V_2} \frac{dV}{V} = N k_B \ln(V_2/V_1). \tag{1.99}$$

Since expansion is isothermic, the reservoir receives as much heat as is lost by the gas due to which change in entropy of the reservoir is $-\Delta S_{\text{gas}}$ so that change in the entropy of universe is zero. Consequently there is no work loss in the process.

Ex. 1.5. Find change in entropy of the gas and that of the environment in Joule's experiment described in Sect. 1.4.1 in which gas expands irreversibly from volume V_1 to V_2. Find also the amount of lost work. Hint: There is no change in temperature and internal energy of the gas in Joule's experiment. Hence change in entropy of the gas in the irreversible process in question is same as that in isothermic reversible expansion from V_1 to V_2 which is given by (1.99). However, since system does not exchange heat with the environment, change in entropy of the environment is zero. Change in entropy of universe in this case being ΔS_{gas}, lost work is $T \Delta S_{\text{gas}}$.

1.5.1 Entropy of Ideal Gas

Let N molecules of an ideal gas be in equilibrium in the state A characterized by (P_A, V_A, T_A). Consider a reversible process which takes it to the equilibrium state B characterized by (P_B, V_B, T_B). The process being reversible is described by the form of the first law in (1.67) which, on invoking the equations of state in (1.80), assumes the form

$$dS = Nk_B \left(c\frac{dU}{U} + \frac{dV}{V} \right). \tag{1.100}$$

On integrating the equation above, we get

$$S = Nk_B\{c\ln(U) + \ln(V)\} + K, \tag{1.101}$$

where K is the integration constant. It plays no role in computing entropy change in going from state A to state B as in that case

$$S_B - S_A = Nk_B\{c\ln(U_B/U_A) + \ln(V_B/V_A)\}. \tag{1.102}$$

Determination of absolute value of entropy in a state would, of course, require knowledge of K. In principle, it may be determined by the condition $S = 0$ at $T = 0$. However, the ideal gas law is not applicable at low temperatures. The integration constant should therefore be determined by other considerations. We will revisit this question in Ex. 2.4.

1.6   Van der Waals Equation of State

The assumption that the gases are ideal cannot describe all relevant experimental observations. The search for the law for non-ideal gases faced several failed attempts till van der Waals derived one in his Ph.D. thesis in 1873 which goes by his name. It won him the Nobel Prize in 1910.

As we saw above, thermodynamics was developed as an independent subject. However, with the discovery of the concept of atom, the connection of thermodynamics with mechanics started getting attention leading to the emergence of what is known as the kinetic theory, outlined in Chap. 3. The attempts to find the equation of state for a real gas were based mainly on the kinetic theory.

The equation of state for a real gas derived by van der Waals was also based on the kinetic theory and the virial theorem. Remarkably, as we will see in Chap. 12, it emerges as the leading order correction to the ideal gas law in statistical mechanics with smallness of the molecular volume compared with the volume occupied per molecule as the smallness parameter. Here we give standard phenomenological derivation of the equation, obtained by introducing following modifications in the ideal gas law:

1. It is assumed that the molecules are not point particles but are hard spheres. As a result, they cannot come closer than a certain distance. This results in each molecule excluding some volume, say b, from the total volume V occupied by the gas. Hence if N is the number of molecules in the gas then the volume available effectively to it is $V - Nb$. It was argued that the ideal gas law should therefore be modified to replace V by $V - Nb$.
2. The molecules attract each other when separated by distances greater than the molecular radius r_0. The force vanishes when separation between them is large. Assuming that the molecules are distributed uniformly, each molecule in the interior is acted upon by forces on all sides resulting in net zero force. However, the layer near a boundary surface has no molecules on its boundary side. The layers near the boundary surfaces therefore experience a net inward force resulting in reduction of pressure. The force on a molecule is proportional to the number density N/V and the number of molecules in the layer next to a bounding surface is also proportional to N/V. This causes net reduction of pressure proportional to $(N/V)^2$ which must be subtracted from the pressure appearing in the ideal gas law.

Under the suggested modifications, the ideal gas law assumes the form

$$P = \frac{Nk_{\mathrm{B}}T}{V - Nb} - \frac{a}{(V/N)^2}, \qquad (1.103)$$

where a is the proportionality constant referred to in the item 2 above. In terms of the specific volume $v = V/N$, (1.103) reads

$$P = \frac{k_B T}{v - b} - \frac{a}{v^2}. \qquad (1.104)$$

This is the equation of state derived by van der Waals. We will discuss its properties in Chap. 12. As mentioned before, in addition to the equation of state relating pressure, volume, and temperature, complete description of thermodynamic properties requires also the knowledge of its internal energy as a function of the state variables. We derived the expression for internal energy of an ideal gas empirically. Based on phenomenological considerations in Sect. 2.14 and on the theory of statistical mechanics in Sect. 12.2, it will be shown that the internal energy per molecule of the van der Waals gas is

$$u = c k_B T - \frac{a}{v}. \qquad (1.105)$$

Summary

The equations of state describing the van der Waals gas are

$$P = \frac{k_B T}{v - b} - \frac{a}{v^2}, \qquad u = c k_B T - \frac{a}{v}, \qquad (1.106)$$

where v is volume per molecule and u is the internal energy per molecule.
In this chapter we presented traditional formulation of thermodynamics. In the next chapter we present the axiomatic approach.

Exercises

Ex. 1.6. (a) Show that the $T - S$ diagram corresponding to the $P - V$ diagram in Fig. 1.10 is given by Fig. 1.11. (b) Show that the heat absorbed in the process is the area bound by the rectangle $ABEF$ and that released in

Fig. 1.11 $T - S$ diagram corresponding to the $P - V$ diagram in Fig. 1.10

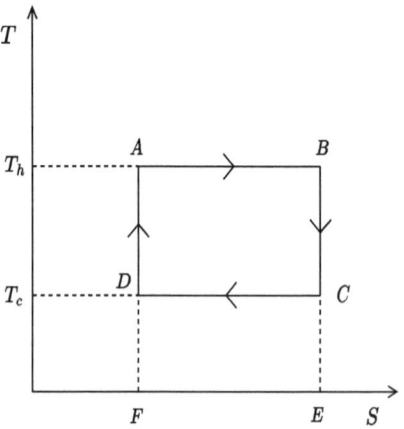

the process is the area bound by the rectangle $CDFE$. (c) Hence find the efficiency of the process and confirm Carnot's formula.

Ex. 1.7. Show that the adiabatic equation of state of van der Waals gas undergoing reversible adiabatic transformation is

$$(v - b)T^c = \text{constant}. \tag{1.107}$$

Hint: Since $\dbar Q = 0$ in an adiabatic process, integrate the equation $\dbar Q = P dv + du = 0$ where P and u are given by (1.106).

Ex. 1.8. Consider the cyclic transformation depicted in Fig. 1.10. Assuming the operating fluid to be van der Waals gas described by the equations of state (1.106), verify Carnot engine formula (1.92). Hint: The amount Q_h of heat absorbed in going from A to B at the fixed temperature T_h and the amount Q_c given out in going from C to D at the fixed temperature T_c are obtained by integrating $\dbar Q = N(du + Pdv)$ with P and u given by (1.106):

$$Q_h = N \int_A^B (du + Pdv) = Nk_B T_h \{\ln(v_B - b) - \ln(v_A - b)\},$$

$$Q_c = -N \int_C^D (du + Pdv) = Nk_B T_c \{\ln(v_C - b) - \ln(v_D - b)\}. \tag{1.108}$$

Along the adiabats BC and DA (1.107) holds

$$(v_B - b)T_h^c = (v_C - b)T_c^c, \quad (v_D - b)T_c^c = (v_A - b)T_h^c. \tag{1.109}$$

Rewrite $(v_B - b)$ in the first equation in (1.108) using the expression for $(v_B - b)$ in the first equation in (1.109), and rewrite $(v_D - b)$ in the second equation in (1.108) using the expression for $(v_D - b)$ in the second equation in (1.109) to show that $Q_h/T_h = Q_c/T_c$.

Ex. 1.9. Verify the Carnot formula (1.15) when the working gas in the cyclic transformation depicted in Fig. 1.10 obeys following equations of state:

$$P = Tf(V) + g(V), \quad U = -\int g(V) + \Phi(T), \tag{1.110}$$

where $f(V)$, $g(V)$ are some functions of volume and $\Phi(T)$ some function of temperature. Hint: From (2.247) we know that the amount of heat absorbed when system undergoes isothermal transformation at temperature T_h from volume V_A to V_B is

$$Q_{AB} = T_h (I(V_B) - I(V_A)), \quad I(V) = \int f(V)dV. \tag{1.111}$$

The $Q_{AB} \equiv Q_h$ is the heat absorbed in the isothermal process from A to B. Also, from (2.240) it follows that under reversible adiabatic transformation from (V_B, T_h) to (V_C, T_c) we have the relation

$$(I(V_C) - I(V_B)) + (J(T_c) - J(T_h)) = 0, \quad J(T) = \int \frac{d\Phi(T)}{T}. \quad (1.112)$$

Similarly

$$\begin{aligned} Q_{CD} &= T_c \left(I(V_C) - I(V_D) \right), \\ (I(V_A) - I(V_D)) &+ (J(T_h) - J(T_c)) = 0. \end{aligned} \quad (1.113)$$

The $Q_{CD} \equiv Q_c$ is the heat lost in the isothermal process from C to D. On comparing (1.112) and the second equation in (1.113) it follows that $I(V_B) - I(V_A) = I(V_C) - I(V_D)$ due to which the comparison of the expression (1.111) for Q_h and the first expression in (1.113) for heat lost Q_c confirms (1.15).

Ex. 1.10. Verify the Carnot engine formula (1.15) when the working gas in the cyclic transformation depicted in Fig. 1.10 obeys arbitrary equations of state. Hint: Since the transformation along AB is at constant temperature T_h and that along CD is at constant temperature T_c heat absorbed along AB and that given out along CD is

$$Q_h = T_h(S_B - S_A), \qquad Q_c = T_c(S_C - S_D), \quad (1.114)$$

where S_A, S_B, S_C, S_D stand for entropy at A, B, C, D. Since the processes along BC and DA are isentropic,

$$S_C = S_B, \qquad S_A = S_D. \quad (1.115)$$

On combining (1.114) and (1.115) follows the Carnot engine formula (1.15).

References

1. R.P. Feynman, *Feynman Lectures on Physics*, vol. 1 (Addisson-Wesely Publications, 1964)
2. I. Müller, *A History of Thermodynamics* (Springer, 2007)
3. G.S. Girolami, J. Chem. Eng. Data **65** 298 (2020)
4. Sadi Carnot *Reflections on the Motive Power of Fire and other papers on the second law of thermodynamics* by É. Clapeyron, R. Clausius edited with an introduction by E. Mendoza (Dover Publications, 1988)

Chapter 2
Fundamentals of Thermodynamics-II

The laws of thermodynamics are empirical. They, however, lead to the concept of entropy, entirely alien to common experience, their other ingredients—temperature, heat, pressure, energy—being intuitive. It is the law regarding change in entropy, along with the law of conservation of energy, which characterize permissible thermodynamic processes. The postulatory approach to thermodynamics adopts entropy as the fundamental entity characterizing thermodynamic state of a system as a function of its energy, volume, and number of particles. The partial derivatives of entropy with respect to energy, volume, and number of particles define the thermodynamic state variables temperature, pressure, and (to be introduced) chemical potential. Since it treats entropy as a function of quantities which characterize a mechanical system as well, the postulatory approach facilitates a natural way of establishing connection between thermodynamics and mechanics if only one could construct mechanical counterpart of thermodynamic entropy. We will construct mechanical counterpart of thermodynamic entropy in different ways in the subsequent chapters to establish connection between thermodynamics and mechanics.

The postulatory approach is due to Callen [Callen]. Before describing it, we summarize essential terminology.

1. Let x be a parameter which has value x_A in system A and x_B in system B. Let the said two systems be combined to form one system. If the value of x in the composite system is $x_A + x_B$ then x is called an *extensive* parameter. For example, volume, number of moles of a molecule are extensive parameters.
2. If a function $f(x_1, x_2, \ldots, x_n)$ of n variables $x_1, x_2, \ldots, x_n$ is such that, for any constant λ,

$$f(\lambda x_1, \lambda x_2, \ldots, \lambda x_n) = \lambda^m f(x_1, x_2, \ldots, x_n), \qquad (2.1)$$

then $f(x_1, x_2, \ldots, x_n)$ is called homogeneous of degree m. Its partial derivatives obey the relation

R. R. Puri, *Modern Thermodynamics and Statistical Mechanics*, Undergraduate Lecture Notes in Physics, https://doi.org/10.1007/978-3-031-54310-4_2

$$\sum_{i=1}^{n} x_i \frac{\partial f}{\partial x_i} = mf. \tag{2.2}$$

This is *Euler's theorem* on homogeneous functions.

A homogeneous function of degree zero is called *intensive*.

3. A system enclosed by walls that keep fixed its volume and mole numbers of all its components, and does not allow it to exchange energy (both heat and work) with its surroundings is called *closed*.[1]
4. A system consisting of two or more subsystems is said to be *composite*.
5. A system which is macroscopically homogeneous, isotropic, electrically neutral, and large enough so that surface effects can be neglected and not acted upon by external forces is called a *simple system*.

Exercises

Ex. 2.1. Prove Euler's theorem on homogeneous functions (2.2). Hint: Due to its definition (2.1), a homogeneous function of degree m in n variables may be expressed in the form

$$f(x_1, x_2, \ldots, x_n)$$
$$= \sum_{i_1, i_2, \ldots, i_n} a_{i_1 i_2, \ldots i_n} x_1^{i_1} x_2^{i_2} \cdots x_n^{i_n} \delta(i_1 + i_2 + \cdots + i_n - m). \tag{2.3}$$

Using the form of $f(x_1, x_2, \ldots, x_n)$ above, the desired relation (2.2) follows immediately.

2.1 Postulates of Thermodynamics

In this approach, the terms work and energy carry the commonly understood meaning. Heat defined as the difference between the increase in energy of a system and work done on it:

$$dQ = dU + dW. \tag{2.4}$$

Note that our notation dW stands for the work done by the system so that the work done on the system is $-dW$.

---

[1] An alternative terminology is to call the systems, named closed here, as isolated. The closed systems in the said alternative terminology are the ones which can exchange only matter with its surroundings.

2.1.1 First Postulate

There exist particular states of simple systems, called equilibrium states, which are macroscopically characterized completely by internal energy U, the volume V, and the numbers $N_1, N_2, \ldots, N_r$ of molecules of its chemical components.

This postulate conceptualizes only the existence of an equilibrium state but provides no means of characterizing and determining it. Those questions are addressed by the second and third postulates.

2.1.2 Second Postulate

There exists a function, called entropy S, of the extensive parameters of any composite system defined for all equilibrium states and having the following property: When an internal constraint in a composite system is removed, the values assumed by the extensive parameters in the absence of that constraint are those that maximize the entropy over the manifold of constrained equilibrium states.

This postulate provides a means for characterizing an equilibrium state in terms of a function called entropy but evaluation of that function would require knowledge of its properties which is the question answered by the third postulate.

2.1.3 Third Postulate

The entropy of a composite system has following properties:

1. *The entropy of a composite system is sum of the entropies of its constituent subsystems.*
2. *Entropy is a single-valued, continuous, and differentiable function of its arguments and is a monotonically increasing function of energy.*

Some consequences of this postulate are

1. Since entropy is assumed to be a function of extensive parameters and since by virtue of first postulate, the extensive parameters characterizing a system are internal energy U, volume V, and the numbers $N_1, N_2, \ldots, N_r$ of molecules of the chemical components constituting it, we have

$$S = S(U, V, \{N_i\}_r), \quad \{N_i\}_r \equiv N_1, N_2, \ldots, N_r. \tag{2.5}$$

We will see that all thermodynamic properties emerge from the knowledge of the functional form of entropy. Hence (2.5) is called the *entropic fundamental equation of thermodynamics*.

2. Since entropy of a composite system is sum of the entropies of its constituent subsystems, it follows that $S(U, V, \{N_i\}_r)$ is a homogeneous function of order one in its independent variables (see Ex. 2.2):

$$S(\lambda U, \lambda V, \{\lambda N_i\}_r). = \lambda S(U, V, \{N_i\}_r) \tag{2.6}$$

3. The property of increasing monotonicity with respect to energy implies

$$\left(\frac{\partial S}{\partial U}\right)_{V, \{N_i\}_r} > 0. \tag{2.7}$$

4. Properties of single valuedness, continuity, differentiability, and monotonicity with respect to U imply that the function $S = S(U, V, \{N_i\}_r)$ can be inverted to express U as a function of $S, V, \{N_i\}_r$:

$$U = U(S, V, \{N_i\}_r). \tag{2.8}$$

This is the inverse of the entropic fundamental equation and is called the *energetic fundamental equation of thermodynamics*.

Exercises

Ex. 2.2. Show that the postulate that the entropy S of a system is sum of the entropies of its subsystems implies (2.6). Hint: Due to additivity of S, $S(U, V, \{N_i\}_r) = S(U/k, V/k, \{N_i/k\}_r) + k$ times $= kS(U/k, V/k, \{N_i/k\}_r)$ so that $S(U/k, V/k, \{N_i/k\}_r) = S(U, V, \{N_i\}_r)/k$.

2.1.4 Fourth Postulate

The entropy of any system vanishes in the state for which

$$\left(\frac{\partial U}{\partial S}\right)_{V, \{N_i\}_r} = 0. \tag{2.9}$$

We will see that this is equivalent with the third law of thermodynamics.

2.1.5 Examining Admissible Forms of $S(U, V, N)$

The postulates restrict the admissible functional forms of $S(U, V, N)$. In order to find whether a given $S(U, V, N)$ is in admissible form, apart from its continuity and differentiability, we need to examine whether it satisfies three conditions: (1) The extensivity condition (2.6), (2) the condition (2.7) of monotonicity of S with respect to U, and (3) the condition (2.9). These conditions restrict the functional form of $S(U, V, N)$. However, additional restrictions arise from the requirements of stability of the equilibrium state which restrict further the form of $S(U, V, N)$:

1. We will see that $(\partial S/\partial V)_{U,N} = P/T$. The stability demands P to be positive [Landau and Lifshitz]. For, positive P means $(\partial S/\partial V)_{U,N} > 0$, i.e. S increases if the body expands. The expansion is not uncontrolled as the stability is achieved by its prevention by the surroundings. On the other hand, $P < 0$ means $(\partial S/\partial V)_{U,N} < 0$, i.e. S would increases even as the body contracts spontaneously leading to instability. The possibility of existence of negative pressure though is not ruled out in metastable states. Since we are concerned only with states of stable equilibrium, we would demand $P > 0$ which is equivalent with demanding

$$\left(\frac{\partial S}{\partial V}\right)_{U,N} > 0. \tag{2.10}$$

2. The stability conditions (2.141)

$$\frac{\partial^2 S}{\partial U^2} < 0, \qquad \frac{\partial^2 S}{\partial V^2} < 0, \qquad \left(\frac{\partial^2 S}{\partial U^2}\right)\left(\frac{\partial^2 S}{\partial V^2}\right) - \left(\frac{\partial^2 S}{\partial U \partial V}\right)^2 > 0$$

derived in Sect. 2.10 also place restrictions on the form of $S(U, V, N)$.

Exercises

Ex. 2.3. Assume that the fundamental entropic equation of some thermodynamic system is given by

$$S(U, V, N) = AU^a V^b N^c, \tag{2.11}$$

where A is a constant. Find the restrictions on the values of the constants a, b, c: (1) arising from the postulates and (2) from the stability conditions. Hint: (1) The extensivity condition (2.6) evidently requires $a + b + c = 1$. The monotonicity condition (2.7) will be satisfied if $a > 0$. The condition (2.9) of fourth postulate demands $a < 1$. The restrictions on the values of a, b, c due to postulates therefore are:

$$a + b + c = 1, \qquad 0 < a < 1. \tag{2.12}$$

(2) The condition (2.10) demands $b > 0$. The stability condition $\partial^2 S/\partial U^2 < 0$ will be satisfied if $a(1 - a) > 0$, i.e. if $0 < a < 1$. This is same as the one arrived at by postulate based considerations. The stability condition $\partial^2 S/\partial V^2 < 0$ will be satisfied if $b(1 - b) > 0$, i.e. if $0 < b < 1$. The stability condition $(\partial^2 S/\partial U^2)(\partial^2 S/\partial V^2) - (\partial^2 S/\partial U \partial V)^2 > 0$ will be sat-

isfied if $a + b < 1$. Since $c = 1 - a - b$, it follows that $0 < c < 1$. Thus $S(U, V, N)$ in (2.11) will be an admissible expression for entropy if

$$a + b + c = 1, \qquad 0 < a < 1, \qquad 0 < b < 1 \qquad 0 < c < 1. \qquad (2.13)$$

Ex. 2.4. We found that the first law of thermodynamics, along with the equations of state of an ideal gas lead to the expression (1.101) for entropy of an ideal gas as the function of U, V with K therein being an unknown constant. If we consider S as a function also of N then that constant must be considered a function of N and (1.101) should be rewritten as

$$S(U, V, N) = N k_B \left\{ c \ln(U) + \ln(V) + f(N) \right\}. \qquad (2.14)$$

Show that $S(U, V, N)$ is extensive if

$$f(N) = -(c + 1)\ln(N) + A, \qquad (2.15)$$

where A is an unknown constant. It cannot be determined by thermodynamic considerations. Its value follows naturally in statistical mechanical formalism. Hint: Show that the extensivity property (2.6) will be obeyed if $f(\lambda N) - f(N) = -(c + 1)\ln(\lambda)$. Let $N = 1$ to get the desired result with $A = f(1)$ as an arbitrary constant. With $f(N)$ given by (2.15), (2.14) may be rewritten as

$$S(U, V, N) = N k_B \left(\ln(u^c v) + A \right), \quad u = U/N, \quad v = V/N. \qquad (2.16)$$

2.1.6 Connection with the Laws of Thermodynamics

The connection of the postulatory approach with the laws of thermodynamics is established as follows.

Invoking (2.8) we have

$$dU = \left(\frac{\partial U}{\partial S} \right)_{V, \{N_i\}} dS + \left(\frac{\partial U}{\partial V} \right)_{S, \{N_i\}} dV + \sum_{j=1}^{r} \left(\frac{\partial U}{\partial N_j} \right)_{S, V, \{N_{i \neq j}\}} dN_j.$$

$$(2.17)$$

Define

$$T = \left(\frac{\partial U}{\partial S}\right)_{V,\{N_i\}} \qquad : \quad \text{Temperature,}$$

$$P = -\left(\frac{\partial U}{\partial V}\right)_{S,\{N_i\}} \qquad : \quad \text{Pressure,}$$

$$\mu_j = \left(\frac{\partial U}{\partial N_j}\right)_{S,V,\{N_{i\neq j}\}} \qquad : \quad \text{Chemical potential of } j\text{th component,}$$

$$(2.18)$$

to rewrite (2.17) as

$$dU = TdS - PdV + \sum_{j=1}^{r} \mu_j dN_j. \tag{2.19}$$

At this stage, the definitions of temperature, pressure and chemical potential given in (2.18) are without justification. Though comparison of (2.19) with the first law of thermodynamics for constant N_j would lead to the identification of P and T with the partial derivatives as in (2.18), their justification within the postulatory framework will be provided in the sections to follow. For now we proceed by assuming correctness of the identifications in (2.18).

The expressions in (2.18) show that T, P, μ_j are intensive parameters. In many situations it is one or more intensive parameters which are experimentally controlled. It is then convenient to choose the relevant controllable intensive parameters as independent variables. In Sect. 2.6 we will carry the task of transforming one or more independent extensive parameter to appropriate intensive parameters as independent variables.

If the number of molecules of each component is fixed then $dN_j = 0$ for all j and (2.19) reduces to (1.67) wherein the first term is identified as the amount of heat absorbed by the system and the second term as the mechanical work done by it. In case the number of molecules changes, the applicable equation is (2.19). We may identify the last term therein as the "chemical work". The chemical potential may be viewed as the increase in energy of the system due to addition of a molecule keeping its entropy and volume constant. To distinguish different kinds of works, we denote the mechanical wok by W_m and the chemical work by W_c. With

$$dW_c \equiv \sum_{j=1}^{r} \mu_j dN_j, \tag{2.20}$$

the (2.19) assumes the form

$$dU = TdS - dW_m + dW_c. \tag{2.21}$$

For further insight into the meaning of chemical potential, including its relation with thermodynamic potentials, see [1, 2].

The relations in (2.18) define intensive variables in terms of the partial derivatives of $U(S, V, \{N_j\}_r)$. We use the relations between partial derivatives derived in the Appendix A to express the intensive variables defined in (2.18) in terms of the partial derivatives of $S(U, V, \{N_j\}_r)$.

1. Invoking (A.8) with $x \rightarrow U$, $y \rightarrow V$, $z \rightarrow S$ therein and by keeping the N_i's fixed in all partial derivatives, we get

$$\left(\frac{\partial S}{\partial U}\right)_{V,\{N_i\}} = \left\{\left(\frac{\partial U}{\partial S}\right)_{V,\{N_i\}}\right\}^{-1} = \frac{1}{T}. \tag{2.22}$$

2. Invoke (A.9) with $x \rightarrow U$, $y \rightarrow V$, $z \rightarrow S$ therein keeping the N_i's constant in all partial derivatives to obtain

$$\left(\frac{\partial U}{\partial S}\right)_{V,\{N_i\}} \left(\frac{\partial S}{\partial V}\right)_{U,\{N_i\}} \left(\frac{\partial V}{\partial U}\right)_{S,\{N_i\}} = -1. \tag{2.23}$$

Recall (2.18) and (2.22) to get

$$\left(\frac{\partial S}{\partial V}\right)_{U,\{N_i\}} = \frac{P}{T}. \tag{2.24}$$

3. Invoke (A.9) with $x \rightarrow U$, $y \rightarrow N_j$, $z \rightarrow S$ therein and keep V, $\{N_{i \neq j}\}$ constant in all partial derivatives, to obtain

$$\left(\frac{\partial U}{\partial S}\right)_{V,\{N_i\}} \left(\frac{\partial N_j}{\partial U}\right)_{S,V,\{N_{i \neq j}\}} \left(\frac{\partial S}{\partial N_j}\right)_{U,V,\{N_{i \neq j}\}} = -1. \tag{2.25}$$

The use of (2.18) and (2.22) in the equation above yields

$$\left(\frac{\partial S}{\partial N_j}\right)_{U,V,\{N_{i \neq j}\}} = -\frac{\mu_j}{T}. \tag{2.26}$$

Summary

The intensive parameters in entropic representation are:

$$\frac{1}{T} = \left(\frac{\partial S}{\partial U}\right)_{V,\{N_i\}}, \quad \frac{P}{T} = \left(\frac{\partial S}{\partial V}\right)_{U,\{N_i\}}, \quad \frac{\mu_j}{T} = -\left(\frac{\partial S}{\partial N_j}\right)_{U,V,\{N_{i \neq j}\}}. \tag{2.27}$$

Next we justify the definitions of various intensive parameters in (2.18).

2.2 Justification of Definitions of Intensive Parameters

2.2.1 Temperature

Consider an isolated cylinder containing a gas. It is partitioned into two parts which are, to begin with, separated by a partition which does not allow flow of heat, work and gas from one part to the other. Let the internal energy, volume and the number of molecules in one part be denoted by U_1, V_1, N_1 and let U_2, V_2, N_2 be the corresponding quantities in the other part. Now, change the partition so that it allows exchange of heat between the two parts but it is still rigid and also does not allow exchange of gas. Since the system is isolated, the total energy $U_T = U_1 + U_2$ is conserved so that

$$dU_T = dU_1 + dU_2 = 0, \qquad dV_1 = dV_2 = 0, \qquad dN_1 = dN_2 = 0. \quad (2.28)$$

The entropy S_T of the system is sum of the entropies S_1 and S_2 of its parts:

$$S = S_1(U_1, V_1, N_1) + S_2(U_2, V_2, N_2). \quad (2.29)$$

Using (2.28) it follows that

$$dS_T = \frac{\partial S_1}{\partial U_1} dU_1 + \frac{\partial S_2}{\partial U_2} dU_2 = \left(\frac{1}{T_1} - \frac{1}{T_2} \right) dU_1. \quad (2.30)$$

The state of equilibrium of the system corresponds to maximum entropy. Hence, at equilibrium $dS_T = 0$, as a result (2.30) yields

$$\left(\frac{1}{T_1} \right)_{\text{equ.}} = \left(\frac{1}{T_2} \right)_{\text{equ.}}. \quad (2.31)$$

This shows that the quantity T which may have different values for two systems when isolated from each other, equalizes when the systems exchange heat. Since the quantity that equalizes on exchange of heat is temperature, the identification of T as temperature suggested in (2.18) is justified.

Further justification is provided by showing that the equilibrium is achieved by flow of heat from the part having higher T to the one of lower T. To that end note that, by postulate, entropy never decreases, it follow that $\Delta S_T \geq 0$. The relation (2.30) then yields

$$\Delta S_T = \left(\frac{1}{T_1} - \frac{1}{T_2} \right) \Delta U_1 \geq 0. \quad (2.32)$$

This shows that if $T_1 \geq T_2$ then we must have $\Delta U_1 \leq 0$. Since ΔU_1 denotes the amount of energy received by the system numbered 1, non-positivity of ΔU_1 implies that system numbered 1 gives out energy which in the present case is heat energy as no other form of energy exchange is allowed by assumption. Since $T_1 \geq T_2$ it follows that heat flows from the system having higher value of T to the one at the lower value. This is consistent with our understanding of temperature.

2.2.2 Pressure

Continuing the consideration of the system discussed in the last subsection, let the partition between the two parts allow, in addition to the exchange of heat, also the exchange of work. In other words, the partition is no longer rigid, it can move. In this case, in addition to the conservation of total internal energy, the condition of conservation of volume also applies so that

$$dU_T = dU_1 + dU_2 = 0, \quad dV = dV_1 + dV_2 = 0, \quad dN_1 = dN_2 = 0. \quad (2.33)$$

Hence

$$
\begin{aligned}
dS_T &= \frac{\partial S_1}{\partial U_1} dU_1 + \frac{\partial S_2}{\partial U_2} dU_2 + \frac{\partial S_1}{\partial V_1} dV_1 + \frac{\partial S_2}{\partial V_2} dV_2 \\
&= \left(\frac{1}{T_1} - \frac{1}{T_2} \right) dU_1 + \left(\frac{P_1}{T_1} - \frac{P_2}{T_2} \right) dV_1,
\end{aligned}
\quad (2.34)
$$

where use has been made of (2.27) in writing the last line. Since $dS_T = 0$ at equilibrium, we see that the equation above yields

$$\left(\frac{1}{T_1} \right)_{equ.} = \left(\frac{1}{T_2} \right)_{equ.}, \quad \left(\frac{P_1}{T_1} \right)_{equ.} = \left(\frac{P_2}{T_2} \right)_{equ.}. \quad (2.35)$$

This shows that, at equilibrium, T and P both equalize:

$$(T_1)_{equ.} = (T_2)_{equ.}, \quad (P_1)_{equ.} = (P_2)_{equ.}. \quad (2.36)$$

Since in the present case we have allowed exchange of heat as well as that of work and, as argued circa (2.31), the equalization of T is due to exchange of heat, equalization of P is due to exchange of work. Since the quantity that equalizes due to exchange of work is pressure, we see that identification of P as pressure suggested in the (2.18) is justified.

We have considered the consequences of allowing exchange of heat alone and the exchange of heat and work together. Allowing exchange of work alone leads to an indeterminate problem (see [Callen] for details).

2.2.3   Chemical Potential

To find the consequences of exchange of matter, consider again the system discussed in last two subsections. Let the partition between the two parts allow, in addition to the exchange of heat, also the exchange of matter but not of work. In other words, the partition is rigid. In this case, in addition to the conservation of total internal energy, the condition of conservation of number of molecules also applies so that

$$dU_T = dU_1 + dU_2 = 0, \quad dN_1 + dN_2 = 0, \quad dV_1 = dV_2 = 0. \qquad (2.37)$$

Hence

$$
\begin{aligned}
dS_T &= \frac{\partial S_1}{\partial U_1} dU_1 + \frac{\partial S_2}{\partial U_2} dU_2 + \frac{\partial S_1}{\partial N_1} dN_1 + \frac{\partial S_2}{\partial N_2} dN_2 \\
&= \left(\frac{1}{T_1} - \frac{1}{T_2} \right) dU_1 - \left(\frac{\mu_1}{T_1} - \frac{\mu_2}{T_2} \right) dN_1,
\end{aligned}
\qquad (2.38)
$$

where use has been made of (2.27) in writing the last line. Since $dS_T = 0$ at equilibrium, we see that the equation above yields

$$\left(\frac{1}{T_1} \right)_{\text{equ.}} = \left(\frac{1}{T_2} \right)_{\text{equ.}}, \qquad \left(\frac{\mu_1}{T_1} \right)_{\text{equ.}} = \left(\frac{\mu_2}{T_2} \right)_{\text{equ.}}. \qquad (2.39)$$

This shows that, at equilibrium, T and μ both equalize:

$$(T_1)_{\text{equ.}} = (T_2)_{\text{equ.}}, \quad (\mu_1)_{\text{equ.}} = (\mu_2)_{\text{equ.}}. \qquad (2.40)$$

Since in the present case we have allowed exchange of heat as well as that of matter and, as argued circa (2.31), the equalization of T is due to exchange of heat, equalization of μ is due to exchange of matter. Since the quantity that equalizes due to exchange of matter is chemical potential, we see that identification of μ as chemical potential suggested in (2.18) is justified.

2.3   Equations of State

The relations in (2.18) express intensive parameters in terms of independent extensive variables $S, V, \{N_j\}$:

$$T = T(S, V, \{N_j\}), \quad P = P(S, V, \{N_j\}), \quad \mu_i = \mu_i(S, V, \{N_j\}). \qquad (2.41)$$

The equations expressing intensive parameters in terms of independent extensive variables are known as the *equations of state*. The equations of state in (2.41) arise in energy representation in which $(S, V, \{N_j\})$ are independent variables. We can instead work in the entropy representation in which $(U, V, \{N_j\})$ are independent variables and the corresponding intensive parameters are defined by (2.27). The equations of state in entropy representation therefore read

$$\frac{1}{T} = \frac{1}{T}\left(U, V, \{N_j\}\right), \quad \frac{P}{T} = \frac{P}{T}\left(U, V, \{N_j\}\right), \quad \frac{\mu_i}{T} = \frac{\mu_i}{T}\left(U, V, \{N_j\}\right).$$
$$(2.42)$$

As mentioned before, thermodynamics does not determine equations of state.

The relations $P = P(V, T)$, $U = U(V, T)$ commonly called equations of state are not in the forms (2.41) or (2.42). The equations of state will be said to be in "proper form" if written as in (2.41) or in (2.42).

Exercises

Ex. 2.5. Consider the fundamental entropic relation (2.11) in which the parameters a, b, c obey (2.13). Find the three equations of state.

Ex. 2.6. Consider two systems, named 1 and 2, described by the fundamental entropic relation (2.11) in which the parameters a, b, c obey the conditions in (2.13). Let $S_i(U_i, V_i, N_i)$ denote entropy of the ith system $(i = 1, 2)$ when isolated from each other. If the systems are allowed to exchange only heat then show that their internal energies when equilibrium is reached are given by

$$U_1 = U\left[1 + \left(\frac{V_2}{V_1}\right)^{b/(1-a)}\left(\frac{N_2}{N_1}\right)^{c/(1-a)}\right]^{-1},$$

$$U_2 = U\left[1 + \left(\frac{V_1}{V_2}\right)^{b/(1-a)}\left(\frac{N_1}{N_2}\right)^{c/(1-a)}\right]^{-1}. \qquad (2.43)$$

Hint: Extremize total entropy $S = S_1(U_1, V_1, N_1) + S_2(U_2, V_2, N_2)$ subject to the constraint $U_1 + U_2 = U$.

Ex. 2.7. Rewrite the equations of state (1.80) for an ideal gas in proper form. Hint: Since second of those equations relates U and T, the proper form of the equations must be entropic form. Indeed, the two equations of state may be rewritten in the entropic forms $P/T = Nk_B/V$, $1/T = cNk_B/U$.

Ex. 2.8. Write the equations of state (1.106) of van der Waals gas in proper form. Hint: The proper form of the energy equation is $1/T = ck_B v/(a + uv)$. Use this to rewrite the P-equation as $P/T = k_B/(v - b) - ack_B/(av + uv^2)$.

2.4 Euler Equation

An important consequence of extensivity of entropy (or equivalently of internal energy) is the so-called Euler equation. To derive it, recall from (2.6) that $S(U, V, \{N_i\}_r)$ is a homogeneous function of $(U, V, \{N_i\}_r)$ of degree one. Hence, due to (2.2) with $m = 1$, we have

$$
S = U \left(\frac{\partial S}{\partial U} \right)_{V, \{N_i\}} + V \left(\frac{\partial S}{\partial V} \right)_{S, \{N_i\}} + \sum_{j=1}^{r} N_j \left(\frac{\partial S}{\partial U} \right)_{U, V, \{N_{i \neq j}\}}. \tag{2.44}
$$

Invoking the relations (2.27) between intensive parameters and the partial derivatives of S, (2.44) assumes the form

$$
S = \frac{U}{T} + \frac{PV}{T} - \sum_{j} \frac{N_j \mu_j}{T}. \tag{2.45}
$$

On rearrangement of its terms, the equation above may be rewritten as

$$
U = ST - PV + \sum_{j} N_j \mu_j. \tag{2.46}
$$

The equation above may also be arrived at by starting with the property of homogeneity of U with respect to the independent variables $S, V, \{N_i\}_r$:

$$
U(\lambda S, \lambda V, \{\lambda N_i\}_r) = \lambda U(S, V, \{N_i\}_r). \tag{2.47}
$$

Equation (2.45) or its equivalent (2.46) is known as the *Euler equation*.

2.5 Gibbs–Duhem Relation

The Gibbs–Duhem relation is the relation between intensive variables. To derive it, note that (2.46) implies

$$
dU = \{T dS - P dV + \sum_{j} \mu_j dN_j\} + \{S dT - V dP + \sum_{j} N_j d\mu_j\}. \tag{2.48}
$$

On recalling (2.19), we see that the left side of the equation above is equal to the first bracketed term on its right side. The second bracketed term must therefore be zero:

$$
S dT - V dP + \sum_{j} N_j d\mu_j = 0. \tag{2.49}
$$

This is known as the *Gibbs–Duhem* relation. It is straightforward to see that, in the entropy representation, the Gibbs–Duhem relation reads

$$U \mathrm{d}\left(\frac{1}{T}\right) + V \mathrm{d}\left(\frac{P}{T}\right) - \sum_j N_j \mathrm{d}\left(\frac{\mu_j}{T}\right) = 0. \tag{2.50}$$

The Gibbs–Duhem relation shows that the intensive parameters are not independent. Thus in a system having r types of chemical components, there are r chemical potentials which along with temperature and pressure constitute $r + 2$ intensive parameters. However, because of the Gibbs–Duhem relation, the number of independent parameters reduces to $r + 1$. The number of independent intensive parameters is known as the number of *thermodynamic degrees of freedom.* We thus see that the number of thermodynamic degrees of freedom of an r-component system is $r + 1$. Accordingly, the number of independent equations of state for single component system is two.

Though we do not illustrate it here, the Gibbs–Duhem relation is useful in determining the fundamental thermodynamic equation from the knowledge of the equations of state.

Exercises

Ex. 2.9. Using the two equations of state of an ideal gas in (1.80), show that its third equation of state is given by ($u = U/N$, $v = V/N$)

$$\frac{\mu}{T} = -k_{\mathrm{B}}\left(\ln(u^c v) + B\right), \tag{2.51}$$

where B is a constant. Hint: Rewrite the said two equations of state in the form of entropic equations of state: $P/T = k_{\mathrm{B}}/v$, $1/T = ck_{\mathrm{B}}/u$ and use them in the entropic Gibbs–Duhem relation (2.50) to get

$$\mathrm{d}\left(\frac{\mu}{T}\right) = k_{\mathrm{B}}\left\{cu\mathrm{d}\left(\frac{1}{u}\right) + v\mathrm{d}\left(\frac{1}{v}\right)\right\}. \tag{2.52}$$

Integration of the equation above yields the desired result (2.51).

Ex. 2.10. Using Euler equation, show that the entropy and chemical potential of the ideal gas are related by

$$\frac{S}{N} + \frac{\mu}{T} = k_{\mathrm{B}}(c + 1). \tag{2.53}$$

Hence show that the expression (2.16) for entropy and expression (2.51) for μ for an ideal gas are consistent with (2.53) if the constants A and B in the said two expression are such that $A - B = c + 1$.

Ex. 2.11. Consider the two equations of state given by (2.242) and (2.243). Show that the third equation of state is ($\Phi(T)/N \rightarrow \Phi(T)$)

$$\frac{\mu}{T} = \int v \left(f'(v) + \frac{g'(v)}{T} \right) dv + \int \Phi(T) d(1/T) + C, \qquad (2.54)$$

where C is a constant and $F'(x) \equiv dF(x)/dx$. Hint: Using the said two equations of state, the entropic Gibbs–Duhem relation (2.50) reads

$$\begin{aligned}
d(\mu/T) &= \left(vg(v) - \int g(v) dv + \Phi(T) \right) d(1/T) \\
&\quad + v \left(f'(v) + \frac{g'(v)}{T} \right) dv \\
&= \left(\int vg'(v) dv \right) d(1/T) + \Phi(T) d(1/T) \\
&\quad + v \left(f'(v) + \frac{g'(v)}{T} \right) dv \\
&= d \left\{ \int v \left(f'(v) + \frac{g'(v)}{T} \right) dv \right\} + \Phi(T) d(1/T). \qquad (2.55)
\end{aligned}$$

Integration of this leads to the desired result (2.54).

2.6 Thermodynamic Potentials

Entropy and energy are functions of extensive variables. However, the control parameters in experiments are often intensive variables: temperature and pressure. It is then
more convenient to work with functions having one or more intensive parameters
as independent variables. In the following, we study the theory of transformation of
the energy function to the functions in which one or more of the extensive variables
characterizing the energy function are replaced by the corresponding intensive variables. We find also the extremum principle in terms of the transformed functions that
enables one to identify the equilibrium state in each case.

Recall that an intensive variable is the derivative of the energy function $U(S, V, N)$
with respect to one of its arguments. The transformation of U to a function in
which an extensive variable characterizing U is replaced by the corresponding intensive variable is mathematically therefore a Legendre transformation, summarized in
Appendix B. For, it is the transformation of a function $f(x)$ of x to the function $G(s)$
of its slope $s = df(x)/dx$ defined by

$$G(s) = f(x(s)) - sx(s), \qquad (2.56)$$

called the Legendre transform of $f(x)$.

We use the concept of Legendre transform to generate from the energy function
$U(S, V, \{N_i\})$ the thermodynamic functions, also called *thermodynamic potentials*,

which are functions of combinations of one or more of the intensive and extensive variables. Their physics origin will be examined in Sect. 2.13. For simplicity, we consider one component systems.

2.6.1 Helmholtz Potential

We know that $T = \partial U(S, V, N)/\partial S$. Hence the Legendre transform of $U(S, V, N)$ with respect to S is the function $F(T, V, N)$ having T in place of S as the variable which, on invoking (2.56), is given by

$$F(T, V, N) = U(S, V, N) - TS. \tag{2.57}$$

The function $F(T, V, S)$ is called *Helmholtz Potential* or *Free Energy*.

The definition (2.57) implies

$$dF(T, V, N) = dU(S, V, N) - T dS - S dT. \tag{2.58}$$

On using (2.19), the equation above assumes the form

$$dF(T, V, N) = -S dT - P dV + \mu dN. \tag{2.59}$$

Hence it follows that

$$S = -\left(\frac{\partial F}{\partial T}\right)_{V,N}, \quad P = -\left(\frac{\partial F}{\partial V}\right)_{N,T}, \quad \mu = \left(\frac{\partial F}{\partial N}\right)_{V,T}. \tag{2.60}$$

Knowing $F(T, V, N)$, the relations above provide a means of evaluating pressure, entropy, and chemical potential.

Let us examine how $F(T, V, N)$ changes in an arbitrary isothermal process. We assume fixed N. From the second law of thermodynamics expressed in the form (1.53) we know that, if T is constant all through the process, then the amount ΔQ of heat absorbed by the system in going from the state A to B along an arbitrary path is such that

$$\Delta Q \leq T \Delta S, \qquad T \text{ constant}, \tag{2.61}$$

where $\Delta S \equiv S(B) - S(A)$ is the difference in the entropy of the state B and that of A. Now, for constant T, (2.58) implies

$$\Delta F(T, V, N) = \Delta U(S, V, N) - T \Delta S \leq \Delta U(S, V, N) - \Delta Q, \tag{2.62}$$

where the inequality is by virtue of (2.61). Since $\Delta T = \Delta N = 0$, the first law gives $\Delta U(S, V, N) - \Delta Q = -P\Delta V$ which on substitution in (2.62) leads to the inequality

$$\Delta F(T, V, N) \leq -P\Delta V. \tag{2.63}$$

For $\Delta V = 0$ this reduces to

$$\Delta F(T, V, N) \leq 0. \tag{2.64}$$

We have thus shown that *Helmholtz free energy never increases in an isothermal process if the volume and the particle number are also kept constant.*

2.6.2 Gibbs Potential

Consider the transformation from the set of variables (S, V, N) characterizing U to the variables (T, P, N) where $T = \partial U/\partial S$, $P = -\partial U/\partial V$. Extending the relation (2.56) defining single variable Legendre transform to two variables, the Legendre transform of $U(S, V, N)$ under $S \to T$, $V \to P$ is

$$G(T, P, N) = U(S, V, N) - TS + PV. \tag{2.65}$$

The function $G(T, P, N)$ is called the *Gibbs potential* or *Gibbs free energy*. Invoking Euler's equation (2.46), it can be seen that $G(T, P, N)$ is expressible in terms of the chemical potential by the relation

$$G(T, P, N) = N\mu. \tag{2.66}$$

On using (2.19) for dU, (2.65) leads to

$$dG(T, P, N) = -SdT + VdP + \mu dN. \tag{2.67}$$

Hence it follows that

$$S = -\left(\frac{\partial G}{\partial T}\right)_{P,N}, \quad V = \left(\frac{\partial G}{\partial P}\right)_{T,N}, \quad \mu = \left(\frac{\partial G}{\partial N}\right)_{T,P}. \tag{2.68}$$

Knowing $G(T, P, N)$, the relations above provide a means of evaluating volume, entropy, and the chemical potential.

By the arguments similar to those leading to (2.64), it can be shown that

$$\Delta G \leq 0, \quad \text{constant } T, P, N. \tag{2.69}$$

It states that *the Gibbs free energy never increases in an isothermal process if pressure and number of moles are also kept constant.*

2.6.3 Enthalpy

Consider the transformation from the set of variables (S, V, N) characterizing U to the variables (S, P, N), where $P = -\partial U/\partial V$. The Legendre transform of $U(S, V, N)$ under the transformation $V \to P$ is

$$H(S, P, N) = U(S, V, N) + PV. \tag{2.70}$$

The function $H(S, P, N)$ is called the *enthalpy.*
   Due to first law, (2.70) gives, for constant N,

$$\Delta H = \Delta Q + V \Delta P. \tag{2.71}$$

This shows that, in an isobaric process, $\Delta H = \Delta Q$, i.e. change in enthalpy in an isobaric process is the heat absorbed when N is constant. Enthalpy is therefore also called *heat function.*
   With dU given by (2.19), the expression for dH reads

$$dH(S, P, N) = TdS + VdP + \mu dN. \tag{2.72}$$

Hence it follows that

$$T = \left(\frac{\partial H}{\partial S}\right)_{P,N}, \quad V = \left(\frac{\partial H}{\partial P}\right)_{S,N}, \quad \mu = \left(\frac{\partial H}{\partial N}\right)_{S,P}. \tag{2.73}$$

Knowing $H(S, P, N)$, the relations above provide a means of evaluating temperature, volume, and the chemical potential.
   By the arguments similar to those leading to (2.64), it can be shown that

$$\Delta H \leq 0, \quad \text{constant } S, P, N. \tag{2.74}$$

This states that *in a process in which S, P, N are kept constant, enthalpy never increases.*

2.6.4 Grand Potential

Consider the transformation from the set of variables (S, V, N) characterizing U to the variables (T, V, μ). The Legendre transform of $U(S, V, N)$ under $(S, V, N) \to (T, V, \mu)$ evidently is

$$\Omega(T, V, \mu) = U(S, V, N) - TS - \mu N. \tag{2.75}$$

The function $\Omega(T, V, \mu)$ is called the the *grand potential*. Due to Euler equation, (2.75) may be rewritten as

$$\Omega = -PV. \tag{2.76}$$

Invoking (2.19) for dU, the expression (2.75) yields

$$d\Omega(T, V, \mu) = -SdT - PdV - Nd\mu. \tag{2.77}$$

It then follows that

$$S = -\left(\frac{\partial \Omega}{\partial T}\right)_{V,\mu}, \quad P = -\left(\frac{\partial \Omega}{\partial V}\right)_{T,\mu}, \quad N = -\left(\frac{\partial \Omega}{\partial \mu}\right)_{T,V}. \tag{2.78}$$

Knowing $\Omega(T, V, \mu)$, the relations above provide a means of evaluating entropy, pressure, and molecular number.

2.7 Massieu Functions

In the foregoing we constructed various thermodynamic potentials as Legendre transforms of the energy function $U(S, V, N)$. Another set of potentials, called *Massieu functions*, can be constructed as Legendre transforms of the entropy function $S(U, V, N)$:

1. Since $(\partial S/\partial U)_{V,N} = 1/T$, the Legendre transform of $S(U, V, N)$ with respect to U would be

$$\Psi(1/T, V, N) = S(U, V, N) - U\frac{1}{T}. \tag{2.79}$$

This is often called the Massieu function. Clearly

$$d\Psi(1/T, V, N) = -Ud\frac{1}{T} + \frac{P}{T}dV - \frac{\mu}{T}dN. \tag{2.80}$$

2. Since $(\partial S/\partial V)_{U,N} = P/T$, the Legendre transform of $S(U, V, N)$ with respect to V would be

$$\Xi(U, P/T, N) = S(U, V, N) - V\frac{P}{T}. \tag{2.81}$$

Its differential form is

$$d\Xi(U, P/T, N) = \frac{1}{T}dU - Vd\frac{P}{T} - \frac{\mu}{T}dN. \tag{2.82}$$

3. Since $(\partial S/\partial N)_{U,V} = -\mu/T$, the Legendre transform of $S(U, V, N)$ with respect to N would be

$$\Theta(U, V, \mu/T) = S(U, V, N) + N\frac{\mu}{T}. \tag{2.83}$$

Its differential form is

$$d\Theta = \frac{1}{T}dU + \frac{P}{T}dV + Nd\frac{\mu}{T}. \tag{2.84}$$

4. The Legendre transform of $S(U, V, N)$ with respect to U and V is

$$\Phi(1/T, P/T, N) = S(U, V, N) - U\frac{1}{T} - V\frac{P}{T}. \tag{2.85}$$

This is often called Planck's function. Its differential form is

$$d\Phi = -Ud\frac{1}{T} - Vd\frac{P}{T} - \frac{\mu}{T}dN. \tag{2.86}$$

5. The Legendre transform of $S(U, V, N)$ with respect to U and N is

$$K(1/T, V, \mu/T) = S(U, V, N) - U\frac{1}{T} + N\frac{\mu}{T}. \tag{2.87}$$

This is often called Kramer function. Its differential form is

$$dK = -Ud\frac{1}{T} + \frac{P}{T}dV + Nd\frac{\mu}{T}. \tag{2.88}$$

6. The Legendre transform of $S(U, V, N)$ with respect to V and N is

$$\Gamma(U, P/T, \mu/T) = S(U, V, N) - V\frac{P}{T} + N\frac{\mu}{T}. \tag{2.89}$$

Its differential form is

$$d\Gamma = \frac{1}{T}dU - Vd\frac{P}{T} + Nd\frac{\mu}{T}. \tag{2.90}$$

Historically, the Massieu functions were introduced in 1869 which was before the advent of the thermodynamic potentials by Gibbs (1873) and Helmholtz (1882). However, it is the thermodynamic potentials which are widely used.

2.8 Maxwell Relations

We know that U and S are state functions. Being their Legendre transforms, so are the thermodynamic potentials and the Massieu functions. We will show that this fact leads to several relations between the derivatives of the state functions and those of the intensive parameters with respect to each other. They prove useful in identifying useful relations between thermodynamic observables.

To find the said relations, consider the function $f(\{x_i\}_r)$ of r independent variables $(x_1, x_2, \ldots, x_r) \equiv (\{x_i\}_r)$. We say that $df(\{x_i\}_r)$ is a *perfect or exact differential* if the integral,

$$I = \int_{A \to B} df, \qquad (2.91)$$

between the points A and B in the space of the said variables is independent of the path of integration. In that case df is expressible as

$$df = \sum_{i=1}^{r} \frac{\partial f}{\partial x_i} dx_i. \qquad (2.92)$$

If the integral (2.91) depends on the path of integration, then $df(\{x_i\}_r)$ is called imperfect or inexact differential.

Now, let it be given that

$$df = \sum_{i=1}^{r} M_i(\{x_i\}_r) dx_i. \qquad (2.93)$$

If df is a perfect differential, the comparison of (2.93) with (2.92) shows that in that case

$$M_i = \frac{\partial f}{\partial x_i}. \qquad (2.94)$$

This clearly implies and

$$\frac{\partial M_i}{\partial x_j} = \frac{\partial M_j}{\partial x_i}. \qquad (2.95)$$

Now let $f(\{x_i\}_r)$ be a function of thermodynamic variables $\{x_i\}_r$. If $f(\{x_i\}_r)$ is a state function, then we know that change in its value in a transformation from the state A to the state B, described by the integral (2.91), is independent of the path along which the variables change. Hence df is a perfect differential if $f(\{x_i\}_r)$ is a state function. Consequently, the coefficients of the dx_i's in the expression of df must obey the relations (2.95).

Since U, S, the thermodynamic potentials and the Massieu functions are state functions, we apply the relations (2.95) to their differential forms.

Several thermodynamic transformations of widespread interest are those which leave unchanged the number of molecules of each kind. If we characterize the thermodynamic properties in terms of $U(S, V)$ then, with N constant,

$$dU(S, V) = TdS - PdV. \tag{2.96}$$

In this case, we can construct three thermodynamic potentials which are the Legendre transforms of $U(S, V)$ one with respect to S, second with respect to V, and the third with respect to both S and V, called the Helmholtz potential, the enthalpy, and the Gibbs potential defined in Sect. 2.6. Since we are assuming N to be constant their differential forms are given by

$$dF(T, V) = -SdT - PdV,$$
$$dH(S, P) = TdS + VdP,$$
$$dG(T, P) = -SdT + VdP. \tag{2.97}$$

On applying the condition (2.95) to (2.96) and to each of the three expressions in (2.97), follow the relations

$$\left(\frac{\partial T}{\partial V}\right)_S = -\left(\frac{\partial P}{\partial S}\right)_V, \quad \left(\frac{\partial S}{\partial V}\right)_T = \left(\frac{\partial P}{\partial T}\right)_V,$$
$$\left(\frac{\partial T}{\partial P}\right)_S = \left(\frac{\partial V}{\partial S}\right)_P, \quad \left(\frac{\partial S}{\partial P}\right)_T = -\left(\frac{\partial V}{\partial T}\right)_P, \tag{2.98}$$

called the *Maxwell relations*.

Similar relations can be derived corresponding to differential forms of various Massieu functions. One such relation that we will need follows from (2.80) reading

$$\left(\frac{\partial U}{\partial V}\right)_T = -\left(\frac{\partial P/T}{\partial 1/T}\right)_V. \tag{2.99}$$

We will refer to the Maxwell relations corresponding to the Massieu functions as entropic Maxwell relations.

Next we show that, along with the relations between the partial derivatives derived in Appendix A, the Maxwell relations are useful in expressing various thermodynamic quantities in terms of the set of three independent thermodynamic observables introduced next.

2.9 Independent Thermodynamic Observables

The first derivatives of energy U (or those of the entropy S) give the intensive quantities T, P, μ. Their second derivatives yield thermodynamic observables. For example, consider a system of fixed number of molecules so that its energy is a function of S, V. The second derivatives of U with respect to S, V are

$$\left(\frac{\partial^2 U}{\partial S^2}\right)_V = \left(\frac{\partial T}{\partial S}\right)_V = \left[\left(\frac{\partial S}{\partial T}\right)_V\right]^{-1} = \frac{T}{C_V}, \qquad (2.100)$$

where

$$C_V = \left(\frac{\partial Q}{\partial T}\right)_V = T\left(\frac{\partial S}{\partial T}\right)_V = \left(\frac{\partial U}{\partial T}\right)_V \qquad (2.101)$$

is the heat capacity at constant volume. Next,

$$\left(\frac{\partial^2 U}{\partial V^2}\right)_S = -\left(\frac{\partial P}{\partial V}\right)_S = -\left[\left(\frac{\partial V}{\partial P}\right)_S\right]^{-1} = \frac{1}{V\kappa_S}, \qquad (2.102)$$

where

$$\kappa_S = -\frac{1}{V}\left(\frac{\partial V}{\partial P}\right)_S \qquad (2.103)$$

is the isentropic or adiabatic compressibility. Finally

$$\frac{\partial^2 U}{\partial V \partial S} = \left(\frac{\partial T}{\partial V}\right)_S = \left[\left(\frac{\partial V}{\partial T}\right)_S\right]^{-1} = \frac{1}{V\alpha_S}, \qquad (2.104)$$

where

$$\alpha_S = \frac{1}{V}\left(\frac{\partial V}{\partial T}\right)_S \qquad (2.105)$$

is the isentropic or adiabatic coefficient of thermal expansion.

The thermodynamic quantities in general are formed by the partial derivatives between two of the variables from the set U, S, V, T, P keeping one of the others a constant. It can be seen that there are 30 such derivatives. Using Maxwell relations and the identities between partial derivatives, it turns out that not all of them are independent. In fact all those derivatives can be expressed in terms of the set of three derivatives defining $(C_V, \kappa_S, \alpha_S)$, called the *fundamental set*. However, the fundamental set involves keeping an extensive variable constant. In practice, it is often more convenient to hold an intensive variable under control. It is therefore useful to adopt a set having independent second partial derivatives of a thermodynamic potential with respect to intensive parameters as the independent set. In view of this, being a function of intensive variables T, P, the Gibbs potential $G(T, P)$ becomes a natural choice for defining independent observables. Its three second derivatives are

$$\left(\frac{\partial^2 G}{\partial T^2}\right)_P = -\left(\frac{\partial S}{\partial T}\right)_P = -\frac{C_P}{T}, \tag{2.106}$$

where

$$C_P = \left(\frac{\partial Q}{\partial T}\right)_P = T\left(\frac{\partial S}{\partial T}\right)_P \tag{2.107}$$

is the specific heat at constant pressure. Next,

$$\left(\frac{\partial^2 G}{\partial P^2}\right)_T = \left(\frac{\partial V}{\partial P}\right)_T = -V\kappa_T, \tag{2.108}$$

where

$$\kappa_T = -\frac{1}{V}\left(\frac{\partial V}{\partial P}\right)_T \tag{2.109}$$

is the isothermal compressibility. Finally

$$\frac{\partial^2 G}{\partial T \partial P} = \left(\frac{\partial V}{\partial T}\right)_P = V\alpha_P, \tag{2.110}$$

where

$$\alpha_P = \frac{1}{V}\left(\frac{\partial V}{\partial T}\right)_P \tag{2.111}$$

is the coefficient of isobaric thermal expansion.

The set $(C_P, \kappa_T, \alpha_P)$ is called the *primary set*. Using the Maxwell relations and the relations between partial derivatives derived in Appendix A, all the derivatives between two of the variables from the set U, S, V, T, P keeping one of the others a constant can be expressed in terms of the primary set $(C_P, \kappa_T, \alpha_P)$.

Complete list of all the second derivatives of U, S, V, T, P in terms of independent second derivatives is compiled in [3], and is reproduced in [4].

In the following, we derive relations between the observables in the primary set $(C_P, \kappa_T, \alpha_P)$ and those in the the fundamental set $(C_V, \kappa_S, \alpha_S)$.

1. To derive the expression for C_V in terms of the primary set, write its definition (2.101) in terms of the Jacobian:

$$C_V = T\frac{\partial(S, V)}{\partial(T, V)} = T\frac{\partial(S, V)}{\partial(T, P)}\frac{\partial(T, P)}{\partial(T, V)}$$

$$= T \det \begin{pmatrix} \left(\dfrac{\partial S}{\partial T}\right)_P & \left(\dfrac{\partial S}{\partial P}\right)_T \\ \left(\dfrac{\partial V}{\partial T}\right)_P & \left(\dfrac{\partial V}{\partial P}\right)_T \end{pmatrix} \left(\frac{\partial P}{\partial V}\right)_T$$

$$= T \det \begin{pmatrix} \left(\dfrac{\partial S}{\partial T}\right)_P & -\left(\dfrac{\partial V}{\partial T}\right)_P \\ \left(\dfrac{\partial V}{\partial T}\right)_P & \left(\dfrac{\partial V}{\partial P}\right)_T \end{pmatrix} \left(\frac{\partial P}{\partial V}\right)_T, \qquad (2.112)$$

where the fourth Maxwell relation in (2.98) has been used in writing the last equation. It then follows that

$$C_V = C_P - \frac{VT}{\kappa_T}\alpha_P^2. \qquad (2.113)$$

This expresses C_V in the fundamental set in terms of the primary set.

2. To express adiabatic compressibility κ_S in terms of the primary set, write its expression (2.103) in terms of the Jacobian:

$$\kappa_S = -\frac{1}{V}\frac{\partial(V, S)}{\partial(P, S)} = -\frac{1}{V}\frac{\partial(V, S)}{\partial(P, T)}\frac{\partial(P, T)}{\partial(P, S)}$$

$$= -\frac{1}{V} \det \begin{pmatrix} \left(\dfrac{\partial V}{\partial P}\right)_T & \left(\dfrac{\partial V}{\partial T}\right)_P \\ \left(\dfrac{\partial S}{\partial P}\right)_T & \left(\dfrac{\partial S}{\partial T}\right)_P \end{pmatrix} \left(\frac{\partial T}{\partial S}\right)_P$$

$$= -\frac{1}{V} \det \begin{pmatrix} \left(\dfrac{\partial V}{\partial P}\right)_T & \left(\dfrac{\partial V}{\partial T}\right)_P \\ -\left(\dfrac{\partial V}{\partial T}\right)_P & \left(\dfrac{\partial S}{\partial T}\right)_P \end{pmatrix} \left(\frac{\partial T}{\partial S}\right)_P, \qquad (2.114)$$

where the fourth Maxwell relation in (2.98) has been used in writing the last equation. It then follows that

$$\kappa_S = \kappa_T - \frac{VT}{C_P}\alpha_P^2. \tag{2.115}$$

This expresses κ_S in the fundamental set in terms of the primary set.

3. To obtain the adiabatic expansion coefficient in terms of the primary set, write its definition (2.105) in terms of the Jacobian:

$$
\begin{aligned}
\alpha_S &= \frac{1}{V}\frac{\partial(V, S)}{\partial(T, S)} = \frac{1}{V}\frac{\partial(V, S)}{\partial(T, P)}\frac{\partial(T, P)}{\partial(T, S)} \\
&= \frac{1}{V}\det
\begin{pmatrix}
\left(\dfrac{\partial V}{\partial T}\right)_P & \left(\dfrac{\partial V}{\partial P}\right)_T \\[2mm]
\left(\dfrac{\partial S}{\partial T}\right)_P & \left(\dfrac{\partial S}{\partial P}\right)_T
\end{pmatrix}
\left(\frac{\partial P}{\partial S}\right)_T \\
&= \frac{1}{V}\left[\left(\frac{\partial V}{\partial T}\right)_P - \left(\frac{\partial S}{\partial T}\right)_P\left(\frac{\partial V}{\partial P}\right)_T\left(\frac{\partial P}{\partial S}\right)_T\right] \\
&= \frac{1}{V}\left[\left(\frac{\partial V}{\partial T}\right)_P + \left(\frac{\partial S}{\partial T}\right)_P\left(\frac{\partial V}{\partial P}\right)_T\left(\frac{\partial T}{\partial V}\right)_P\right], \tag{2.116}
\end{aligned}
$$

where the fourth Maxwell relation in (2.98) has been used in writing the last equation. It then follows that

$$\alpha_S = \alpha_P - \frac{C_P \kappa_T}{VT\alpha_P}. \tag{2.117}$$

This expresses α_S in the fundamental set in terms of the primary set.

Some other useful relations between the observables in the primary and the fundamental sets are

1. The ratio C_P/C_V is given by

$$
\begin{aligned}
\frac{C_P}{C_V} &= \frac{\partial(S, P)}{\partial(T, P)}\frac{\partial(T, V)}{\partial(S, V)} = \frac{\partial(S, P)}{\partial(S, V)}\frac{\partial(T, V)}{\partial(T, P)} = \left(\frac{\partial P}{\partial V}\right)_S\left(\frac{\partial V}{\partial P}\right)_T \\
&= \frac{\kappa_T}{\kappa_S}. \tag{2.118}
\end{aligned}
$$

2. To obtain α_P/α_S, rewrite (2.117) as

$$\alpha_S = \alpha_P\left(1 - \frac{C_P \kappa_T}{VT\alpha_P^2}\right) = -\frac{\alpha_P C_V}{C_P - C_V}, \tag{2.119}$$

where the second equation is due to (2.113). This yields

$$\frac{\alpha_P}{\alpha_S} = 1 - \frac{C_P}{C_V}.$$ (2.120)

Using (2.118), the equation above can be written in terms of κ_T/κ_S:

$$\frac{\alpha_P}{\alpha_S} = 1 - \frac{\kappa_T}{\kappa_S}.$$ (2.121)

3. Using (2.113) to write C_P in (2.119) in terms of C_V, we get the first of the equations below:

$$\alpha_S = -\frac{C_V \kappa_T}{VT\alpha_P} = -\frac{C_P \kappa_S}{VT\alpha_P}.$$ (2.122)

The second equation is due to the use of (2.118) in the first one. For an alternative derivation of (2.122) see Ex. 2.18.

4. A parameter of interest in certain applications is *Grüneisen parameter*:

$$\Gamma = V\left(\frac{\partial P}{\partial U}\right)_V.$$ (2.123)

This can be rewritten as (see Ex. 2.19)

$$\Gamma = -\frac{V}{T}\left(\frac{\partial T}{\partial V}\right)_S = -\frac{1}{T\alpha_S}.$$ (2.124)

Using (2.122) in (2.124) we obtain first of the equations,

$$\Gamma = \frac{V\alpha_P}{C_P\kappa_S} = \frac{V\alpha_P}{C_V\kappa_T}.$$ (2.125)

The second equation is due to the use of (2.118) in the first one. The equation above is a widely used form of Γ.

Exercises

Ex. 2.12. Show that

$$\left(\frac{\partial T}{\partial P}\right)_V = \frac{\kappa_T}{\alpha_P}.$$ (2.126)

Hint: Use the cyclic rule (A.9).

Ex. 2.13. Show that

$$\left(\frac{\partial S}{\partial V}\right)_T = \frac{\alpha_P}{\kappa_T}. \tag{2.127}$$

Hint: Use second of the Maxwell relation in (2.98) and use (2.126).

Ex. 2.14. Show that

$$\left(\frac{\partial S}{\partial V}\right)_P = \frac{C_P}{T V \alpha_P}. \tag{2.128}$$

Hint: Note that $(\partial S/\partial V)_P = (\partial S/\partial T)_P (\partial T/\partial V)_P$.

Ex. 2.15. Show that

$$\left(\frac{\partial S}{\partial P}\right)_V = \frac{C_V}{T \kappa_T \alpha_P}. \tag{2.129}$$

Hint: Note that $(\partial S/\partial P)_V = (\partial S/\partial T)_V (\partial T/\partial P)_V$ and use (2.127).

Ex. 2.16. Show that

$$\left(\frac{\partial S}{\partial P}\right)_T = V \alpha_P. \tag{2.130}$$

Hint: Use fourth of the Maxwell relations in (2.98).

Ex. 2.17. Derive the so-called $T\mathrm{d}S$ equations:

$$\begin{aligned}
T\mathrm{d}S &= C_V \mathrm{d}T + \frac{T\alpha_P}{\kappa_T}\mathrm{d}V, \\
T\mathrm{d}S &= C_P \mathrm{d}T - V T \alpha_P \mathrm{d}P, \\
T\mathrm{d}S &= \frac{C_P}{V \alpha_P}\mathrm{d}V + \frac{C_V \kappa_T}{\alpha_P}\mathrm{d}P.
\end{aligned} \tag{2.131}$$

Since $T\mathrm{d}S$ is the amount of heat absorbed in a reversible process, the $T\mathrm{d}S$ equations are useful in determining the heat absorbed in a reversible process in terms of measurable quantities when the system is described in terms of different combinations of two independent variables from the set of three state variables (P, V, T). Hint: Let $S = S(\alpha, \beta)$ so that $\mathrm{d}S = (\partial S/\partial \alpha)_\beta \mathrm{d}\alpha + (\partial S/\partial \beta)_\alpha \mathrm{d}\beta$ where $(\alpha, \beta) = (T, V)$, $(\alpha, \beta) = (T, P)$, $(\alpha, \beta) = (V, P)$, respectively, in the three equations above. The desired results are obtained by invoking equations (2.127)–(2.130).

Ex. 2.18. Derive (2.122) using the cyclic rule (A.9).

Ex. 2.19. Show that

$$\left(\frac{\partial P}{\partial U}\right)_V = -\frac{1}{T}\left(\frac{\partial T}{\partial V}\right)_S. \tag{2.132}$$

Hint: Note that $(\partial P/\partial U)_V = (\partial P/\partial S)_V (\partial S/\partial U)_V = (\partial P/\partial S)_V / T$. The desired result is obtained by the use of first Maxwell relation in (2.98).

2.10 Stability from Maximum Entropy Principle

In Sect. 2.2, we applied the principle of maximum entropy to find the equilibrium values of the thermodynamic variables when one or the other internal constraint in an isolated system is removed. The condition used for the purpose was the extremum condition $\delta S_T = 0$. However, the said condition does not ensure that the state is of maximum entropy. Whether the extremum is a minimum or a maximum depends on whether the second-order change in the function is positive or negative at the position of the extremum. The condition $\delta S_T = 0$ will correspond to maximum entropy if $\delta^2 S_T < 0$, called the stability condition. Assuming fixed number of molecules, in the following we examine the consequences of the stability condition.

Like in Sect. 2.2, we begin by considering an isolated system consisting of two subsystems separated by a rigid wall which does not allow exchange of heat as well. When exchange of heat is allowed and the partition is made movable, we showed in Sect. 2.2.2 that the condition $\delta S_T = 0$ ($S_T = S_1(U_1, V_1) + S_2(U_2, V_2)$) leads to equalization of temperature and pressure of the subsystem. We examine now the second-order variation $\delta^2 S_T$ in total entropy. It is given by

$$\delta^2 S_T = \frac{1}{2} \sum_{i=1,2} \left[\frac{\partial^2 S_i}{\partial U_i^2} (\delta U_i)^2 + 2\frac{\partial^2 S_i}{\partial U_i \partial V_i} (\delta U_i)(\delta V_i) + \frac{\partial^2 S_i}{\partial V_i^2} (\delta V_i)^2 \right], (2.133)$$

where it is understood that the derivatives are evaluated at the position of the extremum. Using (2.33), the equation above reduces to

$$\delta^2 S_T = \frac{\partial^2 S_1}{\partial U_1^2} (\delta U_1)^2 + 2\frac{\partial^2 S_1}{\partial U_1 \partial V_1} (\delta U_1)(\delta V_1) + \frac{\partial^2 S_1}{\partial V_1^2} (\delta V_1)^2. \qquad (2.134)$$

Doing away with the subscript 1 on various quantities, (2.134) may be rewritten in the following useful form:

$$\delta^2 S_T = \begin{pmatrix} \delta U & \delta V \end{pmatrix} \hat{A} \begin{pmatrix} \delta U \\ \delta V \end{pmatrix}, \qquad (2.135)$$

where $\hat{A}$ is a real symmetric matrix given by

$$\hat{A} = \begin{pmatrix} \dfrac{\partial^2 S}{\partial U^2} & \dfrac{\partial^2 S}{\partial U \partial V} \\[2mm] \dfrac{\partial^2 S}{\partial U \partial V} & \dfrac{\partial^2 S}{\partial V^2} \end{pmatrix} \equiv \begin{pmatrix} a_{11} & a_{12} \\ a_{12} & a_{22} \end{pmatrix}. \tag{2.136}$$

The condition $\delta^2 S_T < 0$ for S_T to be maximum then reads

$$\begin{pmatrix} \delta U & \delta V \end{pmatrix} \hat{A} \begin{pmatrix} \delta U \\ \delta V \end{pmatrix} < 0. \tag{2.137}$$

This will hold if the eigenvalues $\lambda_{1,2}$ of $\hat{A}$, given by

$$\lambda_{1,2} = \frac{1}{2} \left(a_{11} + a_{22} \pm \sqrt{(a_{11} + a_{22})^2 - 4(a_{11}a_{22} - a_{12}^2)} \right), \tag{2.138}$$

are negative. To find conditions on the a_{ij}'s under which $\lambda_{1,2} < 0$, note that, being equal to $(a_{11} - a_{22})^2 + 4a_{12}^2$, the factor under the radical sign is positive. Hence, the roots are real as they should be due to the fact that $\hat{A}$ is real symmetric. The roots will be negative if their sum is negative and the product positive. Since the sum of the roots of a matrix is its trace and their product its determinant, it follows that the roots of $\hat{A}$ in (2.137) will be negative if

$$a_{11} + a_{22} < 0, \qquad a_{11}a_{22} - a_{12}^2 > 0. \tag{2.139}$$

For satisfaction of the second condition above, $a_{11}a_{22}$ should be positive which will be the case if a_{11} and a_{22} have same sign. Hence, due to first condition above, a_{11} and a_{22} should be negative. Thus, the condition for $\hat{A}$ to have real negative eigenvalues is

$$a_{11} < 0, \qquad a_{22} < 0, \qquad a_{11}a_{22} - a_{12}^2 > 0. \tag{2.140}$$

With a_{ij} given by (2.136), it follows that $\delta^2 S_T < 0$ if

$$\left(\frac{\partial^2 S}{\partial U^2} \right)_V < 0, \qquad \left(\frac{\partial^2 S}{\partial V^2} \right)_U < 0,$$
$$\left(\frac{\partial^2 S}{\partial U^2} \right)_V \left(\frac{\partial^2 S}{\partial V^2} \right)_U - \left(\frac{\partial^2 S}{\partial U \partial V} \right)^2 > 0. \tag{2.141}$$

These are the conditions for the equilibrium state to be stable. Let us examine the consequences of the conditions above on the observables of the system.

1. We examine first the consequences of the first of the stability conditions in (2.141). To that end, we have

$$\left(\frac{\partial^2 S}{\partial U^2}\right)_V = \left(\frac{\partial 1/T}{\partial U}\right)_V = -\frac{1}{T^2}\left(\frac{\partial T}{\partial U}\right)_V = -\frac{1}{T^2 C_V}. \quad (2.142)$$

Hence, in order to satisfy the first condition in (2.141) we should have

$$C_V > 0. \quad (2.143)$$

2. Next, we examine the third condition in (2.141). We have

$$\frac{\partial^2 S}{\partial U^2}\frac{\partial^2 S}{\partial V^2} - \left(\frac{\partial^2 S}{\partial U \partial V}\right)^2 = \frac{\partial(\partial S/\partial U, \partial S/\partial V)}{\partial(U, V)} = \frac{\partial(1/T, P/T)}{\partial(U, V)}. \quad (2.144)$$

Now

$$\frac{\partial(1/T, P/T)}{\partial(U, V)} = \frac{\partial(1/T, P/T)}{\partial(T, V)}\frac{\partial(T, V)}{\partial(U, V)} = -\frac{1}{T^2}\left(\frac{\partial P/T}{\partial V}\right)_T\left(\frac{\partial T}{\partial U}\right)_V$$

$$= \frac{1}{V T^3 C_V \kappa_T}. \quad (2.145)$$

On substituting (2.145) in (2.144) and due to (2.143), we see that the last condition in (2.141) shall be satisfied if

$$\kappa_T \equiv -\frac{1}{V}\left(\frac{\partial V}{\partial P}\right)_T > 0. \quad (2.146)$$

This shows that

$$\left(\frac{\partial V}{\partial P}\right)_T < 0, \quad (2.147)$$

i.e. volume always decreases with increase in pressure.

3. We have examined the consequences of the first and the third stability conditions in (2.141). We examine now the remaining, namely, the second condition. We will see that it does not lead to any new restriction on the properties of the observables. We have

$$\left(\frac{\partial^2 S}{\partial V^2}\right)_U = \left(\frac{\partial P/T}{\partial V}\right)_U = \frac{\partial(P/T, U)}{\partial(V, U)}. \quad (2.148)$$

Rewrite the Jacobian above as

$$\frac{\partial(P/T, U)}{\partial(V, U)} = \frac{\partial(P/T, U)}{\partial(V, T)} \frac{\partial(V, T)}{\partial(V, U)}$$

$$= \begin{pmatrix} \left(\dfrac{\partial P/T}{\partial V}\right)_T & \left(\dfrac{\partial P/T}{\partial T}\right)_V \\ \\ \left(\dfrac{\partial U}{\partial V}\right)_T & \left(\dfrac{\partial U}{\partial T}\right)_V \end{pmatrix} \left(\dfrac{\partial T}{\partial U}\right)_V. \qquad (2.149)$$

Invoking the entropic Maxwell relation (2.99) and with

$$\left(\frac{\partial P/T}{\partial V}\right)_T = \frac{1}{T}\left(\frac{\partial P}{\partial V}\right)_T = -\frac{1}{VT\kappa_T}, \qquad (2.150)$$

we obtain

$$\left(\frac{\partial^2 S}{\partial V^2}\right)_U = \frac{\partial(P/T, U)}{\partial(V, U)} = -\left[\frac{1}{VT\kappa_T} + \frac{1}{T^2 C_V}\left(\frac{\partial U}{\partial V}\right)_T^2\right]. \quad (2.151)$$

Since, as a consequence of other two stability conditions, $C_V > 0$, $\kappa_T > 0$, we see that

$$\left(\frac{\partial^2 S}{\partial V^2}\right)_U < 0. \qquad (2.152)$$

The second condition is thus satisfied once the other two are.

We can express the derivative on the right side of (2.151) in terms of the primary set. To that end, recall the first law to get

$$\left(\frac{\partial U}{\partial V}\right)_T = T\left(\frac{\partial S}{\partial V}\right)_T - P. \qquad (2.153)$$

Using the cyclic rule we have

$$\left(\frac{\partial S}{\partial V}\right)_T = -\frac{C_V}{VT\alpha_S}. \qquad (2.154)$$

Substitute this in (2.153) and the resulting equation in (2.151) to obtain

$$\left(\frac{\partial^2 S}{\partial V^2}\right)_U = -\left[\frac{1}{VT\kappa_T} + \frac{1}{T^2 C_V}\left(P + \frac{C_V}{V\alpha_S}\right)^2\right] < 0. \quad (2.155)$$

The adiabatic expansion coefficient α_S is given in terms of the primary set by (2.117).

4. We examined the third conditions in (2.141) without evaluating the mixed second derivative therein separately. We now evaluate the mixed derivative of S and verify that the result obtained by using it is same as the one derived before. To that end, we have

$$\frac{\partial^2 S}{\partial V \partial U} = \left(\frac{\partial 1/T}{\partial V}\right)_U = -\frac{1}{T^2}\left(\frac{\partial T}{\partial V}\right)_U = -\frac{1}{T^2}\frac{\partial(T,U)}{\partial(V,U)}$$
$$= -\frac{1}{T^2}\frac{\partial(T,U)}{\partial(T,V)}\frac{\partial(T,V)}{\partial(V,U)} = \frac{1}{T^2 C_V}\left(\frac{\partial U}{\partial V}\right)_T, \tag{2.156}$$

where $(\partial U/\partial V)_T$ is given by (2.153). On using the equation above, along with (2.142) and (2.151) to evaluate the left side in (2.144), it may be verified that the result so obtained agrees, as it must, with the directly derived result (2.145).

Summary

The extremum of entropy will be a stable maximum if

$$C_V > 0, \qquad \kappa_T > 0. \tag{2.157}$$

Some consequences of the conditions above are

1. It is readily seen that, when applied to the relation (2.113) for $C_P - C_V$, the conditions (2.157) lead to the inequality

$$C_P > C_V > 0. \tag{2.158}$$

This states that the heat capacity at constant pressure is always greater than that at constant volume. This is expected intuitively as entire supplied heat energy is used in raising temperature of the substance when kept at constant volume but part of it is used in the work that system performs as its volume increases when pressure is kept constant.

2. Since, due to (2.158), $C_P/C_V > 0$, the relation (2.118) implies $\kappa_T/\kappa_S > 0$. But, due to (2.157), $\kappa_T > 0$. Hence $\kappa_S > 0$. Consequently, (2.115) leads to the inequality

$$\kappa_T > \kappa_S > 0. \tag{2.159}$$

3. Due to (2.158) and (2.159), the relation (2.122) yields

$$\alpha_P \alpha_S < 0. \tag{2.160}$$

In Sect. 2.11, we show that maximum entropy principle is equivalent with the minimum energy principle.

Exercises

Ex. 2.20. Show that

$$\left(\frac{\partial P/T}{\partial T}\right)_V = -\frac{1}{T^2}\left(P - \frac{T\alpha_P}{\kappa_T}\right). \tag{2.161}$$

Ex. 2.21. Show that

$$\frac{\partial^2 S}{\partial U \partial V} = \frac{1}{C_V}\left(\frac{\partial P/T}{\partial T}\right)_V. \tag{2.162}$$

2.11 Stability from Minimum Energy Principle

We formulated the maximum entropy principle by assuming that total energy, volume, and the number of particles remain unchanged when two subsystems constituting an isolated system are allowed to exchange energy and the partition between them is movable. We consider now the thermodynamic process between two subsystems such that their total entropy and volume remain unchanged. We show that the equilibrium state of such a system corresponds to the state of minimum total energy U_T:

$$U_T = U_1 + U_2, \tag{2.163}$$

where U_i is the internal energy of the ith system ($i = 1, 2$), when

$$S_1 + S_2 = \text{constant}, \qquad V_1 + V_2 = \text{constant}. \tag{2.164}$$

From (2.163), along with the use of (2.164), it follows that

$$\delta U_T = \left(\frac{\partial U_1}{\partial S_1} - \frac{\partial U_2}{\partial S_2}\right)\delta S_1 + \left(\frac{\partial U_1}{\partial V_1} - \frac{\partial U_2}{\partial V_2}\right)\delta V_1$$
$$= (T_1 - T_2)\,\delta S_1 + (P_1 - P_2)\,\delta V_1. \tag{2.165}$$

From maximum entropy principle we know that, at equilibrium, $T_1 = T_2$, $P_1 = P_2$. Hence it follows that, at equilibrium, $\delta U_T = 0$, i.e. the energy is extremum. To establish that the extremum is a minimum, we examine the second variation $\delta^2 U_T$ given by

$$\delta^2 U_{\rm T} = \frac{1}{2} \sum_{i=1,2} \left[\frac{\partial^2 U_i}{\partial S_i^2} (\delta S_i)^2 + 2 \frac{\partial^2 U_i}{\partial S_i \partial V_i} (\delta S_i)(\delta V_i) + \frac{\partial^2 U_i}{\partial V_i^2} (\delta V_i)^2 \right]$$

$$= \frac{\partial^2 U_1}{\partial S_1^2} (\delta S_1)^2 + 2 \frac{\partial^2 U_1}{\partial S_1 \partial V_1} (\delta S_1)(\delta V_1) + \frac{\partial^2 U_1}{\partial V_1^2} (\delta V_1)^2, \qquad (2.166)$$

where in writing the second line we have invoked the equality of corresponding derivatives of two subsystems at equilibrium and (2.164).

Doing away with the subscript 1 on various quantities, the equation above can be rewritten in following useful form:

$$\delta^2 U_{\rm T} = \begin{pmatrix} \delta S & \delta V \end{pmatrix} \hat{B} \begin{pmatrix} \delta S \\ \delta V \end{pmatrix}, \qquad (2.167)$$

where $\hat{B}$ is a real symmetric matrix given by

$$\hat{B} = \begin{pmatrix} \dfrac{\partial^2 U}{\partial S^2} & \dfrac{\partial^2 U}{\partial S \partial V} \\[2mm] \dfrac{\partial^2 U}{\partial S \partial V} & \dfrac{\partial^2 U}{\partial V^2} \end{pmatrix} \equiv \begin{pmatrix} b_{11} & b_{12} \\ b_{12} & b_{22} \end{pmatrix}. \qquad (2.168)$$

The condition $\delta^2 U_{\rm T} > 0$ for $U_{\rm T}$ to be minimum then reads

$$\begin{pmatrix} (\delta S)^2 & (\delta V)^2 \end{pmatrix} \hat{B} \begin{pmatrix} (\delta S)^2 \\ (\delta V)^2 \end{pmatrix} > 0. \qquad (2.169)$$

The condition (2.169) will hold if the eigenvalues $\lambda_{1,2}$ of $\hat{B}$, given by (2.138) (with the a_{ij}'s therein replaced by the b_{ij}'s) are positive. To find the conditions on the b_{ij}'s under which $\lambda_{1,2} > 0$, recall from the discussion circa (2.138) that the roots are real. The roots will be positive if the sum as well as the product of the eigenvalues is positive which will be the case if the trace and the determinant of $\hat{B}$ are positive:

$$b_{11} + b_{22} > 0, \qquad b_{11} b_{22} - b_{12}^2 > 0. \qquad (2.170)$$

Clearly, if the second condition above is to be satisfied, $b_{11} b_{22}$ should be positive which will be the case if b_{11} and b_{22} have same sign which in turn, due to the first condition above, requires b_{11} and b_{22} to be positive. The condition for $\hat{B}$ to have real positive eigenvalues therefore is

$$b_{11} > 0, \qquad b_{22} > 0, \qquad b_{11} b_{22} - b_{12}^2 > 0. \qquad (2.171)$$

With b_{ij} given by (2.168), the equations above read

$$\left(\frac{\partial^2 U}{\partial S^2}\right)_V > 0, \qquad \left(\frac{\partial^2 U}{\partial V^2}\right)_S > 0,$$

$$\left(\frac{\partial^2 U}{\partial S^2}\right)\left(\frac{\partial^2 U}{\partial V^2}\right) - \left(\frac{\partial^2 U}{\partial S \partial V}\right)^2 > 0. \tag{2.172}$$

These are the conditions for $U(S, V)$ to be a minimum at the values of (S, T) at which $\delta U = 0$.

We leave it to the exercises to show that the inequalities in (2.172) are satisfied under the same conditions under which entropy is maximum. This shows that the principle of maximum entropy is equivalent with that of minimum energy.

Next we consider a system interacting with one or more reservoirs and show that the state of equilibrium of the system is determined by some extremum principle on an appropriately chosen Legendre transform of energy or entropy.

Exercises

Ex. 2.22. Show that

$$\left(\frac{\partial^2 U}{\partial S^2}\right)_V = \frac{T}{C_V}. \tag{2.173}$$

Since the positivity of C_V has been shown to hold due to entropy maximum principle, it follows that the first inequality in (2.172) is satisfied.

Ex. 2.23. Show that

$$\left(\frac{\partial^2 U}{\partial V^2}\right)_S = \frac{1}{V \kappa_S}. \tag{2.174}$$

Since the positivity of κ_S has been shown to hold due to entropy maximum principle, it follows that the second inequality in (2.172) is satisfied.

Ex. 2.24. Show that

$$\frac{\partial^2 U}{\partial S \partial V} = -\frac{T \alpha_P}{C_V \kappa_T}. \tag{2.175}$$

Ex. 2.25. Prove the equation below by expressing its left side as a Jacobian

$$\left(\frac{\partial^2 U}{\partial S^2}\right)_V \left(\frac{\partial^2 U}{\partial V^2}\right)_S - \left(\frac{\partial^2 U}{\partial S \partial V}\right)^2 = -\left(\frac{\partial S}{\partial T}\right)_V^{-1} \left(\frac{\partial P}{\partial V}\right)_T$$

$$= \frac{T}{V \kappa_T C_V}. \tag{2.176}$$

Since the positivity of C_V and κ_T has been shown to hold due to entropy maximum principle, it follows that the last inequality in (2.172) is satisfied.

Ex. 2.26. Establish the correctness of (2.176) by evaluating its left side using the expressions (2.173)–(2.175).

2.12 Stability in Terms of Thermodynamic Potentials

In the problems formulated in terms of thermodynamic potentials, it is desirable to know as to how those functions behave at thermodynamic equilibrium, i.e. whether they too possess an extremum at equilibrium. The answer to that question requires evaluation of the function's second partial derivatives with respect to its arguments. Since thermodynamic potentials are Legendre transforms of energy $U(S, V)$ with respect to one or the other variable or both, the second derivatives of the potential in question can be expressed in terms of the second derivatives of $U(S, V)$ using the results derived in Appendix B, and deduce therefrom the nature of the second derivatives of the potential invoking the conditions (2.172) on the second derivatives of $U(S, V)$ at equilibrium. In this section, we apply the said procedure to examine the nature of second-order variation of the Helmholtz and Gibbs potentials and that of enthalpy. We will see that the nature of variation of the potentials is different with respect to extensive and intensive variables.

2.12.1 Helmholtz Potential

The Helmholtz potential $F(T, V)$ is the Legendre transform of $U(S, V)$ with respect to S. Invoking the results derived in Appendix B.1 we have

$$\left(\frac{\partial^2 F(T, V)}{\partial T^2}\right)_V = -\left[\left(\frac{\partial^2 U(S, V)}{\partial S^2}\right)_V\right]^{-1} < 0, \tag{2.177}$$

$$\left(\frac{\partial^2 F(T, V)}{\partial V^2}\right)_T = \left(\frac{\partial^2 U}{\partial S^2}\right)^{-1}\left(\frac{\partial^2 U}{\partial S^2}\frac{\partial^2 U}{\partial V^2} - \left(\frac{\partial^2 U}{\partial S\partial V}\right)^2\right) > 0, \tag{2.178}$$

$$\frac{\partial^2 F(T, V)}{\partial T^2}\frac{\partial^2 F(T, V)}{\partial V^2} - \left(\frac{\partial^2 F(T, V)}{\partial T\partial V}\right)^2$$
$$= -\left(\frac{\partial^2 U}{\partial S^2}\right)^{-1}\left(\frac{\partial^2 U}{\partial V^2}\right) < 0. \tag{2.179}$$

The inequalities in the equations above are due to the stability conditions (2.172) on $U(S, V)$. The mixed second-order derivative of $F(T, V)$ is

$$\frac{\partial^2 F(T, V)}{\partial T \partial V} = \left(\frac{\partial^2 U}{\partial S^2}\right)^{-1} \left(\frac{\partial^2 U}{\partial S \partial V}\right). \tag{2.180}$$

This does not have a definite sign (see (Ex. 2.27)).

We see that $F(T, V)$ is a convex function with respect to its extensive variable V but is concave with respect to its intensive variable T. See Appendix C for the concept of concave and convex functions.

Exercises

Ex. 2.27. Using (2.60) show that

$$\left(\frac{\partial^2 F}{\partial V^2}\right)_T = \frac{1}{V \kappa_T}, \qquad \left(\frac{\partial^2 F}{\partial T^2}\right)_V = -\frac{C_V}{T}$$

$$\frac{\partial^2 F(T, V)}{\partial T \partial V} = -\frac{\alpha_P}{\kappa_T}. \tag{2.181}$$

Show that the sign of the mixed second derivative above is indefinite.

Ex. 2.28. Prove the equation below by expressing its left side as a Jacobian

$$\frac{\partial^2 F(T, V)}{\partial T^2} \frac{\partial^2 F(T, V)}{\partial V^2} - \left(\frac{\partial^2 F(T, V)}{\partial T \partial V}\right)^2 = -\frac{C_P}{T V \kappa_T}. \tag{2.182}$$

Show that it would hold, as it must, if its left side is evaluated using instead the expressions derived in (2.181).

Ex. 2.29. Verify that the expressions in (2.181) and (2.182) satisfy the corresponding equations (2.177)–(2.180) in terms of the derivatives of $U(S, V)$.

2.12.2 Enthalpy

Enthalpy $H(S, P)$ is the Legendre transform of $U(S, V)$ with respect to V. Invoking the results derived in Appendix B.1 we have

$$\left(\frac{\partial^2 H(S, P)}{\partial P^2}\right)_S = -\left[\left(\frac{\partial^2 U(S, V)}{\partial V^2}\right)_S\right]^{-1} < 0, \tag{2.183}$$

$$\left(\frac{\partial^2 H(S, P)}{\partial S^2}\right)_P = \left(\frac{\partial^2 U}{\partial V^2}\right)^{-1} \left(\frac{\partial^2 U}{\partial V^2} \frac{\partial^2 U}{\partial S^2} - \left(\frac{\partial^2 U}{\partial S \partial V}\right)^2\right) > 0, \tag{2.184}$$

$$\frac{\partial^2 H(S, P)}{\partial P^2} \frac{\partial^2 H(S, P)}{\partial S^2} - \left(\frac{\partial^2 H(S, P)}{\partial P \partial S}\right)^2$$

$$= -\left(\frac{\partial^2 U}{\partial V^2}\right)^{-1} \left(\frac{\partial^2 U}{\partial S^2}\right) < 0. \tag{2.185}$$

The inequalities in the equations above are due to the stability conditions on U derived in Sect. 2.11. The mixed second-order derivative of $H(S, P)$ is

$$\frac{\partial^2 H(S, P)}{\partial P \partial S} = \left(\frac{\partial^2 U}{\partial V^2}\right)^{-1} \left(\frac{\partial^2 U}{\partial V \partial S}\right). \tag{2.186}$$

This does not have a definite sign (see (Ex. 2.30)).

We see that $H(S, P)$ is a convex function with respect to its extensive variable S but is concave with respect to its intensive variable P.

Exercises

Ex. 2.30. Using (2.73) show that

$$\left(\frac{\partial^2 H}{\partial P^2}\right)_S = -V \kappa_S, \qquad \left(\frac{\partial^2 H}{\partial S^2}\right)_P = \frac{T}{C_P}$$

$$\frac{\partial^2 H}{\partial S \partial P} = \frac{V T \alpha_P}{C_P}. \tag{2.187}$$

Show that the mixed derivative above does not have a definite sign.

Ex. 2.31. Prove the equation below by expressing its left side as a Jacobian

$$\frac{\partial^2 H(S, P)}{\partial S^2} \frac{\partial^2 H(S, P)}{\partial P^2} - \left(\frac{\partial^2 H(S, P)}{\partial S \partial P}\right)^2 = -\frac{\kappa_T V T}{C_P}. \tag{2.188}$$

Show that it would hold, as it must, if its left side is evaluated using instead the expressions derived in (2.187).

Ex. 2.32. Verify that the expressions in (2.187) and (2.188) satisfy the corresponding equations (2.183)–(2.186) in terms of the derivatives of $U(S, V)$.

2.12.3 Gibbs Potential

Gibbs potential $G(T, P)$ is the Legendre transform of $U(S, V)$ with respect to both of its variables, S and V. Invoking the results derived in Appendix B.2 we have

$$\left(\frac{\partial^2 G(T, P)}{\partial T^2}\right)_P = -\left(\frac{\partial^2 U}{\partial V^2}\right)\left[\frac{\partial^2 U}{\partial S^2}\frac{\partial^2 U}{\partial V^2} - \left(\frac{\partial^2 U}{\partial S \partial V}\right)^2\right]^{-1} < 0, \quad (2.189)$$

$$\left(\frac{\partial^2 G(T, P)}{\partial P^2}\right)_T = -\left(\frac{\partial^2 U}{\partial S^2}\right)\left[\frac{\partial^2 U}{\partial S^2}\frac{\partial^2 U}{\partial V^2} - \left(\frac{\partial^2 U}{\partial S \partial V}\right)^2\right]^{-1} < 0, \quad (2.190)$$

$$\left(\frac{\partial^2 G(T, P)}{\partial T^2}\right)_P \left(\frac{\partial^2 G(T, P)}{\partial P^2}\right)_T - \left(\frac{\partial^2 G(T, P)}{\partial T \partial P}\right)^2$$

$$= \left[\left(\frac{\partial^2 U}{\partial S^2}\right)_V \left(\frac{\partial^2 U}{\partial V^2}\right)_S - \left(\frac{\partial^2 U}{\partial S \partial V}\right)^2\right]^{-1} > 0, \quad (2.191)$$

$$\frac{\partial^2 G(T, P)}{\partial T \partial P} = \left(\frac{\partial^2 U}{\partial S \partial V}\right)\left[\frac{\partial^2 U}{\partial S^2}\frac{\partial^2 U}{\partial V^2} - \left(\frac{\partial^2 U}{\partial S \partial V}\right)^2\right]^{-1} < 0. \quad (2.192)$$

The inequalities in the equations above are due to the stability conditions on U derived in Sect. 2.11. The function $G(T, P)$ is convex with respect to both its variables which are intensive. Also, unlike the mixed second derivative of $F(T, V)$ and that of $H(S, P)$, the mixed second derivative of $G(T, P)$ has definite sign.

Exercises

Ex. 2.33. Using (2.56) show that

$$\left(\frac{\partial^2 G}{\partial P^2}\right)_T = -V\kappa_T, \qquad \left(\frac{\partial^2 G}{\partial T^2}\right)_P = -\frac{C_P}{T}$$

$$\frac{\partial^2 G}{\partial T \partial P} = V\alpha_P. \qquad (2.193)$$

Show that the mixed derivative above does not have a definite sign.

Ex. 2.34. Prove the equation below by expressing its left side as a Jacobian

$$\frac{\partial^2 G}{\partial T^2}\frac{\partial^2 G}{\partial P^2} - \left(\frac{\partial^2 G}{\partial T \partial P}\right)^2 = \frac{V C_V \kappa_T}{T}. \qquad (2.194)$$

Show that it would hold, as it must, if its left side is evaluated using instead the expressions derived in (2.193).

Ex. 2.35. Verify that the expressions in (2.193) and (2.194) satisfy the corresponding equations (2.189)–(2.192) in terms of the derivatives of $U(S, V)$.

2.13 Thermodynamic Potentials: Alternative Formulation

We introduced the thermodynamic potentials as the Legendre transforms of the energy function $U(S, V, N)$. One may ask: to what physical situations they correspond to. In this section, we address that question for a system of fixed number of particles and show that a thermodynamic potential describes a system in equilibrium with a reservoir of heat or pressure or both. We find also the maximum or available work that can be extracted from such a system and introduce the concept of exergy.

2.13.1 System Interacting with Heat Reservoir

Consider a system A interacting with the heat reservoir R maintained at a constant temperature T_0. The two systems exchange only heat while their individual volumes remain unchanged. Let δQ be the amount of heat drawn by A from R and let δW_e be the external work performed by A. By external work we mean the work other than that associated with the change of volume of the system. Since its volume remains constant, the change in the internal energy U of A is given by

$$\delta U = \delta Q - \delta W_e. \tag{2.195}$$

The amount of heat δQ drawn by A is the amount lost by R. Since R is a reservoir at temperature T_0, the heat lost by it is $\delta Q = -T_0 \delta S_R$ where δS_R is change in its entropy. Equation (2.195) then assumes the form

$$\delta U = -T_0 \delta S_R - \delta W_e. \tag{2.196}$$

Let S denote the entropy of A and $S_T = S + S_R$ total entropy of A and R. The equation above can then be rewritten as

$$- T_0 \delta S_T = \delta(F_0 + W_e), \tag{2.197}$$

where

$$F_0 = U - T_0 S. \tag{2.198}$$

The function F_0 is analogous to but not same as the Helmholtz potential because T in F_0 has fixed value T_0, the temperature of the reservoir.

Since the system and the reservoir together are isolated and there is no entropy change associated with the execution of the outside work, $\delta S_T \geq 0$. Consequently (2.197) leads to the inequality

$$\delta W_e \leq -\delta F_0. \tag{2.199}$$

When no external work is performed, the (2.197) reduces to

$$- T_0 \delta S_T = \delta F_0. \tag{2.200}$$

The state of stable thermodynamic equilibrium is reached when $\delta S_T = 0$, $\delta^2 S_T < 0$, i.e. when

$$\delta F_0 = 0, \qquad \delta^2 F_0 > 0. \tag{2.201}$$

Let us examine the conditions above.

Since it has been assumed that the volume does not change, considering it a function of S, V, the definition (2.198) of F_0 yields

$$\delta F_0 = \left[\left(\frac{\partial U}{\partial S} \right)_V - T_0 \right] \delta S. \tag{2.202}$$

This shows that the first condition in (2.201) will be satisfied if equilibrium temperature of the system $T = (\partial U / \partial S)_V$ is same as that of the reservoir.

Next, (2.198) yields

$$\delta^2 F_0 = \frac{1}{2} \left(\frac{\partial^2 U}{\partial S^2} \right)_V (\delta S)^2 = \frac{1}{2} \left(\frac{\partial T}{\partial S} \right)_V (\delta S)^2 = \frac{T_0}{2C_V} (\delta S)^2. \tag{2.203}$$

Since $C_V > 0$, we see that $\delta^2 F_0 > 0$, i.e. the extremum of F_0 is a minimum if the volume of the system is unchanged. See also Ex. 2.36.

Exercises

Ex. 2.36. Considering F_0 a function of (V, T) show that when $\delta V = 0$, the temperature of the system at equilibrium is $T = T_0$ and $\delta^2 F_0 > 0$. Hint: Since $\delta V = 0$,

$$\delta F_0 = \left\{ \left(\frac{\partial U}{\partial T} \right)_V - T_0 \left(\frac{\partial S}{\partial T} \right)_V \right\} \delta T = C_V \left(1 - \frac{T_0}{T} \right) \delta T. \tag{2.204}$$

Hence the extremization condition requires $\delta F_0 = 0$. To prove the second part, show that at $T = T_0$,

$$\delta^2 F_0 = \frac{C_V}{2T_0} (\delta T)^2. \tag{2.205}$$

Verify that the equation above is equivalent with (2.203).

2.13.2 System Interacting with Heat and Pressure Reservoirs

Consider a system A interacting with the reservoir R maintained at a constant temperature T_0 and constant pressure P_0. The two systems exchange heat and also change volume keeping total volume fixed. Let δQ be the amount of heat drawn by A from R. Let δV be the change in the volume of A. Since it occurs at the constant pressure P_0 of the reservoir and if δW_e is the external work performed by A, the change in the internal energy U of A is given by

$$\delta U = \delta Q - P_0 \delta V - \delta W_e. \tag{2.206}$$

With $\delta Q = -T_0 \delta S_R = -T_0 \delta (S_T - S)$, the equation above yields

$$- T_0 \delta S_T = \delta (G_0 + W_e), \tag{2.207}$$

where

$$G_0 = U - T_0 S + P_0 V. \tag{2.208}$$

The function G_0 is analogous to but not the same as the Gibbs potential because T and P in G_0 have fixed values T_0 and P_0, the temperature and pressure of the reservoir.

Since system and reservoir together are isolated and there is no entropy change associated with the execution of the outside work, $\delta S_T \geq 0$. Consequently (2.207) leads to the inequality

$$\delta W_e \leq -\delta G_0. \tag{2.209}$$

When no external work is performed, (2.207) reads

$$- T_0 \delta S_T = \delta G_0. \tag{2.210}$$

The state of stable equilibrium corresponds to $\delta S_T = 0$, $\delta^2 S_T < 0$ leading to, because of (2.210), the following equilibrium conditions on G_0:

$$\delta G_0 = 0, \qquad \delta^2 G_0 > 0. \tag{2.211}$$

Next we examine the conditions (2.211).

With G_0 given by (2.208), considered a function of (V, T), we see by invoking the first law that

$$\left(\frac{\partial G_0}{\partial T} \right)_V = (T - T_0) \left(\frac{\partial S}{\partial T} \right)_V, \tag{2.212}$$

and

$$\left(\frac{\partial G_0}{\partial V}\right)_T = \left(\frac{\partial U}{\partial V}\right)_T - T_0 \left(\frac{\partial S}{\partial V}\right)_T + P_0$$

$$= (T - T_0) \left(\frac{\partial S}{\partial V}\right)_T + (P_0 - P). \tag{2.213}$$

The equations above show that the first condition of equilibrium in (2.211),

$$\delta G_0 = \left(\frac{\partial G_0}{\partial T}\right)_V \delta T + \left(\frac{\partial G_0}{\partial V}\right)_T \delta V = 0, \tag{2.214}$$

will be satisfied when the system temperature T and pressure P are same, respectively, as the reservoir temperature T_0 and pressure P_0.

To examine the stability condition in (2.211), we have

$$\delta^2 G_0 = \frac{1}{2}\left[\left(\frac{\partial^2 G_0}{\partial T^2}\right)_V (\delta T)^2 + 2\frac{\partial^2 G_0}{\partial T \partial V}(\delta T)(\delta V) + \left(\frac{\partial^2 G_0}{\partial V^2}\right)_T (\delta V)^2\right], \tag{2.215}$$

where it is understood that the derivatives are to be evaluated at equilibrium. Using (2.212) and (2.213) we obtain

$$\left(\frac{\partial^2 G_0}{\partial T^2}\right)_V = \left(\frac{\partial S}{\partial T}\right)_V + (T - T_0)\left(\frac{\partial^2 S}{\partial T^2}\right)_V,$$

$$\left(\frac{\partial^2 G_0}{\partial V^2}\right)_T = -\left(\frac{\partial P}{\partial V}\right)_T + (T - T_0)\left(\frac{\partial^2 S}{\partial V^2}\right)_T,$$

$$\frac{\partial^2 G_0}{\partial T \partial V} = (T - T_0)\frac{\partial^2 S}{\partial T \partial V}. \tag{2.216}$$

Whereas the last equation follows effortlessly from (2.212), the use of second Maxwell relation in (2.98) is needed to get the same from (2.213). Substitute in (2.215) the equations above evaluated at the equilibrium values $T = T_0$, $P = P_0$ of temperature and pressure to get

$$\delta^2 G_0 = \frac{1}{2}\left[\frac{C_V}{T}(\delta T)^2 - \left(\frac{\partial P}{\partial V}\right)_T (\delta V)^2\right]. \tag{2.217}$$

We know that $(\partial P/\partial V)_T < 0$. Hence $\delta^2 G_0 > 0$, i.e. G_0 is minimum in the equilibrium state characterized by temperature and pressure equal to those of the reservoir.

In the study of the critical phenomena, we will see that the critical point corresponds to $(\partial P/\partial V)_T = 0$. The expression (2.217) shows that in that case $\delta^2 G_0 = 0$ at the critical point along the isotherm passing through it. The stability in that case is determined by higher order terms in the expansion of G_0 in powers of δT and δV.

In the study of the critical phenomena, we will need terms up to fourth order when $\delta T = 0$. The next two higher order terms when $\delta T = 0$ are

$$\delta^3 G_0 + \delta^4 G_0 = \left[\frac{1}{3!} \left(\frac{\partial^3 G_0}{\partial V^3} \right)_T (\delta V)^3 + \frac{1}{4!} \left(\frac{\partial^4 G_0}{\partial V^4} \right)_T (\delta V)^4 \right]. \quad (2.218)$$

The right side of the equation can be evaluated using (2.216) and the resulting expression substituted in (2.218) when $(T, P) = (T_0, P_0)$ to get

$$\delta^3 G_0 + \delta^4 G_0 = -\frac{(\delta V)^3}{3!} \left(\frac{\partial^2 P}{\partial V^2} \right)_T - \frac{(\delta V)^4}{4!} \left(\frac{\partial^3 P}{\partial V^3} \right)_T. \quad (2.219)$$

We thus have at hand dG_0 up to fourth order in δV along an isotherm passing through the equilibrium point.

2.13.3 System Interacting with Pressure Reservoir

Consider a system interacting with a pressure reservoir maintained at pressure P_0 producing work δW_e. There is no exchange of heat between the system and the reservoir. The change in its internal energy is then given by

$$\delta U = -P_0 \delta V - \delta W_e. \quad (2.220)$$

This implies

$$\delta W_e = -\delta H_0, \quad (2.221)$$

where

$$H_0 = U + P_0 V. \quad (2.222)$$

Since there is no exchange of heat, $\delta U = -P \delta V$ so that

$$\delta H_0 = (P_0 - P) \delta V. \quad (2.223)$$

This shows that $\delta H_0 = 0$, i.e. H_0 has the extremum when pressure of the system is same as that of the reservoir. To examine the nature of the extremum, we evaluate $\delta^2 H_0$:

$$\delta^2 H_0 = -\frac{1}{2} \left(\frac{\partial P}{\partial V} \right)_S (\delta V)^2. \quad (2.224)$$

We know that $(\partial P/\partial V)_S < 0$. Hence $\delta^2 H_0 > 0$. This shows that the extremum of H_0 is a minimum when the system undergoes isentropic transformation at constant pressure.

The function H_0 is analogous to but not the same as the enthalpy because P in H_0 has fixed value P_0, the pressure of the reservoir.

2.13.4 Exergy

We have found expressions for maximum work available from a system interacting with reservoirs of various kinds. The system cannot perform any work when in equilibrium with the reservoirs it is interacting with. The equilibrium state of the system is therefore called the *dead state*.

A quantity of interest in several applications is the amount of available work from a system in a given state as it approaches the dead state. It is called *exergy*. It is the combination of the Greek words "ex" (meaning from) and "ergon" (meaning work). It was coined by Zoran Rant in 1956. See [5] for introduction to the concept of exergy and some examples.

We list below expression for exergy when the system interacts with only a heat reservoir and when it interacts with the heat and pressure reservoirs.

1. Equation (2.199) shows that the maximum available work when the system is in contact with a heat reservoir at temperature T_0 is

$$\delta W_{emax} = -\delta U + T_0 \delta S. \qquad (2.225)$$

   Let U, S denote, respectively, the internal energy and entropy when the system is brought in contact with the reservoir and U_0, S_0 the corresponding values when it attains equilibrium. The maximum possible work output while it attains equilibrium, obtained by integrating (2.225), is

$$X_{EF} = (U - U_0) - (S - S_0)T_0. \qquad (2.226)$$

   The X_{EF} is therefore the exergy for a system undergoing thermodynamic transformation while in contact with a reservoir at temperature T_0.
2. Equation (2.209) shows that the maximum available work when the system is in contact with a heat reservoir at temperature T_0 and a pressure reservoir at pressure P_0 is given by

$$\delta W_{emax} = -\delta U + T_0 \delta S - P_0 \delta V. \qquad (2.227)$$

   Let U, S, V denote, respectively, the internal energy, entropy, and temperature when the system is brought in contact with the reservoir and U_0, S_0, V_0 their values at equilibrium. Integration of (2.225) gives

$$X_{EG} = (U - U_0) - (S - S_0)T_0 + (V - V_0)P_0. \tag{2.228}$$

The X_{EG} is therefore the exergy of the system while it undergoes thermodynamic transformation remaining in contact with the heat reservoir at temperature T_0 and the pressure reservoir at pressure P_0.

Exercises

Ex. 2.37. Show that the exergy of the ideal gas interacting with a heat reservoir at temperature T_0 is given by

$$X_{EF} = Nk_B c \left[(T - T_0) - T_0 \ln(T/T_0) \right], \tag{2.229}$$

where c is the heat capacity per molecule at constant volume.

Ex. 2.38. Show that the exergy of the ideal gas interacting with heat and pressure reservoirs at temperature T_0 and pressure P_0 is given by

$$X_{EG} = Nk_B \left[c(T - T_0) + \left(\frac{P_0}{P}T - T_0 \right) \right.$$
$$\left. + (c + 1)T_0 \ln \left(\frac{T}{T_0} \right) - T_0 \ln \left(\frac{P}{P_0} \right) \right]. \tag{2.230}$$

2.14 Second Equation of State

We know that, for one component system, we require three equations of state of which two are independent. The familiar equation of state is $P = P(V, T)$. We call it *first equation of state*. An additional equation of state is therefore required for complete thermodynamic description. The additional equation of state is often the energy $U = U(V, T)$ as a function of (V, T). We call it the *second equation of state*. Though they are independent, the Maxwell relations lead to consistency condition between them. In this section, we derive the said consistency condition and also the expressions for entropy and chemical potential as functions of (V, T) assuming fixed N. The first equation of state $P = P(V, T)$ is assumed to be known.

1. Recall the entropic Maxwell relation (2.99):

$$\left(\frac{\partial U}{\partial V} \right)_T = T \left(\frac{\partial P}{\partial T} \right)_V - P. \tag{2.231}$$

Formal integration of the equation above gives

$$U(V, T) = \int \left\{ T \left(\frac{\partial P}{\partial T} \right)_V - P \right\} dV + \Phi(T),$$ (2.232)

where the unknown function $\Phi(T)$ cannot be determined by first equation of state. It must be constructed independently. It is not difficult to see that it is related with the heat capacity at constant volume C_V:

$$C_V(V, T) \equiv \left(\frac{\partial U}{\partial T} \right)_V = \int T \left(\frac{\partial^2 P}{\partial T^2} \right)_V dV + \Phi'(T), \quad \Phi'(T) = \frac{d\Phi(T)}{dT}.$$ (2.233)

The function $\Phi(T)$ can thus be constructed from the data of C_V as a function of (V, T). It is also readily seen that

$$\left(\frac{\partial C_V(V, T)}{\partial V} \right)_T = T \left(\frac{\partial^2 P}{\partial T^2} \right)_V.$$ (2.234)

Equation (2.232) is the second equation of state which expresses internal energy U as a function of (V, T). The first term in it is determined by the equation of state $P = P(V, T)$ whereas the second term, namely, $\Phi(T)$, is independent of the first. As shown in (2.233), $\Phi(T)$ may be obtained from the knowledge of the heat capacity $C_V(V, T)$.

2. Knowing the two equations of state, we may compute entropy as a function of (V, T) as follows. The formal integration of second Maxwell relation in (2.98) yields

$$S(V, T) = \int \left(\frac{\partial P}{\partial T} \right)_V dV + F(T),$$ (2.235)

where $F(T)$ is an unknown function. It may be expressed in terms of $\Phi(T)$, the function appearing in the equation of state (2.232) for $U(V, T)$, by using $T(\partial S/\partial T)_V = C_V = (\partial U/\partial T)_V$ so that

$$F'(T) = \frac{\Phi'(T)}{T}.$$ (2.236)

This is the desired relation expressing $F(T)$ in terms of $\Phi(T)$.

3. To evaluate the chemical potential, substitute (2.232) and (2.235) for U and S in Euler's relation (2.46) to get

$$\mu(V, T) = \frac{1}{N} \left\{ \int V \left(\frac{\partial P}{\partial V} \right)_T dV + \Phi(T) - TF(T) \right\}.$$ (2.237)

The same result may be arrived at by integrating the energetic or entropic Gibbs–Duhem relation (see Exs. 2.40 and 2.41).

Next, we address some questions of importance in applications: (a) determining amount of heat absorbed in an isothermal process, (b) determining adiabatic equation of state in a reversible transformation, and (c) determining change in temperature of adiabatically freely expanding gas.

(a) The amount of heat absorbed by the system while going from state $A \equiv (V_A, T_0)$ to the state $B \equiv (V_B, T_0)$ at constant temperature T_0 along a reversible path is

$$Q_{AB} = T_0 \int_A^B dS = T_0 \left[I(V_B, T_0) - I(V_A, T_0) \right], \qquad (2.238)$$

where use has been made of (2.235) with

$$I(V, T) = \int \left(\frac{\partial P}{\partial T} \right)_V dV. \qquad (2.239)$$

Some examples are discussed in Sect. 2.14.1 and in the exercises.

(b) A reversible adiabatic process is one in which entropy does not change. Hence the relation between V and T in such a process is obtained by equating $S(V, T)$ in (2.235) to a constant to get

$$\int \left(\frac{\partial P}{\partial T} \right)_V dV + F(T) = S_0. \qquad (2.240)$$

Some examples are discussed in Sect. 2.14.1 and in the exercises.

(c) Consider a gas expanding freely adiabatically, confined initially to volume V_A at temperature T_A. Its temperature T_B when it is in equilibrium at volume V_B may be found by recalling from (1.69) that its internal energy remains unchanged during such a process. Using (2.232), this implies

$$\Delta \left[\int \left\{ T \left(\frac{\partial P}{\partial T} \right)_V - P \right\} dV \right] + \Delta \Phi(T) = 0. \qquad (2.241)$$

As an illustration, we apply the results derived above to systems described by the equation of state linear in T.

2.14.1   P Linear in T

Let P be of the form

$$P = f(V)T + g(V).\tag{2.242}$$

1. Expressions (2.232), (2.233), respectively, for U, C_V then read

$$U(V, T) = -\int g(V)dV + \Phi(T),\tag{2.243}$$

$$C_V = \frac{d\Phi(T)}{dT},\tag{2.244}$$

   which shows that C_V is independent of V.
2. Invoking (2.235), entropy may be seen to be given by

$$S(V, T) = \int f(V)dV + F(T).\tag{2.245}$$

3. Use of (2.237) leads to the following expression for μ:

$$\mu(V, T) = \frac{1}{N}\left\{\int V\left\{Tf'(V) + g'(V)\right\}dV + \Phi(T) - TF(T)\right\}.\tag{2.246}$$

4. The amount of heat absorbed when system transforms isothermally at temperature T_0 from the state of volume V_A to that of volume V_B, given by (2.238), in the present case is

$$Q_{AB} = T_0\int_{V_A}^{V_B} f(V)dV.\tag{2.247}$$

5. In case the transformation is adiabatic in which the initial state of the system is $A \equiv (V_A, T_A)$ and the final state is $B \equiv (V_B, T_B)$ then the governing equation (2.240) reduces to

$$\int_{V_A}^{V_B} f(V)dV + \int_{T_A}^{T_B} \frac{\Phi'(T)}{T}dT = S_0.\tag{2.248}$$

6. If volume and temperature of the gas initially are (V_A, T_A) and it undergoes free adiabatic expansion to acquire the values (V_B, T_B) then the condition (1.69), with U in the present case given by (2.243), leads to

$$\int_{V_A}^{V_B} g(V) dV = \Phi(T_B) - \Phi(T_A). \tag{2.249}$$

We apply the results derived above to an ideal gas and the van der Waals gas which are special cases of (2.242).

Ideal Gas

Consider first the ideal gas. It is a special case of (2.242) with

$$f(V) = Nk_B/V, \qquad g(V) = 0. \tag{2.250}$$

Evaluated in the following are various thermodynamic quantities.

1. First we need to specify $\Phi(T)$. As discussed in Sect. 1.4.1, the experimental data shows independence of C_V on V as well as on T. We see from (2.244) that the V-independence of C_V emerges even as an outcome of the consistency condition between the equation of state $P = P(V, T)$, and the equation determining internal energy. However, we have to take recourse to experiments or statistical mechanics to establish the T-independence of C_V which in turn means T-independence of $d\Phi(T)/dT$. Either way, it is found that C_V for an ideal gas is given by (1.77). This corresponds to

$$\Phi(T) = Nk_B cT, \qquad c = 1/(\gamma - 1). \tag{2.251}$$

Using (2.236), the function $F(T)$ turns out to be given by

$$F(T) = Nk_B c\ln(T) + K, \tag{2.252}$$

where K is a constant.
2. Using (2.251), the expression (2.243) for internal energy reads

$$U(V, T) = Nk_B cT. \tag{2.253}$$

3. With $F(T)$ as in (2.236), the expression (2.245) for entropy reduces to

$$S(V, T) = Nk_B \left(\ln(V) + c\ln(T) + K \right). \tag{2.254}$$

4. Equation (2.246) for chemical potential yields

$$\mu(V, T) = -k_B T \left(\ln(V) + c\ln(T) + C \right), \tag{2.255}$$

where C is a constant. By writing T in the logarithmic term in the equation above in terms of U, it is not difficult to see that this is same as (2.51) derived by solving the Gibbs–Duhem equation.

5. The formula (2.247) leads to the following expressions for heat absorbed in a reversible isothermal process at temperature T_0:

$$Q_{AB} = Nk_B T_0 \ln(V_B/V_A). \tag{2.256}$$

6. Due to (2.240), the adiabatic equation of state turns out to be given by

$$T V^{\gamma-1} = \text{constant},$$

between V and T in an adiabatic process.

7. Using (2.249) we see that the temperature of an ideal gas does not change on free adiabatic expansion which is in agreement with observations.

van der Waals Gas

The van der Waals gas is a special case of (2.242) with

$$f(v) = \frac{k_B}{v - b}, \quad g(v) = -\frac{a}{v^2}. \tag{2.257}$$

1. We can determine $\Phi(T)$ as follows. Verify that (2.243) in the present case leads to the following expression for internal energy per molecule:

$$u(V, T) = -\frac{a}{v} + \Phi(T)/N. \tag{2.258}$$

On comparing (2.257) and (2.250) we see that the van der Waals gas behaves as the ideal gas in the limit $v \to \infty$. We may therefore expect the expression (2.258) for U for the van der Waals gas to reduce to (2.253) for the ideal gas in the said limit. Since $\Phi(T)$ is independent of V, that would be the case if $\Phi(T)$ in the present case is same as that in (2.251) for the ideal gas. Expression (2.258) then reads

$$u(v, T) = -\frac{a}{v} + ck_B T. \tag{2.259}$$

2. Expression (2.245) for entropy per molecule reads

$$s(v, T) = k_B \{\ln(v - b) + c\ln(T) + K\}. \tag{2.260}$$

3. Equation (2.246) for chemical potential assumes the form

$$\mu(v, T) = -k_B T \left\{ \ln(v - b) - \frac{b}{v - b} + c\ln(T) + C \right\} - \frac{2a}{v}. \tag{2.261}$$

4. Using (2.247), the heat absorbed in a reversible isothermal transformation at temperature T_0 may be seen to be given by

$$Q_{AB} = Nk_B[\ln\{(v_B - b)/(v_A - b)\}]. \tag{2.262}$$

5. The adiabatic equation of state (2.240) reads

$$T(v - b)^{\gamma-1} = \text{constant}. \tag{2.263}$$

6. Using (2.249) we see that the initial and final volumes and temperature in adiabatic free expansion are related by

$$T_B - T_A = \frac{a}{ck_B}\left(\frac{1}{v_B} - \frac{1}{v_A}\right). \tag{2.264}$$

Exercises

Ex. 2.39. If $F(T)$ and $\Phi(T)$ are related by (2.236) then show that

$$\int F(T)dT = FT - \Phi(T), \quad \int \Phi(T)d(1/T) = \frac{1}{T}(\Phi(T) - FT). \tag{2.265}$$

Ex. 2.40. Derive (2.237) by integration of the energetic Gibbs–Duhem relation. Hint: Consider P a function of (V, T) so that $dP = (\partial P/\partial V)_T dV + (\partial P/\partial T)_V dT$. Also, carry the integration in the expression (2.235) for S by parts and show that

$$d\mu = d\left(\int V\left(\frac{\partial P}{\partial V}\right)_T dV\right) - F(T)dT. \tag{2.266}$$

Make use of the first equation in (2.265) to arrive at the desired result.

Ex. 2.41. Derive (2.237) by integration of the entropic Gibbs–Duhem relation. Hint: Follow the procedure analogous to that outlined in the Ex. 2.40 and make use of the second equation in (2.265).

Ex. 2.42. Assuming $P(V, T)$ to be of the form (2.242) which is linear in T and assuming that $\Phi(T)$ is also linear in T: $\Phi(T) = CT$, find change in entropy in free adiabatic expansion when gas expands from equilibrium state (V_A, T_A) to (V_B, T_B). Hint: Use (2.249) to show that T_B is given by

$$T_B = T_A + \frac{1}{C}\int_{V_A}^{V_B} g(V)dV. \tag{2.267}$$

Using (2.245), change in entropy is

$$\Delta S = \int_{V_A}^{V_B} f(V)dV + C\ln(T_B/T_A),$$ (2.268)

where T_B is as in (2.267).

Ex. 2.43. Consider the equation of state having the form

$$P = f_1(T) + g_1(V),$$ (2.269)

with C_V given by

$$C_V = f_3(T)V + f_4(T).$$ (2.270)

(a) Show that the equations above are consistent provided

$$f_3(T) = T\frac{d^2 f_1(T)}{dT^2}.$$ (2.271)

(b) Show that

$$\Phi(T) = \int f_4(T)dT + K,$$

$$U(V, T) = VT^2\frac{d f_1(T)/T}{dT} + \int f_4(T)dT$$

$$- \int g_1(V)dV + K.$$ (2.272)

Ex. 2.44. Derive (2.246) from entropic Gibbs–Duhem relation. Hint: Assuming $P = P(V, T)$, write $dP = (\partial P/\partial V)_T dV + (\partial P/\partial T)_V dT$ in the entropic Gibbs–Duhem relation (2.50) for single component. Invoke (2.232) too.

2.15 Joule–Thomson Process

We conclude the thermodynamics part by discussing an important experiment, called *Joule–Thomson process* which showed that internal energy of ideal gas is independent of volume. The experiment consisted of a thermally insulated cylinder, partitioned into two parts separated by porous plug. Gas in one part is at higher pressure P_1 at volume V_1 from which it is forced to seep through the porous plug to the part at the lower pressure P_2 at volume V_2. The aim is to find change in temperature of the gas, if any, at the end of the process when gas in high-pressure part is pushed completely into the low-pressure one.

Let U_1 be internal energy of the gas in the beginning of the process and U_2 that at the end. The work $P_1 V_1$ is performed on the gas in pushing it through the plug and $P_2 V_2$ the work done by it while expanding against pressure P_2 to occupy volume V_2. The porous plug slows down the speed of gas molecules to practically zero. Since there is no exchange of heat with the surrounding, conservation of energy implies

$$U_2 = U_1 + P_1 V_1 - P_2 V_2. \tag{2.273}$$

This shows that $U_2 + P_2 V_2 = U_1 + P_1 V_1$, i.e. enthalpy $H \equiv U + PV$ is conserved during the process. The change in temperature due to change in pressure is then determined by the equation

$$dT = \left(\frac{\partial T}{\partial P} \right)_H dP. \tag{2.274}$$

We need to convert the partial derivative in the equation above to measurable quantities. To that end, use to cyclic rule to show that

$$\left(\frac{\partial T}{\partial P} \right)_H = -\frac{(\partial H/\partial P)_T}{(\partial H/\partial T)_P}. \tag{2.275}$$

From the expression (2.72) for dH we have

$$\left(\frac{\partial H}{\partial T} \right)_P = T \left(\frac{\partial S}{\partial T} \right)_P = C_P. \tag{2.276}$$

This determines the denominator in (2.275). The numerator therein can be evaluated as follows:

$$\left(\frac{\partial H}{\partial P} \right)_T = T \left(\frac{\partial S}{\partial P} \right)_T + V = -\left(\frac{\partial V}{\partial T} \right)_P + V, \tag{2.277}$$

where second equation is due to fourth Maxwell relation in (2.98). Substitute (2.276) and (2.277) in (2.275) to get

$$\left(\frac{\partial T}{\partial P} \right)_H = \frac{1}{C_P} \left\{ T \left(\frac{\partial V}{\partial T} \right)_P - V \right\}. \tag{2.278}$$

This can also be written as

$$\left(\frac{\partial T}{\partial P} \right)_H = \frac{V}{C_P} (T \alpha_P - 1), \tag{2.279}$$

where α_P is the coefficient of isobaric thermal expansion.

Equation (2.278) shows that, for the ideal gas,

$$\left(\frac{\partial T}{\partial P}\right)_H = 0, \quad \text{ideal gas.} \tag{2.280}$$

No change in temperature is predicted and none was observed, when the operating gas in the process is the ideal gas.

Change in temperature is expected to be observed for non-ideal gases. Equation (2.278) shows that temperature will decrease as the pressure decreases if the right side therein is positive and temperature will increase as the pressure decreases if the right side therein is negative. The temperature T_{inv} at which the right side changes sign is called the *inversion temperature*. It is the solution of the equation

$$T\left(\frac{\partial V}{\partial T}\right)_P - V = 0. \tag{2.281}$$

Let us estimate the inversion temperature for the van der Waals gas. Using its equation of state (1.104), (2.281) for the van der Waals gas leads to following expression for the inversion temperature:

$$T_{inv} = \frac{2a(v-b)^2}{k_B b v^2}. \tag{2.282}$$

Substitute this in the equation of state (1.104) to get

$$Pv^2 - \frac{2av}{b} + 3a = 0. \tag{2.283}$$

This is solved by

$$v = \frac{a}{bP}\left(1 \pm \sqrt{1 - \frac{3b^2 P}{a}}\right). \tag{2.284}$$

Substitute this in (2.282) to obtain

$$T_{inv} = \frac{2a}{9bk_B}\left(2 \pm \sqrt{1 - \frac{3b^2 P}{a}}\right)^2. \tag{2.285}$$

This shows that there will not be any inversion temperature if $P > a/3b^2$ and that there will be two inversion temperatures if $P < a/3b^2$. Between the two inversion points, $(\partial T/\partial P)_H > 0$ whereas $(\partial T/\partial P)_H < 0$ outside the inversion points. Thus the gas will cool as it is pushed from higher to lower pressure between the two inversion points, and heat outside of them. For low pressures such that $bP/a \ll 1$, the higher inversion temperature is given by

$$T_{\text{inv}} \approx \frac{2a}{bk_B}.\tag{2.286}$$

As argued in [Landau and Lifshitz], the lower inversion temperature for small P may not occur due to condensation of gas into liquid at the pressures and temperatures in question.

References

1. G. Cook, R.H. Dickerson, Am. J. Phys. **63**, 737 (1995)
2. R. Baierlein, Am. J. Phys. **69**, 483 (2001)
3. F.D. Stacey, Geophys. Surv. **3**, 175 (1977)
4. D.L. Anderson, *Theory of the Earth* (Blackwell Scientific Publications, 1989). http://resolver.caltech.edu/CaltechBook:1989.001
5. M.J. Moran, H.N. Shapiri, D.D. Boettner, M.B. Bailey, *Fundamentals of Engineering Thermodynamics* (Wiley, 2018)

Chapter 3
Kinetic Theory

Heat has for long been attributed to random molecular motion or agitation. The attempts to find quantitative relation between heat and random molecular motion led to the kinetic theory of gases, a forerunner of statistical mechanics. It describes the state of a gas of molecules in terms of distribution functions of its molecular positions and momenta in phase space, their time evolution being governed by the laws of mechanics. Linking mechanical description with thermodynamics is fundamentally challenging for one because the mechanical evolution is time-reversal symmetric, whereas the macroscopic bodies evolve irreversibly in time. The arrow of time can emerge in the mechanical description evidently under certain assumptions. Another is identifying thermodynamic quantities with mechanical ones. In this chapter, we describe how those ends are achieved by the kinetic theory.

3.1 Early Kinetic Theory

For a comprehensive account of early kinetic theory, see Truesdell [1].

The history of the kinetic theory may be traced back to the treatise on hydrodynamics by Daniel Bernoulli published in 1738. He considered gas contained in a cylinder fitted with a movable piston and modeled the gas as an elastic fluid consisting of spherical balls moving rapidly with speed u. He calculated change in pressure as the piston moved and showed that, at constant temperature, pressure is (i) inversely proportional to volume and (ii) directly proportional to the square of the speed u of the particles. However, Bernoulli's work was ignored in then prevailing firm belief in the caloric theory of heat.

The interest in Bernoulli's model was revived circa 1820 by John Herapath. Herapath took momentum as the measure of temperature. This hypothesis, like Bernoulli's approach, though led him to the ideal gas equation, but was found inconsistent with several other known results. J. P. Joule, based on experimental evidence, established in 1847 that heat was not a fluid but a form of energy and found the relationship

© The Author(s), under exclusive license to Springer Nature Switzerland AG 2024
R. R. Puri, *Modern Thermodynamics and Statistical Mechanics*, Undergraduate Lecture Notes in Physics, https://doi.org/10.1007/978-3-031-54310-4_3

between heat and kinetic energy. Using the theory of Herapath, he derived relationship between heat and kinetic energy in 1851.

The next important development was due to John James Waterston in 1851. His work, however, remained unrecognized. The concepts similar to Waterston's were advanced by Karl Krönig in 1856. He assumed that a particle constituting gas moves uniformly in a straight line until it changes direction on elastic collision with another particle or with the walls of the container. It led Rudolf Clausius to introduce the concept of mean free path which is the average distance traveled by a sphere between two successive collisions and showed that the mean free path l is given by

$$l = \frac{3}{4n\pi d^2},$$

(3.1)

where n is the number of molecules per unit volume and d is the diameter of the sphere modeling the molecule. There was no way of ascertaining the value of the mean free path using the formula above as experimental measurement of the quantities in it was out of question. It was his desire to link the length of mean free path with observables that prompted Maxwell to formulate his now well-known theory in [2].

As observed by Truesdell [1], early kinetic theory culminated in Maxwell's said paper. It will therefore be appropriate to begin discussion of the kinetic theory with Maxwell's paper [2].

3.2 Maxwell Distribution

Maxwell abandoned the assumption of uniform motion and introduced the concept of random distribution of velocity among the molecules and went on to derive the probability function describing distribution of the velocity among molecules. In fact, as can be seen from the following excerpt from his paper [2], it is (3.1) that Maxwell had in mind while deriving the said distribution function:

Mr. Clausius has determined the mean length of path in terms of the average distance of particles and distance between the centers of two particles when collision takes place. We have at present no means of ascertaining either of these distances, but certain phenomena, such as the internal friction of gases, the conduction of heat through a gas, and the diffusion of one gas through another, seem to indicate the possibility of determining accurately the mean length of path which a particle describes between two successive collisions. In order to lay foundation of such investigations on strict mechanical principles, I shall demonstrate the laws of motion of an indefinite number of small, hard, and perfectly elastic spheres acting on one another only during impact.

Furthermore, he states, *Instead of saying that the particles are hard, spherical, and elastic, we may if we please say that the particles are centers of force, of which the action is insensible except at a certain small distance, when it appears suddenly as a repulsive force of very great intensity. It is evident that either assumption will lead to the same result. For the sake of avoiding the repetition of a long phrase about these repulsive forces, I shall proceed upon the assumption of perfectly spherical bodies.*

Maxwell abandoned the concept of uniform motion and assumed that if a large number of particles are moving in a perfectly elastic vessel then collisions would take place among them altering their velocities such that, after large number of collisions, the kinetic energy will be divided among them according to some regular law. He derived the formula for average number of particles whose velocities lie in the interval between $\mathbf{v}$ and $\mathbf{v} + d\mathbf{v}$ as follows.

To derive the said law, one starts with the assumption that the velocity of particles is a random variable whose distribution is governed by some probability distribution function and that there is no correlation between the motion of different particles. The probability distribution function of all particles then is the product of distribution function of each particle, same for every particle. Let $F(\mathbf{v}) \equiv F(v_x, v_y, v_z)$ be the single-particle distribution function so that $F(\mathbf{v})dv_x dv_y dv_z$ is the probability that the velocity of the particle lies in the interval $(\mathbf{v}, \mathbf{v} + d\mathbf{v})$. The distribution function $F(\mathbf{v})$ is thus the probability per unit volume of the velocity space, or the *probability density*. Maxwell observed that the existence of velocity v_x does not in any way affect that of the velocities v_y or v_z since these are all at right angles and independent. One can therefore write $F(\mathbf{v})$ as the product of same function $f(v_\mu)$ in three directions: $F(v_x, v_y, v_z) = f(v_x)f(v_y)f(v_z)$. Hence the probability that the velocity of a particle lies in $(\mathbf{v}, \mathbf{v} + d\mathbf{v})$ is given by

$$F(\mathbf{v})d^3\mathbf{v} = f(v_x)f(v_y)f(v_z)d^3\mathbf{v}, \qquad d^3\mathbf{v} = dv_x dv_y dv_z. \tag{3.2}$$

Normalization of probability to unity demands

$$\int F(\mathbf{v})d^3\mathbf{v} = 1, \quad -\infty \le v_x, v_y, v_z \le \infty. \tag{3.3}$$

Accordingly, if N is the number of molecules in the gas then the number per unit volume in $(\mathbf{v}, \mathbf{v} + d\mathbf{v})$, denoted by $n(\mathbf{v})$, will be $NF(\mathbf{v})$ so that the number of molecules in the said velocity interval would be

$$n(\mathbf{v})d^3\mathbf{v} = NF(\mathbf{v})d^3\mathbf{v}. \tag{3.4}$$

The problem at hand is to determine $f(v_\mu)$. Maxwell determined it using simple symmetry arguments. Since choice of the direction of the coordinate axes is arbitrary, the probability $F(\mathbf{v})d^3 v$ should depend only on the magnitude of $\mathbf{v}$, which means the following equation should hold:

$$f(v_x)f(v_y)f(v_z) = \phi(v_x^2 + v_x^2 + v_x^2). \tag{3.5}$$

This functional equation is solved by

$$f(w) = B \exp(-Aw^2), \tag{3.6}$$

where A, B are constants. Using the form (3.6) of $f(w)$ in (3.2), we get

$$F(\mathbf{v}) = C \exp\{-A(v_x^2 + v_y^2 + v_z^2)\}, \tag{3.7}$$

where C is determined by the normalization condition (3.3):

$$C \int \exp\{-A(v_x^2 + v_y^2 + v_z^2)\} \, d^3\mathbf{v} = 1. \tag{3.8}$$

Convergence of the integral requires $A > 0$. The integral above may be expressed as a product of three identical integrals and evaluated invoking the identity (H.7) to obtain

$$\int \exp\{-A(v_x^2 + v_y^2 + v_z^2)\} \, d^3\mathbf{v} = \left(\int_{-\infty}^{\infty} \exp(-Aw^2) dw \right)^3 = \left(\frac{\pi}{A} \right)^{3/2}. \tag{3.9}$$

On substituting this in (3.8) follows the expression for C:

$$C = \left(\frac{A}{\pi} \right)^{3/2}. \tag{3.10}$$

Hence the velocity distribution function $F(\mathbf{v})$ reads

$$F(\mathbf{v}) = \left(\frac{A}{\pi} \right)^{3/2} \exp(-Av^2). \tag{3.11}$$

It is called the *Maxwell distribution*. Accordingly, recalling (3.4), the number of molecules in $(\mathbf{v}, \mathbf{v} + d\mathbf{v})$ will be

$$dN(\mathbf{v}) \equiv n(\mathbf{v}) d^3\mathbf{v} = N \left(\frac{A}{\pi} \right)^{3/2} \exp(-Av^2) d^3\mathbf{v}. \tag{3.12}$$

The question that remains to be answered is: what is A? The answer to it is provided by (3.31) which expresses A in terms of temperature and the mass of the molecule. We will see that it is arrived at by deriving the equation of state by using Maxwell's distribution and comparing it with the well-known ideal gas equation of state.

The test of correctness of (3.11) lies in its ability to predict correctly the values of the observables. Let $G(\mathbf{v})$, a function of $\mathbf{v}$, be an observable. Since $\mathbf{v}$ is a random variable whose distribution is governed by $F(\mathbf{v})$, the observed value of $G(\mathbf{v})$ is its average over $F(\mathbf{v})$, denoted by $\langle G(\mathbf{v}) \rangle$ and given by

$$\langle G(\mathbf{v})\rangle = \int G(\mathbf{v})F(\mathbf{v})\,d^3\mathbf{v}. \tag{3.13}$$

In particular, like $F(\mathbf{v})$, let $G(\mathbf{v})$ also be a function of v alone. The integrand in (3.13) in that case is independent of the angle. It can then be evaluated conveniently in spherical polar coordinates (v, θ, ϕ) $(0 \le v \le \infty, 0 \le \theta \le \pi, 0 \le \phi \le 2\pi)$ with $d^3v = v^2 \sin(\theta)dvd\theta d\phi$. The integration over the angles can be performed to obtain

$$\langle G(v)\rangle = 4\pi \int_0^{\infty} G(v)F(v)v^2 dv. \tag{3.14}$$

Using the identity (H.8) it is straightforward to see that

$$\langle v\rangle = \frac{2}{\sqrt{\pi A}}, \qquad \langle v^2\rangle = \frac{3}{2A}. \tag{3.15}$$

On eliminating A between the equations above we obtain

$$\langle v^2\rangle = (3\pi/8)\langle v\rangle^2. \tag{3.16}$$

Also, due to equivalence of different Cartesian components of the velocity,

$$\langle v_x^2\rangle = \langle v_y^2\rangle = \langle v_z^2\rangle = \frac{\langle v^2\rangle}{3} = \frac{1}{2A}. \tag{3.17}$$

Using his velocity distribution, Maxwell derived a number of results some of which of our interest are summarized below:

1. Consider a system consisting of N particles of one kind, called the system I, and N' of another, called the system II. The distribution of the velocities in system I is described by (3.11), and that in II by

$$F_{II}(\mathbf{v}) = \left(\frac{B}{\pi}\right)^{3/2} \exp(-Bv^2). \tag{3.18}$$

The problem addressed by Maxwell is: What is the number of pairs of particles, one each from the two systems, whose relative velocity is $\mathbf{V}$?

To find the said number, note that the probability of a molecule of system I to have velocity in the interval $(\mathbf{v}, \mathbf{v} + d\mathbf{v})$ is $F(\mathbf{v})d^3v$ and for given $\mathbf{v}$, the probability of a molecule of system II to have velocity in the interval $(\mathbf{v} + \mathbf{V}, \mathbf{v} + \mathbf{V} + d\mathbf{V})$ is $F_{II}(\mathbf{v} + \mathbf{V})d^3\mathbf{V}$. Hence the probability of having a pair of molecules, one from the system I to have velocity in the vicinity of $\mathbf{v}$ and another from II to have velocity in the vicinity of $\mathbf{v} + \mathbf{V}$ is

$$S(\mathbf{v}, \mathbf{v} + \mathbf{V})d^3vd^3V = F(\mathbf{v})F_{II}(\mathbf{v} + \mathbf{V})d^3vd^3\mathbf{V}. \tag{3.19}$$

The number of pairs having relative velocity in the vicinity of $\mathbf{V}$ is then given by integrating the expression above over the velocity $\mathbf{v}$:

$$R(\mathbf{V})\mathrm{d}^3 V = NN' \left[\int F(\mathbf{v})F'(\mathbf{v}+\mathbf{V})\mathrm{d}^3 v \right] \mathrm{d}^3 \mathbf{V}. \tag{3.20}$$

We leave it as an exercise to show that

$$R(\mathbf{V}) = NN' \left(\frac{AB}{\pi(A+B)} \right)^{3/2} \exp\{-ABV^2/(A+B)\}. \tag{3.21}$$

The distribution function for the relative velocities is thus of the same form as that for the velocities.

Having shown, as above, that the distribution of relative velocities follows his law, Maxwell deduced that even the distribution of the compounded velocities of two systems follows the same law. For, if $\mathbf{v}_1$ is the velocity of a molecule in system I and $\mathbf{v}_2$ that of a molecule in system II, the result derived above shows that their relative velocity $\mathbf{V} = \mathbf{v}_2 - \mathbf{v}_1$ is distributed according to Maxwell distribution. Since reversing direction of velocity does not change distribution, on changing $\mathbf{v}_1$ to $-\mathbf{v}_1$, the distribution does not change. Consequently, since on changing $\mathbf{v}_1 \rightarrow -\mathbf{v}_1$, $\mathbf{V} = \mathbf{v}_2 - \mathbf{v}_1 \rightarrow \mathbf{v}_2 + \mathbf{v}_1$ the distribution of $\mathbf{V}$ does not change. Hence the compounded velocity of particles of two systems is also distributed according to Maxwell distribution.

2. If two systems of particles move in the same vessel then their mean kinetic energy will become same. We summarize below key points of Maxwell's derivation for historical reason though it is known to be erroneous.

    Let system I consist of molecules of mass m_1 and II those of mass m_2 having distribution of velocities described, respectively, by (3.11) and (3.18). Maxwell considered collision between a molecules of system I moving with velocity $\mathbf{v}_1$ with a molecule of system II moving with velocity $\mathbf{v}_2$ such that $\mathbf{v}_1 \cdot \mathbf{v}_2 = 0$. Denoting by $\mathbf{v}_1'$ and $\mathbf{v}_2'$ the particle velocities after collision, Maxwell proved that

$$m_1 \langle v_1' \rangle^2 - m_2 \langle v_2' \rangle^2 = \left(\frac{m_1 - m_2}{m_1 + m_2} \right)^2 \left(m_1 \langle v_1 \rangle^2 - m_2 \langle v_2 \rangle^2 \right). \tag{3.22}$$

This shows that the quantity $m_1 \langle v_1 \rangle^2 - m_2 \langle v_2 \rangle^2$ reduces after each collision by same factor and would become zero after a large number of collisions so that, after a large number of collisions,

$$m_1 \langle v_1 \rangle^2 = m_2 \langle v_2 \rangle^2. \tag{3.23}$$

The equation above may be written in terms of average kinetic energy of the molecules by recalling (3.16):

$$\frac{1}{2}m_1\langle v_1^2\rangle = \frac{1}{2}m_2\langle v_2^2\rangle. \tag{3.24}$$

This shows that two different kinds of molecules in a gas attain same kinetic energy after a large number of collisions. Same argument when extended to mixture of several different kinds of molecules would prove that all the molecules in the mixture acquire same average kinetic energy.

The derivation outlined above considers the velocities of the colliding particles to be orthogonal and hence is not general. Secondly, if the role of the initial and final velocities is interchanged, the collision will result in increase in the difference in the kinetic energy of the particles. Hence Maxwell's proof of the proposition that the difference in the kinetic energy of the colliding particles of two gases decreases is untenable.

3. The formalism outlined above does not provide any link between Maxwell's distribution and thermodynamics. That link is established by deriving equation of state using Maxwell's distribution. The said derivation proceeds by showing that the pressure on the walls of the vessel of volume V, assumed to be perfectly reflecting, containing uniformly distributed gas of N molecules of mass m is given by

$$P = \frac{mN}{3V}\langle v^2\rangle = \frac{2N}{3V}u, \tag{3.25}$$

where u in the second equation is the average kinetic energy per molecule:

$$u = \frac{m}{2}\langle v^2\rangle. \tag{3.26}$$

Deviating from the proof given by Maxwell, we derive the desired result in a simpler way as follows. Consider an area ΔA on the wall of the vessel of volume V containing N molecules of a gas distributed homogeneously. Choose the coordinate system such that the said area is perpendicular to the x-axis. The pressure is the force per unit area in the direction perpendicular to the surface. In the present case, it is caused by the impact of the molecules moving in positive x-direction. The number of molecules moving in the positive x-direction and hitting ΔA in time Δt are those which are contained in the cylinder of length $v_{+x}\Delta t$ with ΔA as its base, where v_{+x} is the speed of a molecule in the positive x-direction. The volume of the said cylinder is $\Delta V = v_{+x}(\Delta t)(\Delta A)$ so that the number of molecules in it is $\Delta V(N/V) = v_{+x}(\Delta t)(\Delta A)(N/V)$. Now, each molecule strikes the wall with momentum mv_{+x} and, since the wall is assumed to be perfectly reflecting, it reflects back with same momentum so that $2mv_{+x}$ is the change in momentum due to each molecular strike. Consequently, total change in momentum in time Δt is $2mv_{+x}^2(\Delta t)(\Delta A)(N/V)$. Hence the force exerted on the area ΔA is $F_x = 2mv_{+x}^2(\Delta A)(N/V)$ so that average force on the wall per unit area, i.e. pressure is given by

$$P = \frac{2mN}{V} \langle v_{+x}^2 \rangle. \tag{3.27}$$

We know that $\langle v_x^2 \rangle = \langle v^2 \rangle/3$ and $\langle v_{+x}^2 \rangle = \langle v_x^2 \rangle/2$. On substituting these relations in (3.27) we get the desired result (3.25).

Consider two gases at same temperature, pressure and having same volume, one consisting of N_1 molecules of mass m_1 and the other consisting of N_2 molecules of mass m_2. Using (3.25) it follows that in that case

$$m_1 N_1 \langle v_1^2 \rangle = m_2 N_2 \langle v_2^2 \rangle. \tag{3.28}$$

If we invoke Avogadro's hypothesis according to which equal volumes of gases at same temperature and pressure contain same number of molecules then in the equation above $N_1 = N_2 \equiv N$ so that

$$m_1 \langle v_1^2 \rangle = m_2 \langle v_2^2 \rangle \equiv m \langle v^2 \rangle. \tag{3.29}$$

This means that the kinetic energy per particle is same for all gases at same temperature and pressure. Hence, under assumed ideal conditions, all gases obey (3.25) with same u. The said equation will be the same as the ideal gas law if

$$u = \frac{3}{2} k_B T. \tag{3.30}$$

The relation above between the average kinetic energy per molecule and temperature for any ideal gas establishes the desired link between microscopic description of the gas with its thermodynamic one.

If the velocity distribution is assumed to be described by Maxwell's distribution (3.11) then we know that $\langle v^2 \rangle$ is given by (3.15). On combining (3.15) with (3.30), the value of A in Maxwell's distribution formula turns out to be given by

$$A = \frac{m}{2k_B T}. \tag{3.31}$$

The Maxwell distribution formula (3.11) then reads

$$F(\mathbf{v}) = \left(\frac{m}{2\pi k_B T} \right)^{3/2} \exp\{-mv^2/(2k_B T)\}. \tag{3.32}$$

This is the well-known form of Maxwell distribution.

The relation (3.29) above has been arrived at as a consequence of Avogadro's hypothesis applied to (3.28) arising due to equality of pressure of two kinds of molecules constituting a gas. The relation (3.29) is same as the relation (3.24) which according to Maxwell's derivation is the result of a large number of collisions. Maxwell assumed its validity and applied it to (3.28) to show that $N_1 = N_2$ and thus claimed to prove Avogadro's hypothesis. However, as we have remarked before, Maxwell's derivation of (3.24) is erroneous and therefore his claim of proving Avogadro's hypothesis does not stand. It is known that (3.29) cannot be established within the framework of the kinetic theory and therefore proof of Avogadro's hypothesis cannot come from it (see [1]).

Generalization of Maxwell's concept of velocity distribution function to the phase space distribution function. Introduced next, and its evolution according to the laws of mechanics forms the basis of the kinetic theory.

Exercises

Ex. 3.1. Derive (3.21) from (3.20). Hint: Express $\mathbf{v} + \mathbf{V}$ in Cartesian components to write $(\mathbf{v} + \mathbf{V})^2 = (v_x + V_x)^2 + (v_y + V_y)^2 + (v_z + V_z)^2$ and carry the integration in (3.20) using (H.7).

Ex. 3.2. Show using (3.21) that the mean relative speed is the square root of the sum of the squares of the mean speeds of the two systems.

Ex. 3.3. A molecule is moving with velocity $\mathbf{v}$ in a gas of N molecules described by Maxwell's law. (1) Show that the number of molecules having speed in the interval $(w, w + dw)$ with respect to the moving molecule is

$$n(w)dw = \frac{Nw}{v}\sqrt{\frac{A}{\pi}}\left[\exp\{-A(v-w)^2\} - \exp\{-A(v+w)^2\}\right]dw.$$

$$(3.33)$$

(2) Show that the integral of (3.33) over w is N, as it should be. This is one of Maxwell's propositions in [2]. Hint: (1) The velocity of the molecule moving with velocity $\mathbf{w}$ with respect to the molecule moving with velocity $\mathbf{v}$ is $\mathbf{u} = \mathbf{v} - \mathbf{w}$. The number of molecules in the interval $(\mathbf{u}, \mathbf{u} + d\mathbf{u})$ is

$$n(\mathbf{u})d^3u = N\left(\frac{A}{\pi}\right)^{3/2}\exp(-A\mathbf{u}\cdot\mathbf{u})d^3u. \qquad (3.34)$$

We have $\mathbf{u}\cdot\mathbf{u} = v^2 + w^2 - 2vw\cos(\theta)$ where θ is the angle between $\mathbf{v}$ and $\mathbf{w}$ and, since $\mathbf{v}$ is fixed, $d^3u = d^3w = w^2\sin(\theta)d\theta d\phi dw$. Integrate (3.34) over θ and ϕ to get the desired result. (2) The said integral can be carried by proving the identity

$$I = \int_0^\infty x\left(\exp\{-\alpha(x-a)^2\} - \exp\{\alpha(x+a)^2\}\right)dx = a\sqrt{\frac{\pi}{\alpha}}. \quad (3.35)$$

This can be proved by changing the variable of integration in the second integral in (3.35) to $y = -x$ and showing that

$$I = \int_{-\infty}^\infty x\exp\{-\alpha(x-a)^2\}dx, \qquad (3.36)$$

from which follows the desired result on changing the variable x of integration to $y = x - a$. Interestingly, assuming the integral of (3.33) over w to be N, Maxwell states *whence the following mathematical result* which is the one in (3.35).

3.3 Phase Space Distribution Function

The kinetic theory treats the gas as a collection of large number N of molecules modeled as point particles moving under the influence of mutual interaction and possibly some external force. The walls of the container are assumed to be perfectly reflecting. In what follows we will assume the gas as consisting of molecules of same kind. The mechanical state of a molecule may be described by a point in the six-dimensional space formed by three components of its position and corresponding three components of its momentum as the axes, called *single-particle phase space* or μ-space. We denote by $(\mathbf{r}_i, \mathbf{p}_i)$ the phase space coordinates of the ith molecule. The volume element around the point $(\mathbf{r}_i, \mathbf{p}_i)$ in the said phase space will be denoted by

$$d\mu_i = d^3\mathbf{r}_i d^3\mathbf{p}_i. \tag{3.37}$$

The mechanical state of a molecule is described by a point in the μ-space and the state of N molecules is represented by N points in it.

Alternatively, the state of the system of N molecules may be described as a point in the $6N$-dimensional phase space, called the Γ space, formed by $3N$ position components and $3N$ momentum components of N molecules. We denote a point in the Γ space by $\{\mathbf{r}_i, \mathbf{p}_i\}_N$:

$$\{\mathbf{r}_i, \mathbf{p}_i\}_N \equiv (\mathbf{r}_1, \mathbf{r}_2, \ldots, \mathbf{r}_N; \mathbf{p}_1, \mathbf{p}_2, \ldots, \mathbf{p}_N), \tag{3.38}$$

where $\mathbf{r}_i$ denotes the position and $\mathbf{p}_i$ the momentum of the ith $(i = 1, 2, \ldots, N)$ molecule. The volume element around the point $\{\mathbf{r}_i, \mathbf{p}_i\}_N$ will be denoted by

$$d\tau_N = \prod_{i=1}^{N} d^3\mathbf{r}_i d^3\mathbf{p}_i \equiv \prod_{i=1}^{N} d\mu_i. \tag{3.39}$$

In what follows, unless stated otherwise, we will describe the state of the system in the $6N$-dimensional Γ space.

As the molecules move under mutual and external forces, the point in the phase space traverses a trajectory. That trajectory can, in principle, be determined by solving coupled equations of motion for each of the $6N$ variables $(\{\mathbf{r}_i, \mathbf{p}_i\}_N)$ if the intermolecular potential, the external forces, and the initial position and momenta of each molecule are known. Since N for macroscopic systems is very large ($N \sim 10^{20}$ or more), it is not possible to specify the initial position and momentum of each and every molecule and not possible to solve the equations even when the potentials are known. However, we have seen that the macroscopic properties are characterized in terms of a small number of thermodynamic variables. The knowledge of the position and momentum of each molecule is therefore unnecessary and appropriate small number of microscopic functions should be constructed instead to relate with the thermodynamic quantities.

To that end, let us assume that no external force is applied on the gas so that the change of position of its representative point in the Γ space and hence its change of state is caused only by collisions between molecules. The time between collisions is generally of the order of 10^{-8} sec., whereas the time scale of observation is much larger. We can then say that the observed value of a dynamical quantity is the average of that quantity over the positions and momenta of the molecules during the time of observation.

To evaluate the said average, we consider all possible molecular states subject to the macroscopic conditions on the system. We call those states the admissible *microstates*. Let $\mathcal{N}$ be the number of the said states. Since each state is represented by a point in the Γ space, all the admissible microstates will be represented by $\mathcal{N}$ points. Let $d\mathcal{N}_i$ be the number of points in the volume element $d\tau_N$ around $\{\mathbf{r}_i, \mathbf{p}_i\}_N$ so that $d\mathcal{N}_i/\mathcal{N}$ stands for the probability that a state lies in $d\tau_N$ at $\{\mathbf{r}_i, \mathbf{p}_i\}_N$. The probability of a point to be in $d\tau_N$ around $\{\mathbf{r}_i, \mathbf{p}_i\}_N$ per unit phase space volume, i.e. the probability density at $\{\mathbf{r}_i, \mathbf{p}_i\}_N$ is $(1/\mathcal{N})(d\mathcal{N}_i/d\tau_N)$. We assume that the numbers $\mathcal{N}, d\mathcal{N}_i$ are large enough to enable one to consider $d\mathcal{N}_i/\mathcal{N}$ a continuous function of $\{\mathbf{r}_i, \mathbf{p}_i\}_N$ and define,

$$f_N\left(\{\mathbf{r}_i, \mathbf{p}_i\}_N; t\right) = \frac{1}{\mathcal{N}} \frac{d\mathcal{N}_i}{d\tau_N}, \tag{3.40}$$

called the *phase space probability distribution function* or simply the *distribution function*. Due to the definition (3.40), the quantity

$$dw = f_N\left(\{\mathbf{r}_i, \mathbf{p}_i\}_N; t\right) d\tau_N \tag{3.41}$$

is same as the probability $d\mathcal{N}_i/\mathcal{N}$ that a state lies in $d\tau_N$ at $\{\mathbf{r}_i, \mathbf{p}_i\}_N$. Its defining relation (3.40) shows that $f_N\left(\{\mathbf{r}_i, \mathbf{p}_i\}_N; t\right)$ obeys the desired normalization condition for a probability distribution:

$$\int f_N\left(\{\mathbf{r}_i, \mathbf{p}_i\}_N; t\right) d\tau_N = 1, \tag{3.42}$$

where the integration is over the entire phase space. The number of states in $d\tau_N$ at $\{\mathbf{r}_i, \mathbf{p}_i\}_N$ is

$$d\mathcal{N}_i = \mathcal{N} f_N\left(\{\mathbf{r}_i, \mathbf{p}_i\}_N; t\right) d\tau_N. \tag{3.43}$$

The average of the function $A(\{\mathbf{r}_i, \mathbf{p}_i\}_N)$ of coordinates and momenta of N particles, denoted by $\langle A(\{\mathbf{r}_i, \mathbf{p}_i\}_N; t)\rangle$, is given by

$$\langle A(\{\mathbf{r}_i, \mathbf{p}_i\}_N; t)\rangle = \int A(\{\mathbf{r}_i, \mathbf{p}_i\}_N) f_N(\{\mathbf{r}_i, \mathbf{p}_i\}_N; t) d\tau_N. \tag{3.44}$$

The thermodynamic properties are expressed in terms of the averages of suitably identified observables.

The time evolution of $f_N(\{\mathbf{r}_i, \mathbf{p}_i\}_N; t)$ is governed by Liouville's equation derived next.

3.3.1 Liouville's Theorem

Proved next, Liouville's theorem states that

$$\frac{d f_N}{dt} = 0. \tag{3.45}$$

Equation (3.45) is called *Liouville's equation*. For convenience we have suppressed and will keep doing so wherever there is no ambiguity, the argument of f_N.

To prove (3.45), it is convenient to rewrite the position variables as $\{q_i\}_{3N} \equiv (q_1, q_2, \ldots, q_{3N})$ and corresponding momenta as $\{p_i\}_{3N} \equiv (p_1, p_2, \ldots, p_{3N})$. The sets $\{q_i\}$ and $\{p_i\}$ are canonically conjugate:

$$\{q_i, p_j\} = \delta_{ij}, \tag{3.46}$$

where $\{A, B\}$ denotes the Poisson bracket of A, B defined by

$$\{A, B\} = \sum_{i=1}^{3N} \left[\frac{\partial A}{\partial q_i} \frac{\partial B}{\partial p_i} - \frac{\partial A}{\partial p_i} \frac{\partial B}{\partial q_i} \right]. \tag{3.47}$$

The distribution function in changed notation is written as

$$f_N(\{q_i, p_i\}_{3N}; t) \equiv f_N(\{\mathbf{r}_i, \mathbf{p}_i\}_N; t)), \quad \{q_i, p_i\}_n \equiv (\{q_i\}_n, \{p_i\}_n). \tag{3.48}$$

The phase space volume element in changed notation reads

$$d\tau_N = \prod_{i=1}^{3N} dq_i dp_i. \tag{3.49}$$

We determine time evolution of f_N assuming that the molecular evolution is governed by the Hamiltonian $H_N \equiv H_N(\{q_i, p_i\}_{3N})$.

We know that the evolution of the position and the momentum under the action of H_N is governed by Hamilton's equations

$$\dot{q}_i = \frac{\partial H_N}{\partial p_i}, \qquad \dot{p}_i = -\frac{\partial H_N}{\partial q_i}, \tag{3.50}$$

where "dot" on a symbol denotes derivative of that symbol with respect to time. Hence, in time δt, the $\{q_i, p_i\}$ evolve to $\{q_i', p_i'\}$ where

$$q_i' = q_i + \dot{q}_i \delta t, \qquad p_i' = p_i + \dot{p}_i \delta t. \tag{3.51}$$

As a consequence, all the points within the volume element $d\tau_N$ at $\{q_i, p_i\}_{3N}$ evolve to occupy the volume element $d\tau_N'$ at $\{q_i', p_i'\}_{3N}$ and $f_N(\{q_i, p_i\}; t)$ evolves to $f_N(\{q_i', p_i'\}; t + \delta t)$. Since the number of points in a volume element is given by (3.43), the fact that there are same number of points in $d\tau_N$ and $d\tau_N'$ leads to the relation

$$f_N(\{q_i', p_i'\}_{3N}; t + \delta t) d\tau_N' = f_N(\{q_i, p_i\}_{3N}; t) d\tau_N. \tag{3.52}$$

We will show that $d\tau_N' = d\tau_N$. To that end, recall the following results (see Ex. 3.4): (1) Under Hamiltonian evolution, canonically conjugate variables evolve to canonically conjugate variables, i.e. the time evolution is a canonical transformation and (2) the phase space volume element is invariant under canonical transformation. Since transformation $d\tau_N \rightarrow d\tau_N'$ is caused by Hamiltonian evolution, due to (1), it is a canonical transformation. Hence, being related by a canonical transformation, due to (2), $d\tau_N = d\tau_N'$. Consequently (3.52) reduces to

$$f_N(\{q_i', p_i'\}_{3N}; t + \delta t) = f_N(\{q_i, p_i\}_{3N}; t). \tag{3.53}$$

Due to (3.51), the equation above implics

$$\sum_{i=1}^{3N} \left[\frac{\partial f_N}{\partial q_i} \dot{q}_i + \frac{\partial f_N}{\partial p_i} \dot{p}_i \right] + \frac{\partial f_N}{\partial t} = 0. \tag{3.54}$$

Since

$$\frac{d f_N}{dt} = \sum_{i=1}^{3N} \left[\frac{\partial f_N}{\partial q_i} \dot{q}_i + \frac{\partial f_N}{\partial p_i} \dot{p}_i \right] + \frac{\partial f_N}{\partial t}, \tag{3.55}$$

(3.54) is same as the form (3.45) of Liouville's equation.

Using Hamilton's equations (3.50), (3.54) can be rewritten as

$$\frac{\partial f_N}{\partial t} = \sum_{i=1}^{3N} \left[\frac{\partial H_N}{\partial q_i} \frac{\partial f_N}{\partial p_i} - \frac{\partial H_N}{\partial p_i} \frac{\partial f_N}{\partial q_i} \right] \equiv \{H_N,\ f_N\}. \tag{3.56}$$

Equation (3.56) is another form of Liouville's theorem. In terms of $\{\mathbf{r}_i, \mathbf{p}_i\}$, (3.56) assumes the form

$$\frac{\partial f_N}{\partial t} = \sum_{i=1}^{N} \left(\nabla_{\mathbf{r}_i} H_N \cdot \nabla_{\mathbf{p}_i} f_N - \nabla_{\mathbf{p}_i} H_N \cdot \nabla_{\mathbf{r}_i} f_N \right), \tag{3.57}$$

where

$$\begin{aligned}
\nabla_{\mathbf{r}_i} A &= \frac{\partial A}{\partial x_i} \mathbf{e}_x + \frac{\partial A}{\partial y_i} \mathbf{e}_y + \frac{\partial A}{\partial z_i} \mathbf{e}_z, \\
\nabla_{\mathbf{p}_i} A &= \frac{\partial A}{\partial p_{x_i}} \mathbf{e}_x + \frac{\partial A}{\partial p_{y_i}} \mathbf{e}_y + \frac{\partial A}{\partial p_{z_i}} \mathbf{e}_z,
\end{aligned} \tag{3.58}$$

the $\mathbf{e}_x, \mathbf{e}_y, \mathbf{e}_z$ being unit Cartesian vectors.

The Liouville equation determines time evolution of the distribution function of all particles. However, several physical properties are determined in terms of reduced distribution functions introduced next.

3.3.2 Reduced Distribution Functions

The distribution function $f_N(\{\mathbf{r}_i, \mathbf{p}_i\}_N; t)$ is the probability that N molecules are in the vicinity of the point $\{\mathbf{r}_i, \mathbf{p}_i\}_N$ in the phase space. However, we will see that for the study of thermodynamic properties we need to know the probability distribution function of a small number of particles called reduced distribution functions.

Consider the question: what is the probability density for the kth molecule to be in the vicinity of $(\mathbf{r}, \mathbf{p})$ in phase space irrespective of the phase space position of other molecules? It is evidently given by integrating over the phase space coordinates of all the molecules while keeping that of the kth one fixed at $(\mathbf{r}, \mathbf{p})$:

$$\tilde{f}_{1k}(\mathbf{r}, \mathbf{p}; t) = \langle \delta^{(3)}(\mathbf{r}_k - \mathbf{r}) \delta^{(3)}(\mathbf{p}_k - \mathbf{p}) \rangle, \tag{3.59}$$

where average is over $f_N(\{\mathbf{r}_i, \mathbf{p}_i\}_N; t)$ so that

$$\begin{aligned}
\tilde{f}_{1k}(\mathbf{r}, \mathbf{p}; t) &= \int f_N(\{\mathbf{r}_i, \mathbf{p}_i\}_N; t) \delta^{(3)}(\mathbf{r}_k - \mathbf{r}) \delta^{(3)}(\mathbf{p}_k - \mathbf{p}) \prod_{i=1}^{N} d\mu_i \\
&= \int f_N(\{\mathbf{r}_i, \mathbf{p}_i\}_N; t) \prod_{i \neq k}^{N} d\mu_i, \quad (\mathbf{r}_k, \mathbf{p}_k) = (\mathbf{r}, \mathbf{p}). \tag{3.60}
\end{aligned}$$

Since the molecules are identical and the system is assumed to be homogeneous, $f_N(\{\mathbf{r}_i, \mathbf{p}_i\}_N; t)$ is symmetric under the exchange of molecules. Consequently the average in (3.60) is independent of the label k identifying the molecule, and can be rewritten in the following form taking $k = 1$,

$$\tilde{f}_1(\mathbf{r}_1, \mathbf{p}_1; t) = \int f_N(\{\mathbf{r}_i, \mathbf{p}_i\}; t) \prod_{i=2}^{N} \mathrm{d}\mu_i. \tag{3.61}$$

Because of the normalization condition (3.42) on f_N, it follows that

$$\int \tilde{f}_1(\mathbf{r}, \mathbf{p}; t)\, \mathrm{d}^3 r \mathrm{d}^3 p = 1. \tag{3.62}$$

The function $\tilde{f}_1(\mathbf{r}, \mathbf{p}; t)$ defined in (3.59) is the probability density for finding the kth molecule, i.e. a particular molecule in the vicinity of $(\mathbf{r}, \mathbf{p})$. The probability density for any of the N molecule to be in the vicinity of $(\mathbf{r}, \mathbf{p})$ is obtained by summing (3.59) over all k:

$$f_1(\mathbf{r}, \mathbf{p}; t) = \sum_{k=1}^{N} \langle \delta^{(3)}(\mathbf{r}_k - \mathbf{r})\delta^{(3)}(\mathbf{p}_k - \mathbf{p}) \rangle. \tag{3.63}$$

Hence

$$f_1(\mathbf{r}, \mathbf{p}; t) = N \tilde{f}_1(\mathbf{r}, \mathbf{p}; t). \tag{3.64}$$

Due to (3.62) the normalization condition on $f_1(\mathbf{r}, \mathbf{p}; t)$ is

$$\int f_1(\mathbf{r}, \mathbf{p}; t)\, \mathrm{d}^3 r \mathrm{d}^3 p = N. \tag{3.65}$$

The function $f_1(\mathbf{r}, \mathbf{p}; t)$ is called *reduced one-particle distribution function* or simply *one-particle distribution function*. Evidently, the average number of molecules in the volume element $\mathrm{d}^3 r \mathrm{d}^3 p$ is

$$\mathrm{d}n(\mathbf{r}, \mathbf{p}) = f_1(\mathbf{r}, \mathbf{p}; t)\mathrm{d}^3 r \mathrm{d}^3 p. \tag{3.66}$$

We will see the usefulness of $f_1(\mathbf{r}, \mathbf{p}; t)$ in evaluating averages of single molecule observables.

We know that the distribution function $f_N(\{\mathbf{r}_i, \mathbf{p}_i\}_N; t)$ determines, by means of (3.44), the average of any function $A(\{\mathbf{r}_i, \mathbf{p}_i\}_N)$ of phase space coordinates. However, several physical properties of interest are averages of functions which depend on the phase space coordinates of single molecule. For example, the kinetic energy $p_k^2/2m$ of the kth molecule. It is a function only of the momentum of the kth molecule. The average of the function $A_1(\mathbf{r}_k, \mathbf{p}_k)$ of the phase space coordinates of the kth molecule may be evaluated as follows:

$$
\begin{aligned}
\langle A_1(\mathbf{r}_k, \mathbf{p}_k; t) \rangle &= \int A_1(\mathbf{r}_k, \mathbf{p}_k) f_N(\{\mathbf{r}_i, \mathbf{p}_i\}_N; t) \prod_{i=1}^{N} d\mu_i \\
&= \int A_1(\mathbf{r}_k, \mathbf{p}_k) \left[\int f_N(\{\mathbf{r}_i, \mathbf{p}_i\}_N; t) \prod_{i \neq k}^{N} d\mu_i \right] d\mu_k \\
&= \int A_1(\mathbf{r}_k, \mathbf{p}_k) \tilde{f}_1(\mathbf{r}_k, \mathbf{p}_k; t) d\mu_k,
\end{aligned}
\tag{3.67}
$$

where $\tilde{f}_1(\mathbf{r}, \mathbf{p}; t)$ is as in (3.61). Hence the average of the sum of $A_1(\mathbf{r}_k, \mathbf{p}_k)$ over k is given by

$$
\sum_{k=1}^{N} \langle A_1(\mathbf{r}_k, \mathbf{p}_k; t) \rangle = N \langle A_1(\mathbf{r}, \mathbf{p}; t) \rangle = \int A_1(\mathbf{r}, \mathbf{p}) f_1(\mathbf{r}, \mathbf{p}; t) \, d^3r d^3p, \tag{3.68}
$$

where $f_1(\mathbf{r}, \mathbf{p}; t)$ is as in (3.64).

Similarly, to evaluate the average of the function $A_2(\mathbf{r}_j, \mathbf{r}_k, \mathbf{p}_j, \mathbf{p}_k)$ dependent on the phase space coordinates of two molecules, we define joint probability of finding jth molecule in the vicinity of $(\mathbf{r}, \mathbf{p})$ and simultaneously the kth molecule in the vicinity of $(\mathbf{r}', \mathbf{p}')$:

$$
\begin{aligned}
&\tilde{f}_{2jk}(\mathbf{r}, \mathbf{r}', \mathbf{p}, \mathbf{p}'; t) \\
&= \sum_{j \neq k=1}^{N} \sum_{k=1}^{N} \langle \delta^{(3)}(\mathbf{r}_j - \mathbf{r}) \delta^{(3)}(\mathbf{r}_k - \mathbf{r}') \delta^{(3)}(\mathbf{p}_j - \mathbf{p}) \delta^{(3)}(\mathbf{p}_k - \mathbf{p}') \rangle,
\end{aligned}
\tag{3.69}
$$

where the average is with respect to $f_N(\{\mathbf{r}_i, \mathbf{p}_i\}_N; t)$. Due to symmetry of $f_N(\{\mathbf{r}_i, \mathbf{p}_i\}_N; t)$ under exchange of molecules, $\tilde{f}_{2jk}$ is independent of j, k. We therefore let $j = 1, k = 2$ to rewrite (3.69) as

$$
\tilde{f}_2(\mathbf{r}_1, \mathbf{r}_2, \mathbf{p}_1, \mathbf{p}_2; t) = \int f_N(\{\mathbf{r}_i, \mathbf{p}_i\}_N; t) \prod_{i=3}^{N} d\mu_i. \tag{3.70}
$$

The average of $A_2(\mathbf{r}_j, \mathbf{r}_k, \mathbf{p}_j, \mathbf{p}_k)$ is given by

$$
\begin{aligned}
&\langle A_2(\mathbf{r}_j, \mathbf{r}_k, \mathbf{p}_j, \mathbf{p}_k; t) \rangle \\
&= \int A_2(\mathbf{r}_j, \mathbf{r}_k, \mathbf{p}_j, \mathbf{p}_k) f_N(\{\mathbf{r}_i, \mathbf{p}_i\}_N; t) \prod_{i=1}^{N} d\mu_i
\end{aligned}
$$

$$= \int A_2(\mathbf{r}_j, \mathbf{r}_k, \mathbf{p}_j, \mathbf{p}_k) \left[f_N(\{\mathbf{r}_i, \mathbf{p}_i\}_N; t) \prod_{i \neq j,k}^{N} d\mu_i \right] d\mu_j d\mu_k$$

$$= \int A_2(\mathbf{r}_j, \mathbf{r}_k, \mathbf{p}_j, \mathbf{p}_k) \tilde{f}_2(\mathbf{r}_j, \mathbf{r}_k, \mathbf{p}_j, \mathbf{p}_k; t) d\mu_j d\mu_k, \qquad (3.71)$$

where

$$\tilde{f}_2(\mathbf{r}_j, \mathbf{r}_k, \mathbf{p}_j, \mathbf{p}_k; t) = \int f_N(\{\mathbf{r}_i, \mathbf{p}_i\}_N; t) \prod_{i \neq j,k}^{N} d\mu_i. \qquad (3.72)$$

It then follows that

$$\sum_{j \neq k=1}^{N} \sum_{k=1}^{N} \langle A_2(\mathbf{r}_j, \mathbf{r}_k, \mathbf{p}_j, \mathbf{p}_k; t) \rangle = N(N-1) \langle A_2(\mathbf{r}_1, \mathbf{r}_2, \mathbf{p}_1, \mathbf{p}_2; t) \rangle$$

$$= \int A_2(\mathbf{r}_1, \mathbf{r}_2, \mathbf{p}_1, \mathbf{p}_2) f_2(\mathbf{r}_1, \mathbf{r}_2, \mathbf{p}_1, \mathbf{p}_2; t) d\mu_1 d\mu_2, \qquad (3.73)$$

where $N(N-1)$ is the result of the sum over $j \neq k$. That sum being the number of pairs of different indices j, k that can be chosen from the set of N numbers and

$$f_2(\mathbf{r}_1, \mathbf{r}_2, \mathbf{p}_1, \mathbf{p}_2; t) = N(N-1) \tilde{f}_2(\mathbf{r}_1, \mathbf{r}_2, \mathbf{p}_1, \mathbf{p}_2; t) \qquad (3.74)$$

is called *reduced two-particle distribution function* or simply *two-particle distribution function*, or *two-particle correlation function*.

In general, due to symmetry of f_N under the exchange of molecules, the average of a function of s distinct phase space points, $A_s(\mathbf{r}_{i_1}, \dots \mathbf{r}_{i_s}, \mathbf{p}_{i_1}, \dots \mathbf{p}_{i_s})$ ($i_1 \neq i_2 \neq \dots \neq i_s$) is independent of the labels $i_1, \dots, i_s$ so that

$$\tilde{f}_s(\mathbf{r}_{i_1}, \dots, \mathbf{r}_{i_s}, \mathbf{p}_{i_1}, \dots, \mathbf{p}_{i_s}; t) \equiv \tilde{f}_s(\{\mathbf{r}_i, \mathbf{p}_i\}_s; t)$$

$$= \int f_N(\{\mathbf{r}_i, \mathbf{p}_i\}_N; t) \prod_{i=s+1}^{N} d\mu_i. \qquad (3.75)$$

As a consequence of this, we have

$$\langle A_s(\mathbf{r}_{i_1}, \dots \mathbf{r}_{i_s}, \mathbf{p}_{i_1}, \dots \mathbf{p}_{i_s}; t) \rangle = \int A_s(\{\mathbf{r}_i, \mathbf{p}_i\}_s) \tilde{f}_s(\{\mathbf{r}_i, \mathbf{p}_i\}_s; t) \prod_{i=1}^{s} d\mu_i$$

$$\equiv \langle A_s(\{\mathbf{r}_i, \mathbf{p}_i\}_s; t) \rangle. \qquad (3.76)$$

This leads to

$$
\sum_{i_1 \neq i_2 \neq \ldots \neq i_s = 1}^{N} \left\langle A_s(\mathbf{r}_{i_1}, \ldots \mathbf{r}_{i_s}, \mathbf{p}_{i_1}, \ldots \mathbf{p}_{i_s}) \right\rangle
$$
$$
= \frac{N!}{(N-s)!} \langle A_s(\{\mathbf{r}_i, \mathbf{p}_i\}_s; t) \rangle
$$
$$
= \int A_s(\{\mathbf{r}_i, \mathbf{p}_i\}_s) f_s(\{\mathbf{r}_i, \mathbf{p}_i\}_s; t) \prod_{i=1}^{s} \mathrm{d}\mu_i, \tag{3.77}
$$

where $N!/(N-s)!$ is the result of the sum over $i_1 \neq i_2 \neq \ldots \neq i_s$. That sum being the number of ways of choosing s different numbers $i_1, i_2, \ldots, i_s$ from the set of N numbers and

$$
f_s(\{\mathbf{r}_i, \mathbf{p}_i\}_s; t) = \frac{N!}{(N-s)!} \tilde{f}_s(\{\mathbf{r}_i, \mathbf{p}_i\}_s; t) \tag{3.78}
$$

is called *reduced s-particle distribution function* or simply *s-particle distribution function* or *s-particle correlation function*. It stands for the probability density of finding s molecules, one each in the vicinity of $(\mathbf{r}_{i_1}, \mathbf{p}_{i_1})$, $(\mathbf{r}_{i_2}, \mathbf{p}_{i_2})$, $\ldots$, $(\mathbf{r}_{i_s}, \mathbf{p}_{i_s})$, irrespective of the phase space position of the remaining molecules.

The task of the kinetic theory is to determine $f_s(\{\mathbf{r}_i, \mathbf{p}_i\}_s; t)$ by constructing and solving the equation obeyed by it. We will see that several thermodynamic quantities can be expressed as averages involving one-particle distribution function $f_1(\mathbf{r}, \mathbf{p}; t)$. By working in the single-particle six-dimensional phase space, we first outline the method to derive under certain approximations a closed-form equation for $f_1(\mathbf{r}, \mathbf{p}; t)$, the so-called Boltzmann equation. We then work in the N-particle phase space and derive a hierarchy of equations, known as the BBGKY hierarchy, which is an integro-differential equation determining time evolution of $f_s(\{\mathbf{r}_i, \mathbf{p}_i\}_s; t)$ in terms of $f_{s+1}(\{\mathbf{r}_i, \mathbf{p}_i\}_{s+1}; t)$, followed by its reduction to the Boltzmann equation.

Exercises

Ex. 3.4. Show that the phase space volume element is invariant under a canonical transformation. Hint: Let the system be described by n generalized coordinates $\{q_i\}_n \equiv (q_1, q_2, \ldots, q_n)$ and momenta $\{p_i\}_n \equiv (p_1, p_2, \ldots, p_n)$. Let $\tilde{\xi}$ be the column constituted by $2n$ elements $\xi_m = q_m$, $\xi_{m+n} = p_m$, $m = 1, 2, \ldots, n$. Consider the transformation to a new set of coordinates and momenta $(\{Q_i\}_n, \{P_i\}_n)$ defined by invertible relations

$$
Q_i = Q_i(\{q_i\}, \{p_i\}, t), \qquad P_i = P_i(\{q_i\}, \{p_i\}, t). \tag{3.79}
$$

Let $\tilde{\eta}$ be the column having $2n$ elements $\eta_m = Q_m$, $\eta_{m+n} = P_m$, $m = 1, 2, \ldots, n$. The transformation (3.79) may then be written as

$$\eta_i = \eta_i(\{\xi_i\}), \qquad i = 1, 2, \ldots, 2n. \tag{3.80}$$

Let $\hat{M}$ denote the $2n \times 2n$ matrix formed by M_{ij} as its elements:

$$M_{ij} = \frac{\partial \eta_i}{\partial \xi_j}. \tag{3.81}$$

It is known that the transformation (3.79) will be canonical if the following condition holds:

$$\hat{M}\hat{J}\hat{M}^{\mathrm{T}} = \hat{M}^{\mathrm{T}}\hat{J}\hat{M} = \hat{J}, \tag{3.82}$$

where M^{T} is the transpose of $\hat{M}$ and $\hat{J}$ is the $2n \times 2n$ anti-symmetric matrix defined by

$$\hat{J} = \begin{pmatrix} \hat{O}_n & \hat{I}_n \\ -\hat{I}_n & \hat{O}_n \end{pmatrix}, \tag{3.83}$$

with $\hat{O}_n$ denoting $n \times n$ null matrix and $\hat{I}_n$ the $n \times n$ identity matrix. The volume element $d\tau$ in the phase space constituted by $\{\xi_i\} \equiv (\{q_i\}, \{p_i\})$ is

$$d\tau = \prod_{i=1}^{n} dq_i dp_i. \equiv \prod_{i=1}^{2n} d\xi_i. \tag{3.84}$$

Under the transformation (3.80), $d\tau \to d\tau'$ where

$$d\tau' = \prod_{i=1}^{n} dQ_i dP_i. \equiv \prod_{i=1}^{2n} d\eta_i = \mathcal{J} d\Gamma, \tag{3.85}$$

the $\mathcal{J}$ being the Jacobian of transformation:

$$\mathcal{J} = \left| \det(\hat{M}) \right|, \tag{3.86}$$

and $\hat{M}$ is as in (3.81). The determinant of (3.82) is

$$\{\det(\hat{M})\}^2 \det(\hat{J}) = \det(\hat{J}). \tag{3.87}$$

Since $\det(\hat{J}) \neq 0$, it follows that $\left| \det(\hat{M}) \right| = 1$. Hence the desired result: $d\tau' = d\tau$.

Ex. 3.5. Show that the time evolution governed by Hamiltonian is a canonical transformation. Hint: This may be proved by invoking the property of invariance of Poisson bracket under canonical transformation. Using Hamilton's equations written in terms of the Poisson bracket ($\dot{q} = \{q, \ H\}, \dot{p} = \{p, \ H\}$), we have

$$q(t + \delta t) = q(t) + \{q, \ H\}\delta t, \quad p(t + \delta t) = p(t) + \{p, \ H\}\delta t. \quad (3.88)$$

From the equations above it follows that

$$
\begin{aligned}
&\{q(t + \delta t), \ p(t + \delta t)\} \\
&= \{q(t), \ p(t)\} + [\{q, \ \{p, \ H\}\} + \{\{q, \ H\}, \ p\}] \, \delta t. \quad (3.89)
\end{aligned}
$$

The term multiplying δt in the equation above may be rewritten as

$$\{q, \ \{p, \ H\}\} + \{\{q, \ H\}, \ p\} = \{q, \ \{p, \ H\}\} + \{p, \ \{H, \ q\}\}. \quad (3.90)$$

Invoking Jacobi's identity:

$$\{A, \ \{B, \ C\}\} + \{C, \ \{A, \ B\}\} + \{B, \ \{C, \ A\}\} = 0, \quad (3.91)$$

we have

$$\{q, \ \{p, \ H\}\} + \{H, \ \{q, \ p\}\} + \{p, \ \{H, \ q\}\} = 0, \quad (3.92)$$

which on combining with (3.90) yields

$$\{q, \ \{p, \ H\}\} + \{\{q, \ H\}, \ p\} = -\{H, \ \{q, \ p\}\} = 0, \quad (3.93)$$

where last equation is due to the fact that $\{q, \ p\} = 1$. Substitution of (3.93) in (3.89) gives

$$\{q(t + \delta t), \ p(t + \delta t)\} = \{q(t), \ p(t)\} = 1. \quad (3.94)$$

Thus the Poisson bracket between position and momentum at time $t + \delta t$ is same as that at time t, i.e. the Poisson bracket is invariant under time evolution. This establishes the desired result, namely, the time evolution governed by Hamiltonian is a canonical transformation.

3.4 Boltzmann Equation: Single-Particle Phase Space Approach

Consider the single-particle six-dimensional phase space. The state of motion of each molecule is described by a point in this space and that of N molecules by N points.

To begin with, we consider the gas of non-interacting molecules. The molecules may, however, be acted upon by external potential $U(\mathbf{r})$. The Hamiltonian governing the evolution of the system is then given by

$$H_N(\{\mathbf{r}_i, \mathbf{p}_i\}_N) \equiv \sum_{i=1}^{N} h(\mathbf{r}_i, \mathbf{p}_i), \qquad h(\mathbf{r}_i, \mathbf{p}_i) = \frac{1}{2m} p_i^2 + U(\mathbf{r}_i). \qquad (3.95)$$

Under H_N, each particle evolves independent of the other. Starting from Liouville's equation, we have shown in Sect. 3.6 that the one-particle distribution function obeys the equation

$$\frac{\partial f_1(\mathbf{r}, \mathbf{p}; t)}{\partial t} = \nabla_\mathbf{r} h \cdot \nabla_\mathbf{p} f_1 - \nabla_\mathbf{p} h \cdot \nabla_\mathbf{r} f_1 \equiv \{ h(\mathbf{r}, \mathbf{p}), \; f_1(\mathbf{r}, \mathbf{p}; t) \}. \qquad (3.96)$$

This determines time evolution of $f_1(\mathbf{r}, \mathbf{p}; t)$ when there is no inter-molecular interaction.

The inter-molecular interaction in the present formalism is included assuming that it causes scattering of molecules. The theory based on actual form of inter-molecular interaction will be outlined in Sect. 3.6. Due to scattering, the molecules may enter or leave a phase space volume element. The number of molecules in a phase space volume element then will no longer be constant and $d f_1/dt$ will no longer be zero. The effect of inter-molecular interaction may therefore be accounted for by including in $d f_1/dt$ the contribution from scattering and rewriting it as

$$\frac{d f_1(\mathbf{r}, \mathbf{p}; t)}{dt} = \left(\frac{\partial f_1(\mathbf{r}, \mathbf{p}; t)}{\partial t} \right)_{\text{coll}}. \qquad (3.97)$$

Consequently, due to collisions, (3.96) changes to

$$\frac{\partial f_1}{\partial t} - \{ h(\mathbf{r}, \mathbf{p}), \; f_1(\mathbf{r}, \mathbf{p}; t) \} = \left(\frac{\partial f_1(\mathbf{r}, \mathbf{p}; t)}{\partial t} \right)_{\text{coll}}. \qquad (3.98)$$

The Poisson bracket term, which describes evolution of non-interacting molecules in the equation above, is called the *streaming term*.

The collision term may be evaluated if the gas is sufficiently dilute to enable one to treat collisions only as binary. Assuming that to be the case, consider collision between a molecule of momentum $\mathbf{p}'$ and another of momentum $\mathbf{p}_1'$, both in the vicinity of $\mathbf{r}$. Let their momenta after scattering be $\mathbf{p}$ and $\mathbf{p}_1$, respectively. Assume the collision to be elastic so that total momentum and kinetic energy of the molecules before and after scattering is same:

$$\mathbf{p} + \mathbf{p}_1 = \mathbf{p}' + \mathbf{p}_1', \qquad p^2 + p_1^2 = p'^2 + p_1'^2. \tag{3.99}$$

We denote this scattering process by $(\mathbf{p}', \mathbf{p}_1') \rightarrow (\mathbf{p}, \mathbf{p}_1)$ and its reverse by $(\mathbf{p}, \mathbf{p}_1) \rightarrow (\mathbf{p}', \mathbf{p}_1')$.

Knowing the inter-molecular force, one can in principle evaluate using the microscopic theory the rate of transition $W(\mathbf{p}, \mathbf{p}_1 | \mathbf{p}', \mathbf{p}_1')$ of the process $(\mathbf{p}', \mathbf{p}_1') \rightarrow (\mathbf{p}, \mathbf{p}_1)$. Our interest is in determining in terms of the said transition rate, the rate at which the number of molecules of given momenta change due to scattering. To that end, note that the number of pairs of molecules in the vicinity of $\mathbf{r}$, one of which has momentum in the vicinity of $\mathbf{p}'$ and another has that in the vicinity of $\mathbf{p}_1'$, is $F(\mathbf{r}, \mathbf{p}', \mathbf{p}_1'; t) d^3 \mathbf{p}' d^3 \mathbf{p}_1'$ where $F(\mathbf{r}, \mathbf{p}', \mathbf{p}_1'; t) \equiv f_2(\mathbf{r}, \mathbf{r}, \mathbf{p}', \mathbf{p}_1'; t)$. Hence the rate at which the process $(\mathbf{p}', \mathbf{p}_1') \rightarrow (\mathbf{p}, \mathbf{p}_1)$ generates a molecule in the vicinity of $\mathbf{p}$ and another in the vicinity of $\mathbf{p}_1$ is

$$R_+(\mathbf{p}, \mathbf{p}_1) d^3 \mathbf{p}' d^3 \mathbf{p}_1' d^3 \mathbf{p} d^3 \mathbf{p}_1$$
$$= W(\mathbf{p}, \mathbf{p}_1 | \mathbf{p}', \mathbf{p}_1') F(\mathbf{r}, \mathbf{p}', \mathbf{p}_1'; t) \delta^{(4)}(P - P') d^3 \mathbf{p}' d^3 \mathbf{p}_1' d^3 \mathbf{p} d^3 \mathbf{p}_1, \tag{3.100}$$

where, with $\mathbf{P} = \mathbf{p} + \mathbf{p}_1$, $E = (p^2 + p_1^2)/2m$, the delta-function

$$\delta^{(4)}(P - P') \equiv \delta^{(3)}(\mathbf{P} - \mathbf{P}') \delta(E - E') \tag{3.101}$$

restricts the values of the momenta only to those which are allowed by the conservation conditions (3.99). Hence the rate at which the molecules emerge with momentum in the vicinity of $\mathbf{p}$ is

$$\left(\frac{\partial f_1(\mathbf{r}, \mathbf{p}; t)}{\partial t} \right)_+ = \int R_+(\mathbf{p}, \mathbf{p}_1) d^3 \mathbf{p}' d^3 \mathbf{p}_1' d^3 \mathbf{p}_1. \tag{3.102}$$

Similarly, the rate of the reverse scattering $(\mathbf{p}, \mathbf{p}_1) \rightarrow (\mathbf{p}', \mathbf{p}_1')$ is

$$R_-(\mathbf{p}, \mathbf{p}_1) d^3 \mathbf{p}' d^3 \mathbf{p}_1' d^3 \mathbf{p} d^3 \mathbf{p}_1$$
$$= W(\mathbf{p}', \mathbf{p}_1' | \mathbf{p}, \mathbf{p}_1) F(\mathbf{r}, \mathbf{p}, \mathbf{p}_1; t) \delta^{(4)}(P' - P) d^3 \mathbf{p}' d^3 \mathbf{p}_1' d^3 \mathbf{p} d^3 \mathbf{p}_1, \tag{3.103}$$

so that the rate at which molecules having momentum in the vicinity of $\mathbf{p}$ are lost is

$$\left(\frac{\partial f_1(\mathbf{r}, \mathbf{p}; t)}{\partial t} \right)_- = \int R_-(\mathbf{p}, \mathbf{p}_1) d^3 \mathbf{p}' d^3 \mathbf{p}_1' d^3 \mathbf{p}_1. \tag{3.104}$$

Net rate of change of the number of molecules having momentum in the vicinity of $\mathbf{p}$ due to collisions therefore is

$$\left(\frac{\partial f_1(\mathbf{r}, \mathbf{p}; t)}{\partial t}\right)_{\text{coll}} = \left(\frac{\partial f_1(\mathbf{r}, \mathbf{p}; t)}{\partial t}\right)_{+} - \left(\frac{\partial f_1(\mathbf{r}, \mathbf{p}; t)}{\partial t}\right)_{-}. \tag{3.105}$$

It is known that, due to time-reversal symmetry,

$$W(\mathbf{p}', \mathbf{p}_1'|\mathbf{p}, \mathbf{p}_1) = W(\mathbf{p}, \mathbf{p}_1|\mathbf{p}', \mathbf{p}_1'). \tag{3.106}$$

Substitution of (3.102) and (3.104) in (3.105), along with the use of (3.106), yields the *collision integral*,

$$\left(\frac{\partial f_1(\mathbf{r}, \mathbf{p}; t)}{\partial t}\right)_{\text{coll}} = \int \Big[\Big(F(\mathbf{r}, \mathbf{p}', \mathbf{p}_1'; t) - F(\mathbf{r}, \mathbf{p}, \mathbf{p}_1; t)\Big)$$
$$\times W(\mathbf{p}, \mathbf{p}_1|\mathbf{p}', \mathbf{p}_1')\delta^{(4)}(P - P')\Big]d^3\mathbf{p}'d^3\mathbf{p}_1'd^3\mathbf{p}_1. \tag{3.107}$$

The expression above is exact for a sufficiently dilute gas. However, it is not useful as it stands because it has in it the unknown function $F(\mathbf{r}, \mathbf{p}, \mathbf{p}_1; t)$ which characterizes correlation between the momenta of two molecules.

To get a closed-form equation, Boltzmann approximated the correlation function $F(\mathbf{r}, \mathbf{p}, \mathbf{p}_1; t)$ by the product of single-particle distribution functions:

$$F(\mathbf{r}, \mathbf{p}, \mathbf{p}_1; t) \approx f_1(\mathbf{r}, \mathbf{p}; t)f_1(\mathbf{r}, \mathbf{p}_1; t). \tag{3.108}$$

This is called the assumption of *molecular chaos*. Equation (3.107) then reads

$$\left(\frac{\partial f_1(\mathbf{r}, \mathbf{p}; t)}{\partial t}\right)_{\text{coll}} = \int \Big[\Big(f_1(\mathbf{p}')f_1(\mathbf{p}_1') - f_1(\mathbf{p})f_1(\mathbf{p}_1)\Big)$$
$$\times W(\mathbf{p}, \mathbf{p}_1|\mathbf{p}', \mathbf{p}_1')\delta^{(4)}(P - P')\Big]d^3\mathbf{p}'d^3\mathbf{p}_1'd^3\mathbf{p}_1, \tag{3.109}$$

where for ease of writing we have suppressed $\mathbf{r}$ and t in the argument of f_1 under the integral above and written $f_1(\mathbf{p}) \equiv f_1(\mathbf{r}, \mathbf{p}; t)$. On substituting (3.109) in (3.98) we obtain closed-form equation for $f_1(\mathbf{p})$:

$$\frac{\partial f_1}{\partial t} - \{h(\mathbf{r}, \mathbf{p}), \ f_1(\mathbf{p})\}$$
$$= \int \Big[\Big(f_1(\mathbf{p}')f_1(\mathbf{p}_1') - f_1(\mathbf{p})f_1(\mathbf{p}_1)\Big)$$
$$\times W(\mathbf{p}, \mathbf{p}_1|\mathbf{p}', \mathbf{p}_1')\delta^{(4)}(P - P')\Big]d^3\mathbf{p}'d^3\mathbf{p}_1'd^3\mathbf{p}_1. \tag{3.110}$$

The non-linear integro-differential (3.110) is called the *Boltzmann transport equation* or simply the *Boltzmann equation*. A useful way of writing the scattering integral

(3.109) is to express the transition rate in terms of the scattering cross section. Using the said relation, derived in (3.125), (3.110) assumes the form

$$\frac{\partial f_1}{\partial t} + \frac{\mathbf{p}}{m} \cdot \nabla_{\mathbf{r}} f_1 - \nabla_{\mathbf{r}} U(\mathbf{r}) \cdot \nabla_{\mathbf{p}} f_1$$
$$= \frac{1}{m} \int |\mathbf{p} - \mathbf{p}_1| \Big(f_1(\mathbf{p}') f_1(\mathbf{p}_1') - f_1(\mathbf{p}) f_1(\mathbf{p}_1) \Big) \frac{d\sigma}{d\Omega} d\Omega d^3 \mathbf{p}_1, \quad (3.111)$$

where $d\sigma/d\Omega$ is the differential scattering cross section. Note that $(\mathbf{p}', \mathbf{p}_1')$ in the equation above are determined in terms of $(\mathbf{p}, \mathbf{p}_1)$ by conservation of energy and momentum.

In Sect. 3.6, we will derive (3.111) making explicit use of inter-particle potential. To that end, we recall first some relevant elements of the theory of elastic scattering.

3.5 Scattering

Consider elastic collision between two particles, one initially stationary at the origin of the coordinate system, called the target, and the other moving toward it, starting far away from it with velocity $\mathbf{v}$. Assuming the interaction to vanish at large separation between the particles, the trajectory of the incident particle shall be straight line in the direction of $\mathbf{v}$ when it is far away from the target. Let us take the line drawn parallel to $\mathbf{v}$ and passing through the target as the z-axis and the direction of $\mathbf{v}$ as the positive z-direction (see Fig. 3.1). Choose two directions orthogonal to each other and to the z-direction as the other two coordinate axes, x and y, of the Cartesian system. We denote the distance of the incoming particle from the z-axis when it is far away from the target, called the *impact parameter*, by b. It is the closest distance between the incident particle and the target in the absence of interaction between them. Draw a circle of radius b passing through the particle with the center of the circle on the z-axis and its plane perpendicular to it. The azimuthal angle ϕ that the line joining the particle with the center of the circle makes with the x-axis denotes the particle position on the circle.

Fig. 3.1 Schematic depiction of scattering

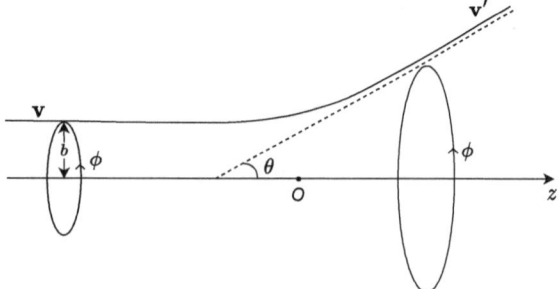

The asymptotic position of the incident particle before it comes close to the target is specified by the impact parameter b and the azimuthal angle ϕ.

Due to its interaction with the target, the trajectory of the incident particle deflects and the particle scattered as it approaches the target. Since we are assuming no interaction when the separation between the particles is large, asymptotic trajectory of the scattered particle is a straight line. The asymptotic position of the scattered particle is described by the angles (θ, ϕ) where θ is the angle that the asymptotic trajectory of the scattered particle makes with the z-axis and ϕ is its azimuthal angle, same as the incident azimuthal angle if the potential is central which is assumed to be the case. The asymptotic position of the particle after scattering is related with its asymptotic position before scattering and its energy. Knowing the potential, (θ, ϕ) can be calculated using the equations of motion.

Consider now the situation wherein, not one, but a stream of particles, all having initial velocity $\mathbf{v}$, is incident on a stationary target. The flux I of particles, defined as the number of particles passing per unit time per unit area perpendicular to its direction of propagation, is evidently

$$I = |\mathbf{v}|n_{\text{inc}}(\mathbf{r}, \mathbf{v}), \qquad (3.112)$$

where $n_{\text{inc}}(\mathbf{r}, \mathbf{v})$ is number of particles per unit volume in the incident beam.

Consider circles of radii b and $b + db$ with their planes perpendicular to the z-axis and their common center on that axis. Consider the region between those circles bound by the parts of the radii whose azimuthal angles are ϕ and $\phi + d\phi$. The area $d\sigma(b, \phi)$ of the said region evidently is

$$d\sigma(b, \phi) = b\, db\, d\phi, \qquad (3.113)$$

and the number of particles crossing it per unit time is $I d\sigma(b, \phi)$. All those particles are scattered within the solid angle $d\Omega(\theta, \phi)$ at (θ, ϕ) where

$$d\Omega(\theta, \phi) = \sin(\theta)d\theta d\phi. \qquad (3.114)$$

Let the number of particles scattered per unit time per unit solid angle surrounding (θ, ϕ), called the *scattered flux*, be $N_s(\theta, \phi)$ so that the number of particles scattered into $d\Omega(\theta, \phi)$ is $N_s(\theta, \phi)d\Omega(\theta, \phi)$. Since the particles scattered into $d\Omega$ are those which were in the area $d\sigma(b, \phi)$ in the incident beam, we must have

$$N_s(\theta, \phi)d\Omega(\theta, \phi) = I d\sigma(b, \phi). \qquad (3.115)$$

The ratio $N_s(\theta, \phi)/I$ of the scattered flux to the incident flux I is called the *differential scattering cross section* or simply the *differential cross section*. The expression (3.115) shows that

$$\frac{N_s(\theta, \phi)}{I} = \frac{d\sigma}{d\Omega}. \qquad (3.116)$$

The differential cross section is thus given by $d\sigma/d\Omega$. It has the units of area.

We can also define *total scattering cross section* σ_T:

$$\sigma_T = \int \frac{d\sigma}{d\Omega}\, d\Omega. \tag{3.117}$$

It has the units of area.

We can generalize the considerations above to the case of the target not initially stationary but moving initially with velocity $\mathbf{v}_1$. That end is achieved easily by working in the rest frame of the target. The expression (3.112) for the incident flux then gets changed to

$$I = |\mathbf{v} - \mathbf{v}_1| n_{\text{inc}}(\mathbf{r}, \mathbf{v}). \tag{3.118}$$

Hence, invoking (3.115) with I given by (3.118), the number of particles scattered from $d\sigma$ into $d\Omega$ is

$$N_s(\theta, \phi)d\Omega = |\mathbf{v} - \mathbf{v}_1| n_{\text{inc}}(\mathbf{r}, \mathbf{v})d\sigma. \tag{3.119}$$

The results above may be generalized to the case of incident particles having distribution of velocities described by the phase space distribution function $f_1(\mathbf{r}, \mathbf{p}; t)$ so that $n_{\text{inc}}(\mathbf{r}, \mathbf{v}) = f_1(\mathbf{r}, \mathbf{p}; t)d^3\mathbf{p}$ yielding

$$I = |\mathbf{v} - \mathbf{v}_1| f_1(\mathbf{r}, \mathbf{p}; t)d^3\mathbf{p}, \tag{3.120}$$

and

$$N_s(\theta, \phi)d\Omega = |\mathbf{v} - \mathbf{v}_1| f_1(\mathbf{r}, \mathbf{p}; t)d^3\mathbf{p}d\sigma. \tag{3.121}$$

Next, consider a dilute gas having a distribution of molecular velocities. The gas is assumed sufficiently dilute so that only binary collisions are important. Consider collisions between molecules of velocity $\mathbf{v}$ and $\mathbf{v}_1$. The (3.121) gives the rate of scattering of molecules of velocity $\mathbf{v} = \mathbf{p}/m$ in the vicinity of $\mathbf{r}$ into $d\Omega$ if there is a molecule with velocity $\mathbf{v}_1$ also in the vicinity of $\mathbf{r}$. In the present case, the probability that a molecule in the vicinity of $\mathbf{r}$ has velocity in the vicinity of $\mathbf{v}_1 = \mathbf{p}_1/m$ is $f_1(\mathbf{r}, \mathbf{p}_1; t)d^3\mathbf{r}d^3\mathbf{p}$. Hence the rate of scattering into $d\Omega$ in the present case is given by multiplying (3.121) by the said probability:

$$N_s(\theta, \phi)d\Omega = |\mathbf{v} - \mathbf{v}_1| f_1(\mathbf{r}, \mathbf{p}; t) f_1(\mathbf{r}, \mathbf{p}_1; t)d^3\mathbf{r}d^3\mathbf{p}d^3\mathbf{p}_1 d\sigma. \tag{3.122}$$

The assumption that the distribution of the velocities of the target and the incident particles is independent is inherent in the derivation of the expression above and is behind the product form $f_1(\mathbf{r}, \mathbf{p}; t) f_1(\mathbf{r}, \mathbf{p}_1; t)$ of the distribution functions. However, the two velocities need not be distributed independently. To account for that possibility, replace $f_1(\mathbf{r}, \mathbf{p}; t) f_1(\mathbf{r}, \mathbf{p}_1; t)$ by two-particle distribution function

$f_2(\mathbf{r}, \mathbf{r}, \mathbf{p}, \mathbf{p}_1; t) \equiv F(\mathbf{r}, \mathbf{p}, \mathbf{p}_1; t)$ to rewrite (3.122) as

$$N_s(\theta, \phi)\mathrm{d}\Omega = |\mathbf{v} - \mathbf{v}_1| F(\mathbf{r}, \mathbf{p}, \mathbf{p}_1; t)\mathrm{d}^3 r\mathrm{d}^3 p\mathrm{d}^3 p_1 \mathrm{d}\sigma. \tag{3.123}$$

Now, if the molecules of momenta in the vicinities of $\mathbf{p}$ and $\mathbf{p}_1$ collide and emerge as having momenta in the vicinities of $\mathbf{p}'$ and $\mathbf{p}'_1$ then, in terms of the transition rate, the number of molecule in the vicinities of $\mathbf{p}'$ and $\mathbf{p}'_1$ is

$$N_s(\theta, \phi)\mathrm{d}\Omega = W(\mathbf{p}', \mathbf{p}'_1 | \mathbf{p}, \mathbf{p}_1) F(\mathbf{r}, \mathbf{p}, \mathbf{p}_1; t)\delta^{(4)}(P - P')$$
$$\times \mathrm{d}^3 r\mathrm{d}^3 p\mathrm{d}^3 p_1 \mathrm{d}^3 p'\mathrm{d}^3 p'_1. \tag{3.124}$$

On equating (3.123) and (3.124) follows the relation,

$$|\mathbf{v} - \mathbf{v}_1|\frac{\mathrm{d}\sigma}{\mathrm{d}\Omega}\mathrm{d}\Omega = W(\mathbf{p}', \mathbf{p}'_1 | \mathbf{p}, \mathbf{p}_1)\delta^{(4)}(P - P')\mathrm{d}^3 p'\mathrm{d}^3 p'_1, \tag{3.125}$$

relating transition rate with the scattering cross section.

We derive next the Boltzmann equation working in N-particles phase space.

Exercises

Ex. 3.6. Let a particle of mass m_1 moving with velocity $\mathbf{v}_1$ collide elastically with another one of mass m_2 moving with velocity $\mathbf{v}_2$. Show that their relative velocity before and after collision has same magnitude. In other words, the relative velocity in binary collision is rotated without change in its magnitude. Since relative velocity and relative momentum are proportional to each other when $m_1 = m_2$, we say that, the relative momentum of equal mass particles gets rotated without change in its magnitude in elastic collision. Hint: Let $\mathbf{R} = (m_1\mathbf{r}_1 + m_2\mathbf{r}_2)/M$ $(M = m_1 + m_2)$ denote the position of the center of mass of the particles and $\mathbf{r} - \mathbf{r}_1 - \mathbf{r}_2$ their relative position so that

$$\mathbf{r}_1 = \mathbf{R} + \frac{m_2}{M}\mathbf{r}, \qquad \mathbf{r}_2 = \mathbf{R} - \frac{m_1}{M}\mathbf{r}. \tag{3.126}$$

Total momentum and kinetic energy before collision are given by

$$\mathbf{P} = m_1\dot{\mathbf{r}}_1 + m_2\dot{\mathbf{r}}_2 \equiv M\dot{\mathbf{R}}, \quad T = \frac{1}{2M}P^2 + \frac{\mu}{2}\dot{v}^2, \tag{3.127}$$

where $\mu = m_1 m_2/M$ is the reduced mass and $\mathbf{v} = \dot{\mathbf{r}}_1 - \dot{\mathbf{r}}_2$ is relative velocity. The kinetic energy after collision is given by

$$T' = \frac{1}{2M}P'^2 + \frac{\mu}{2}\dot{v}'^2. \tag{3.128}$$

Since momentum and kinetic energy remain constant in elastic collision, (the prime on a symbol stands for its value after collision),

$$\mathbf{P}' = \mathbf{P}, \qquad T' = T, \tag{3.129}$$

it follows that

$$\mathbf{v}^2 = \mathbf{v}'^2. \tag{3.130}$$

Hence the desired result.

3.6 BBGKY Hierarchy

Starting from Liouville's equation for the phase space evolution of the distribution function $f_N(\{\mathbf{r}_i, \mathbf{p}_i\}_N; t)$ of all N molecules, in this section, we derive equation governing the evolution of s-particle distribution function $f_s(\{\mathbf{r}_i, \mathbf{p}_i\}_s; t)$. We will see that the evolution equation for f_s is coupled with f_{s+1}. We will use the hierarchy of the said equations to derive the Boltzmann equation under appropriate approximations.

To that end, we consider to begin with the gas of non-interacting molecules. The molecules may, however, be acted upon by an external force. The Hamiltonian H_N governing the evolution of the system is then given by (3.95). Substitute H_N given therein in Liouville's equation (3.56) and integrate over all $\mathbf{r}_i, \mathbf{p}_i$, excluding $i = 1$, to get

$$\frac{\partial \tilde{f}_1(\mathbf{r}_1, \mathbf{p}_1; t)}{\partial t} = \sum_{i=1}^{N} \int \{h(\mathbf{r}_i, \mathbf{p}_i), \ f_N\} \prod_{k \neq 1} d\mu_k. \tag{3.131}$$

Consider the $i = 1$ term in the equation above. It involves the Poisson bracket,

$$\begin{aligned} &\{h(\mathbf{r}_1, \mathbf{p}_1), \ f_N\} \\ &= \left[\nabla_{\mathbf{r}_1} h(\mathbf{r}_1, \mathbf{p}_1) \cdot \nabla_{\mathbf{p}_1} f_N - \nabla_{\mathbf{p}_1} h(\mathbf{r}_1, \mathbf{p}_1) \cdot \nabla_{\mathbf{r}_1} f_N \right]. \end{aligned} \tag{3.132}$$

Since h in the equation above and the derivatives therein are independent of the variables of integration, we can shift the integration entirely on the function f_N as that is the only function in the integrand in (3.131) depending on the variables of integration to obtain

$$\int \{h(\mathbf{r}_1, \mathbf{p}_1), \ f_N\} \prod_{k \neq 1} d\mu_k = \left\{ h(\mathbf{r}_1, \mathbf{p}_1), \ \int f_N \prod_{k \neq 1} d\mu_k \right\}$$

$$= \{h(\mathbf{r}_1, \mathbf{p}_1), \ \tilde{f}_1(\mathbf{r}_1, \mathbf{p}_1)\}. \tag{3.133}$$

Consider now a term in (3.131) for $i \neq 1$:

$$\int \{h(\mathbf{r}_i, \mathbf{p}_i), \ f_N\} \prod_{k \neq 1} d\mu_k$$

$$= \int \left[\nabla_{\mathbf{r}_i} h \cdot \nabla_{\mathbf{p}_i} f_N - \nabla_{\mathbf{p}_i} h \cdot \nabla_{\mathbf{r}_i} f_N \right] d\mu_i \prod_{k \neq 1, i} d\mu_k. \qquad (3.134)$$

Invoking the identity

$$\int_V \mathbf{F} \cdot \nabla \psi \, dV = \int_S \psi \mathbf{F} \cdot \mathbf{n} \, dS - \int_V \psi \nabla \cdot \mathbf{F} \, dV, \qquad (3.135)$$

where S is the surface enclosing the volume V and $\mathbf{n}$ is the unit vector normal to the surface element dS. Application of (3.135) to the integration over the momentum space volume in the first term in (3.134) gives

$$\int \nabla_{\mathbf{r}_i} h \cdot \nabla_{\mathbf{p}_i} f_N \, d^3 \mathbf{p}_i$$

$$= \int_S f_N \nabla_{\mathbf{r}_i} h \cdot \mathbf{n} \, dS - \int f_N \nabla_{\mathbf{p}_i} \cdot \nabla_{\mathbf{r}_i} h \, d^3 \mathbf{p}_i. \qquad (3.136)$$

Since integration is over entire phase space, the surface S is at infinity and hence the value of f_N in the first term in (3.136) is for $\mathbf{p}_i$ at infinity. The contribution from the surface integral in (3.136) will be zero if it is assumed that $f_N \to 0$ when $|\mathbf{p}_i| \to \infty$. Consequently (3.136) reduces to

$$\int \nabla_{\mathbf{r}_i} h \cdot \nabla_{\mathbf{p}_i} f_N \, d^3 \mathbf{p}_i = - \int f_N \nabla_{\mathbf{p}_i} \cdot \nabla_{\mathbf{r}_i} h \, d^3 \mathbf{p}_i. \qquad (3.137)$$

Evaluating the second term on the right-hand side in (3.134) in similar way by assuming $f_N \to 0$ as $|\mathbf{r}_i| \to \infty$, we get

$$\int \nabla_{\mathbf{p}_i} h \cdot \nabla_{\mathbf{r}_i} f_N \, d^3 \mathbf{r}_i = - \int f_N \nabla_{\mathbf{r}_i} \cdot \nabla_{\mathbf{p}_i} h \, d^3 \mathbf{r}_i. \qquad (3.138)$$

The right sides of (3.137) and (3.138) are same. Hence, on substituting (3.137) and (3.138) in (3.134) it follows that

$$\int \{h(\mathbf{r}_i, \mathbf{p}_i), \ f_N\} \prod_{k \neq 1} d\mu_k = 0, \qquad i \neq 1. \qquad (3.139)$$

Substitute (3.133) and (3.139) in (3.131) to get

$$\frac{\partial \tilde{f}_1(\mathbf{r}, \mathbf{p})}{\partial t} = \{h(\mathbf{r}, \mathbf{p}), \ \tilde{f}_1(\mathbf{r}, \mathbf{p})\}. \tag{3.140}$$

Multiply the equation above by N and use (3.64) to transform it to the equation for $f_1(\mathbf{r}, \mathbf{p}; t)$:

$$\frac{\partial f_1(\mathbf{r}, \mathbf{p}; t)}{\partial t} - \{h(\mathbf{r}, \mathbf{p}), \ f_1(\mathbf{r}, \mathbf{p}; t)\} = 0. \tag{3.141}$$

This is the equation for computing single-particle distribution function in the absence of mutual interaction between molecules.

We can similarly derive the equation for f_s corresponding to H_N given by (3.95) by integrating Liouville's (3.56) over $\mathbf{r}_i$, $\mathbf{p}_i$ for $i = s+1, s+2, \dots, N$ to get

$$\frac{\partial \tilde{f}_s(\{\mathbf{r}_i, \mathbf{p}_i\}_s; t)}{\partial t} = \sum_{i=1}^{N} \int \{h(\mathbf{r}_i, \mathbf{p}_i), \ f_N\} \prod_{k \geq s+1} d\mu_k. \tag{3.142}$$

Consider the Poisson bracket for an $i \leq s$ term in the equation above:

$$\{h(\mathbf{r}_i, \mathbf{p}_i), \ f_N\}$$
$$= \left[\nabla_{\mathbf{r}_i} h(\mathbf{r}_i, \mathbf{p}_i) \cdot \nabla_{\mathbf{p}_i} f_N - \nabla_{\mathbf{p}_i} h(\mathbf{r}_i, \mathbf{p}_i) \cdot \nabla_{\mathbf{r}_i} f_N \right], \quad i \leq s. \tag{3.143}$$

Since h in the equation above and the derivatives are independent of the variables of integration, we can write

$$\int \{h(\mathbf{r}_i, \mathbf{p}_i), \ f_N\} \prod_{k \geq s+1} d\mu_k$$
$$= \left\{ h(\mathbf{r}_i, \mathbf{p}_i), \ \int f_N \prod_{k \geq s+1} d\mu_k \right\}$$
$$= \{h(\mathbf{r}_i, \mathbf{p}_i), \ \tilde{f}_s(\{\mathbf{r}_i, \mathbf{p}_i\}_s; t)\}, \quad i \leq s. \tag{3.144}$$

Consider next the Poisson bracket for an $i \geq s+1$ term in (3.142). Following the steps leading to (3.139), it is straightforward to show that the contribution due to such terms vanishes:

$$\int \{h(\mathbf{r}_i, \mathbf{p}_i), \ f_N\} \prod_{k \geq s+1} d\mu_k = 0, \quad i \geq s+1. \tag{3.145}$$

Substitute (3.144) and (3.145) in (3.142) to get the following equation for $\tilde{f}_s$:

$$\frac{\partial \tilde{f}_s(\{\mathbf{r}_i, \mathbf{p}_i\}_s; t)}{\partial t} = \sum_{i=1}^{s} \{h(\mathbf{r}_i, \mathbf{p}_i), \ \tilde{f}_s\}. \tag{3.146}$$

Multiply the equation above by $N!/(N-s)!$ and use (3.78) to transform it to the equation for f_s:

$$\frac{\partial f_s(\{\mathbf{r}_i, \mathbf{p}_i\}_s; t)}{\partial t} = \sum_{i=1}^{s} \{h(\mathbf{r}_i, \mathbf{p}_i), f_s\}. \tag{3.147}$$

This is the closed-form equation for the s-particle distribution function in the absence of interaction between the molecules.

Next let us assume that the molecules interact mutually. Assuming the mutual interaction to be two body and dependent only on the distance between molecules, the Hamiltonian of the system may be written as

$$H_N = \sum_{i=1}^{N} h(\mathbf{r}_i, \mathbf{p}_i) + \frac{1}{2} \sum_{i\neq j=1}^{N} W(|\mathbf{r}_i - \mathbf{r}_j|), \tag{3.148}$$

where $h(\mathbf{r}_i, \mathbf{p}_i)$ is as in (3.95) and $W(|\mathbf{r}_i - \mathbf{r}_j|)$ is the interaction potential between the ith and the jth molecule. It will turn out to be convenient to write the double sum in (3.148) as

$$\sum_{i\neq j=1}^{N} W(|\mathbf{r}_i - \mathbf{r}_j|) = \left(\sum_{i\neq j=1}^{s} + \sum_{i\neq j=s+1}^{N} + 2\sum_{i=1}^{s} \sum_{j=s+1}^{N} \right) W(|\mathbf{r}_i - \mathbf{r}_j|). \tag{3.149}$$

This equation is the result of breaking the sum over i, j form 1 to N into two parts, 1 to s plus $s+1$ to N, and writing the double sum over i, j as

$$\left[\sum_{i=1}^{s} \sum_{j=s+1}^{N} + \sum_{i=s+1}^{N} \sum_{j=1}^{s} \right] W(|\mathbf{r}_i - \mathbf{r}_j|) = 2\sum_{i=1}^{s} \sum_{j=s+1}^{N} W(|\mathbf{r}_i - \mathbf{r}_j|), \tag{3.150}$$

obtained by interchanging i and j in the second sum and noting that $W(|\mathbf{r}_i - \mathbf{r}_j|)$ is unchanged under the exchange of i and j.

To derive equation for f_s we integrate the Liouville equation (3.56) over $s+1, s+2, \ldots, N$. Equation (3.146) gives the result of integration when $W = 0$. Denoting the contribution from the mutual interaction term by I_s,

$$I_s = \frac{1}{2} \sum_{i\neq j=1}^{N} \int \{W(|\mathbf{r}_i - \mathbf{r}_j|), f_N\} \prod_{k\geq s+1} d^3\mathbf{r}_k d^3\mathbf{p}_k, \tag{3.151}$$

the equation for $\tilde{f}_s$ assumes the form

$$\frac{\partial \tilde{f}_s(\{\mathbf{r}_i, \mathbf{p}_i\}_s; t)}{\partial t} = \sum_{i=1}^{s} \{h(\mathbf{r}_i, \mathbf{p}_i), \tilde{f}_s\} + I_s. \tag{3.152}$$

Since $W(|\mathbf{r}_i - \mathbf{r}_j|)$ is independent of momenta, we have

$$\{W(|\mathbf{r}_i - \mathbf{r}_j|), f_N\} = \nabla_{\mathbf{r}_i} W(|\mathbf{r}_i - \mathbf{r}_j|) \cdot \nabla_{\mathbf{p}_i} f_N + \nabla_{\mathbf{r}_j} W(|\mathbf{r}_i - \mathbf{r}_j|) \cdot \nabla_{\mathbf{p}_j} f_N. \tag{3.153}$$

We evaluate I_s by using the identity

$$\int \mathbf{F}(\mathbf{r}) \cdot \nabla_{\mathbf{p}} f(\mathbf{r}, \mathbf{p}) \, d^3\mathbf{p} = -\int f(\mathbf{r}, \mathbf{p}) \nabla_{\mathbf{p}} \cdot \mathbf{F}(\mathbf{r}) \, d^3\mathbf{p} = 0, \tag{3.154}$$

arrived at by invoking the identity (3.135) and the assumption that the surface terms do not contribute.

To evaluate I_s, use the breakup of the sum as in (3.149) to write I_s in (3.151) as

$$I_s = I_{s1} + I_{s2} + I_{s3}, \tag{3.155}$$

where

$$I_{s1} = \frac{1}{2} \sum_{i \neq j=1}^{s} \int \{W(|\mathbf{r}_i - \mathbf{r}_j|), f_N\} \prod_{k \geq s+1} d\mu_k, \tag{3.156}$$

$$I_{s2} = \frac{1}{2} \sum_{i \neq j=s+1}^{N} \int \{W(|\mathbf{r}_i - \mathbf{r}_j|), f_N\} \prod_{k \geq s+1} d\mu_k, \tag{3.157}$$

$$I_{s3} = \sum_{i=1}^{s} \sum_{j=s+1}^{N} \int \{W(|\mathbf{r}_i - \mathbf{r}_j|), f_N\} \prod_{k \geq s+1} d\mu_k. \tag{3.158}$$

The I_{sk} can be evaluated as follows.

1. Consider I_{s1}. The summation over i, j in it is from 1 to s whereas the integration is over the phase space coordinates $\mathbf{r}_k, \mathbf{p}_k$ for $k \geq s+1$. The integration in (3.156) can therefore be performed only over f_N to obtain

$$I_{s1} = \frac{1}{2} \sum_{i \neq j=1}^{s} \{W(|\mathbf{r}_i - \mathbf{r}_j|), \tilde{f}_s\}. \tag{3.159}$$

2. To evaluate I_{s2} we write the commutator in it using (3.153) so that

$$I_{s2} = \frac{1}{2} \sum_{i \neq j = s+1}^{N} \int \left[\nabla_{\mathbf{r}_i} W(|\mathbf{r}_i - \mathbf{r}_j|) \cdot \nabla_{\mathbf{p}_i} f_N \right.$$
$$\left. + \nabla_{\mathbf{r}_j} W(|\mathbf{r}_i - \mathbf{r}_j|) \cdot \nabla_{\mathbf{p}_j} f_N \right] \cdot \prod_{k \geq s+1} d\mu_k \qquad (3.160)$$

Since i, j are both greater than $s + 1$, the momenta with respect to which derivatives in (3.160) are carried are same as those which are integrated. Hence, invoking (3.154)

$$I_{s2} = 0. \qquad (3.161)$$

3. Using (3.153) (3.158) for I_{s3} reads

$$I_{s3} = \sum_{i=1}^{s} \sum_{j=s+1}^{N} \int \left[\nabla_{\mathbf{r}_i} W(|\mathbf{r}_i - \mathbf{r}_j|) \cdot \nabla_{\mathbf{p}_i} f_N \right.$$
$$\left. + \nabla_{\mathbf{r}_j} W(|\mathbf{r}_i - \mathbf{r}_j|) \cdot \nabla_{\mathbf{p}_j} f_N \right] \prod_{k \geq s+1} d\mu_k. \qquad (3.162)$$

Note that the momenta with respect to which the derivatives in the second term in (3.162) are carried are same as the ones which are integrated. Hence the contribution from the second term vanishes due to (3.154) reducing I_{s3} to

$$I_{s3} = \sum_{i=1}^{s} \sum_{j=s+1}^{N} \int \nabla_{\mathbf{r}_i} W(|\mathbf{r}_i - \mathbf{r}_j|) \cdot \nabla_{\mathbf{p}_i} f_N \prod_{k \geq s+1} d\mu_k. \qquad (3.163)$$

This may be rewritten as

$$I_{s3} = \sum_{i=1}^{s} \sum_{j=s+1}^{N} \int \nabla_{\mathbf{r}_i} W(|\mathbf{r}_i - \mathbf{r}_j|) \cdot \nabla_{\mathbf{p}_i} f_N \prod_{k \geq s+1} d\mu_k$$
$$= \sum_{i=1}^{s} \sum_{j=s+1}^{N} \int \left[\nabla_{\mathbf{r}_i} W(|\mathbf{r}_i - \mathbf{r}_j|) \cdot \nabla_{\mathbf{p}_i} \int f_N \prod_{k \geq s+1, k \neq j} d\mu_k \right] d\mu_j.$$
$$\qquad (3.164)$$

The integration of f_N in the equation above is over all phase space variables except over $(\mathbf{r}_l, \mathbf{p}_l)$ ($l = 1, 2, \ldots s$) and $l = j$ ($j \geq s + 1$). Hence

$$\int f_N(\{\mathbf{r}_i, \mathbf{p}_i\}_N; t) \prod_{k \geq s+1, k \neq j} d\mu_k = \tilde{f}_{s+1}(\{\mathbf{r}_i, \mathbf{p}_i\}_s, \mathbf{r}_j, \mathbf{p}_j; t). \qquad (3.165)$$

Substitute this in (3.164) and use the symmetry of f_N under exchange of particles to get

$$I_{s3} = \sum_{i=1}^{s} \sum_{j=s+1}^{N} \int \nabla_{\mathbf{r}_i} W(|\mathbf{r}_i - \mathbf{r}_j|) \cdot \nabla_{\mathbf{p}_i} \tilde{f}_{s+1}(\{\mathbf{r}_i, \mathbf{p}_i\}_s, \mathbf{r}_j, \mathbf{p}_j; t) \, \mathrm{d}\mu_j$$

$$= (N - s) \sum_{i=1}^{s} \int \nabla_{\mathbf{r}_i} W(|\mathbf{r}_i - \mathbf{r}_{s+1}|) \cdot \nabla_{\mathbf{p}_i} \tilde{f}_{s+1} \, \mathrm{d}\mu_{s+1}. \qquad (3.166)$$

Substitute (3.159), (3.161), and (3.166) in (3.155) to get

$$I_s = \frac{1}{2} \sum_{i \neq j=1}^{s} \left\{ W(|\mathbf{r}_i - \mathbf{r}_j|), \ \tilde{f}_s \right\}$$

$$+ (N - s) \sum_{i=1}^{s} \int \nabla_{\mathbf{r}_i} W(|\mathbf{r}_i - \mathbf{r}_{s+1}|) \cdot \nabla_{\mathbf{p}_i} \tilde{f}_{s+1}(\{\mathbf{r}_i, \mathbf{p}_i\}_{s+1}; t) \mathrm{d}\mu_{s+1}.$$

$$(3.167)$$

Substitute (3.167) in (3.152), multiply the resulting equation by $N!/(N - s)!$ and invoke the relation (3.78) to arrive at the following equation for the s-particle distribution function:

$$\frac{\partial f_s}{\partial t} - \left\{ H_s, \ f_s \right\} = \sum_{i=1}^{s} \int \nabla_{\mathbf{r}_i} W(|\mathbf{r}_i - \mathbf{r}_{s+1}|) \cdot \nabla_{\mathbf{p}_i} f_{s+1} \mathrm{d}\mu_{s+1}, \qquad (3.168)$$

where

$$H_s = \sum_{i=1}^{s} h(\mathbf{r}_i, \mathbf{p}_i) + \frac{1}{2} \sum_{i \neq j=1}^{s} W(|\mathbf{r}_i - \mathbf{r}_j|). \qquad (3.169)$$

Consider the equation for f_1. Since $s = 1$, there is no $i \neq j$ term in the sum over i, j in the expression (3.169) for H_s. Hence the interaction term will not appear in H_s in the equation for f_1 obtained from (3.168) by taking $s = 1$:

$$\frac{\partial f_1}{\partial t} - \left\{ h(\mathbf{r}, \mathbf{p}), \ f_1 \right\} = \int \nabla_{\mathbf{r}} W(|\mathbf{r} - \mathbf{r}_1|) \cdot \nabla_{\mathbf{p}} f_2 \, \mathrm{d}\mu_1. \qquad (3.170)$$

The equation for f_1 is thus different from the equations for f_s ($s \geq 2$) in that it does not have the commutator with the interaction potential term in it.

The chain of equations (3.169) and (3.170), relating the evolution of f_s ($s = 1, 2, \ldots, N$) with f_{s+1}, is called the *Bogoliubov–Born–Green–Kirkwood–Yvon* (BBGKY) hierarchy. The determination of f_1 thus requires solving N equations, i.e. as many equations as the number of particles! It is therefore not of help till the chain of equations is terminated at an early step.

In the following we describe conditions under which the BBGKY hierarchy can be terminated at the equation for f_2 to obtain the Boltzmann equation.

3.7 Boltzmann Equation from BBGKY Hierarchy

The Boltzmann equation (3.110) is a closed-form equation for $f_1(\mathbf{r}, \mathbf{p}; t)$. On the other hand, the equation (3.170) for $f_1(\mathbf{r}, \mathbf{p}; t)$ is the first in the BBGKY hierarchy. It is coupled with f_2 whose evolution is governed by (3.168) corresponding to $s = 2$:

$$\frac{\partial f_2}{\partial t} - \left\{ H_2, \ f_2 \right\} = \sum_{i=1}^{2} \int \nabla_{\mathbf{r}_i} W(|\mathbf{r}_i - \mathbf{r}_3|) \cdot \nabla_{\mathbf{p}_i} f_3 \, d\mu_3. \qquad (3.171)$$

The equation for f_2 above is coupled with f_3. By considering relative magnitude of various terms in (3.171) under certain assumptions, we will show that the integral on the right containing f_3 can be ignored. To that end note that each term in the equation for f_s has dimensions of f_s/time.

The explicit expression of the streaming term in (3.171) is

$$\left\{ H_2, \ f_2 \right\} = \sum_{i=1,2} \left[\nabla_{\mathbf{r}_i} U(\mathbf{r}_i) \cdot \nabla_{\mathbf{p}_i} f_2 - \frac{1}{m} \mathbf{p}_i \cdot \nabla_{\mathbf{r}_i} f_2 \right.$$
$$\left. + \nabla_{\mathbf{r}_i} W(|\mathbf{r}_1 - \mathbf{r}_2|) \cdot \nabla_{\mathbf{p}_i} f_2 \right]. \qquad (3.172)$$

For simplicity we assume $U(\mathbf{r}) = 0$ and estimate relative magnitude of various terms in (3.171) based on the following considerations: (1) We consider such inter-molecular potentials which are short ranged in the sense that they are appreciable over distances $r_0 \sim 10^{-8}$ cm, i.e. on the atomic scale. This will exclude potentials such as Coulomb potential which would require different treatment. (2) Typical speed v of the molecules in a gas at room temperature is $\sim 10^4$ cm s^{-1}.

Since, as observed before, each term in the equation for f_s has dimensions of f_s/time, we determine in the following the time scale on which each term makes dominant contribution.

1. Consider the terms involving inter-molecular potential $W(|\mathbf{r}|)$ in (3.172):

$$I_c = \nabla_{\mathbf{r}_i} W(|\mathbf{r}_1 - \mathbf{r}_2|) \cdot \nabla_{\mathbf{p}_i} f_2, \quad i = 1, 2. \qquad (3.173)$$

Clearly, dominant contribution to I_c comes from distances over which inter-molecular potential $W(|\mathbf{r}|)$ varies appreciably. Due to the assumptions listed circa (3.172), the distance over which $W(|\mathbf{r}|)$ varies appreciably is its range $r_0 \sim 10^{-8}$ cm and the molecules move with typical speed $v \sim 10^4$ cm s^{-1}. Hence dominant contribution to I_c is on the time scale τ_c where

$$\tau_c \sim r_0/v \sim 10^{-12} \text{ sec.} \tag{3.174}$$

Evidently τ_c is the time that the particle takes to traverse the interaction region or the *duration of collision*.

2. We now estimate the time scale on which the integral on the right side of (3.171) makes dominant contribution. Note that whereas the left side of (3.171) has f_2, the integral on its right has f_3. To estimate relative magnitude of the f_2 and f_3 terms in the equation, we recall the defining relation:

$$\tilde{f}_2(\mathbf{r}_1, \mathbf{r}_2, \mathbf{p}_1, \mathbf{p}_2; t) = \int \tilde{f}_3(\mathbf{r}_1, \mathbf{r}_2, \mathbf{r}_3, \mathbf{p}_1, \mathbf{p}_2, \mathbf{p}_3; t) \, d\mu_3 \implies$$

$$\int f_3 \, d\mu_3 = (N-2) f_2 \approx N f_2. \tag{3.175}$$

Since f_2 is independent of $\mathbf{r}_3$, we can write

$$f_2 = \frac{1}{V} \int f_2 d^3 \mathbf{r}_3. \tag{3.176}$$

The expression (3.175) may then be rewritten as

$$\int f_3 \, d^3 \mathbf{r}_3 d^3 \mathbf{p}_3 = \frac{N}{V} \int f_2 d^3 \mathbf{r}_3. \tag{3.177}$$

Using this we can approximate the integral in (3.171) by

$$I_{\text{int}} = \int \nabla_{\mathbf{r}_i} W(|\mathbf{r}_i - \mathbf{r}_3|) \cdot \nabla_{\mathbf{p}_i} f_3 \, d^3 \mathbf{r}_3 d^3 \mathbf{p}_3$$

$$\sim \frac{N}{V} \int \nabla_{\mathbf{r}_i} W(|\mathbf{r}_i - \mathbf{r}_3|) \cdot \nabla_{\mathbf{p}_i} f_2 d^3 \mathbf{r}_3, \quad i = 1, 2. \tag{3.178}$$

Due to the assumptions listed below (3.172), the volume integral above is restricted to r_0^3. Recalling also the definition (3.173) of I_c, we can write

$$I_{\text{int}} \sim \frac{N}{V} r_0^3 I_c. \tag{3.179}$$

Since dominant contribution to I_c is on the time scale τ_c, it follows that the time scale τ_{int} on which I_{int} is dominant given by

$$\tau_{\text{int}} \sim \frac{V}{N} r_0^{-3} \tau_c \sim 10^4 \tau_c, \tag{3.180}$$

where we have taken $N/V \approx 10^{20}$ cm^{-3}. This shows that

$$\tau_{\text{int}} >> \tau_c. \tag{3.181}$$

We thus see that the contribution from the integral term in (3.171) is much smaller than that from the inter-molecular potential term in $\{H_2, \ f_2\}$ and can be neglected thereby reducing (3.171) to the form

$$\frac{\partial f_2}{\partial t} - \left\{ H_2, \ f_2 \right\} = 0. \tag{3.182}$$

Since the dominant term in it is $\sim 1/\tau_c$, the f_2 reaches equilibrium on the time scale of τ_c. We show next that the time scale on which f_1 evolves is $\sim \tau_{\text{int}}$ which, due to (3.181) is much longer than τ_c.

To that end, recall equation (3.170) for f_1, rewritten as (for $U(\mathbf{r}) = 0$)

$$\frac{\partial f_1(\mathbf{r}, \mathbf{p}; t)}{\partial t} - \frac{1}{m} \mathbf{p} \cdot \nabla_{\mathbf{r}} f_1 = \left(\frac{\partial f_1}{\partial t} \right)_{\text{coll}}, \tag{3.183}$$

where

$$\left(\frac{\partial f_1}{\partial t} \right)_{\text{coll}} = \int \nabla_{\mathbf{r}} W(|\mathbf{r} - \mathbf{r}_1|) \cdot \nabla_{\mathbf{p}} f_2(\mathbf{r}, \mathbf{r}_1, \mathbf{p}, \mathbf{p}_1; t) \, d\mu_1. \tag{3.184}$$

By using the same arguments as were used for estimating the integral in the equation for f_2, it can be seen that the contribution of the integral term is $1/\tau_{\text{int}}$ given by (3.180). Hence, due to the absence of inter-molecular interaction in the streaming term which amounts to absence of $1/\tau_c$ contribution, the dominant contribution to (3.183) comes from the integral term. The time evolution of f_1 is therefore on the time scale of τ_{int}. Since $\tau_{\text{int}} >> \tau_c$ where τ_c is the scale of evolution of f_2, it follows that f_2 reaches equilibrium on the time scale τ_c much shorter than the scale τ_{int} on which f_1 would do so. Hence, for $t >> \tau_c$ we can take $\partial f_2 / \partial t = 0$ reducing (3.182) to ($U(\mathbf{r}) = 0$)

$$\sum_{i=1,2} \left[\frac{1}{m} \mathbf{p}_i \cdot \nabla_{\mathbf{r}_i} f_2 - \nabla_{\mathbf{r}_i} W(|\mathbf{r}_1 - \mathbf{r}_2|) \cdot \nabla_{\mathbf{p}_i} f_2 \right] = 0. \tag{3.185}$$

Since $\nabla_{\mathbf{r}_2} W(|\mathbf{r}_1 - \mathbf{r}_2|) = -\nabla_{\mathbf{r}_1} W(|\mathbf{r}_1 - \mathbf{r}_2|)$, we can rewrite (3.185) as

$$\frac{1}{m} \sum_{i=1,2} \mathbf{p}_i \cdot \nabla_{\mathbf{r}_i} f_2 = \nabla_{\mathbf{r}_1} W(|\mathbf{r}_1 - \mathbf{r}_2|) \cdot \left(\nabla_{\mathbf{p}_1} - \nabla_{\mathbf{p}_2} \right) f_2. \tag{3.186}$$

We use (3.186) to evaluate the collision term (3.184) by rewriting it as

$$\left(\frac{\partial f_1}{\partial t}\right)_{\text{coll}} = \int \nabla_{\mathbf{r}} W(|\mathbf{r} - \mathbf{r}_1|) \cdot \left(\nabla_{\mathbf{p}} - \nabla_{\mathbf{p}_1}\right) f_2 \, d\mu_1. \qquad (3.187)$$

This is obtained by adding $-\nabla_{\mathbf{p}_1} f_2$ under the integral in (3.184) as its contribution, by virtue of (3.154), is zero. On using (3.186), (3.187) reads

$$\left(\frac{\partial f_1(\mathbf{r}, \mathbf{p}; t)}{\partial t}\right)_{\text{coll}} = \frac{1}{m} \int \left(\mathbf{p} \cdot \nabla_{\mathbf{r}} + \mathbf{p}_1 \cdot \nabla_{\mathbf{r}_1}\right) f_2 \, d\mu_1. \qquad (3.188)$$

It will not be appropriate to apply the assumption of molecular chaos to express f_2 as the product of two $f_1's$. For, the integration over space in (3.188) involves not only the scattering region but also the region in which the particles interact where the assumption of molecular chaos is unlikely to hold. We therefore need to separate the said two regions.

Invoking the assumption that the range of $W(|\mathbf{r}_i - \mathbf{r}_j|)$ is r_0, we can assume $W(|\tilde{\mathbf{r}}|) = 0$ for $|\tilde{\mathbf{r}}| > |\mathbf{r}_0|$ and thereby restrict integration in (3.188) to the region bound by $|\mathbf{r}_0|$:

$$\left(\frac{\partial f_1(\mathbf{r}, \mathbf{p}; t)}{\partial t}\right)_{\text{coll}} = \frac{1}{m} \int\limits_{|\mathbf{r}_1| < |\mathbf{r}_0|} \left(\mathbf{p} \cdot \nabla_{\mathbf{r}} + \mathbf{p}_1 \cdot \nabla_{\mathbf{r}_1}\right) f_2 \, d\mu_1. \qquad (3.189)$$

To evaluate the integral above, it is useful to work in the center of mass coordinate system:

$$\mathbf{R} = \frac{\mathbf{r} + \mathbf{r}_1}{2}, \quad \tilde{\mathbf{r}} = \mathbf{r} - \mathbf{r}_1, \quad \mathbf{V} = \dot{\mathbf{R}}, \quad \tilde{\mathbf{v}} = \dot{\tilde{\mathbf{r}}}. \qquad (3.190)$$

The integrand in (3.189) in the transformed variables reads

$$\left(\mathbf{p} \cdot \nabla_{\mathbf{r}} + \mathbf{p}_1 \cdot \nabla_{\mathbf{r}_1}\right) f_2 = \left(\mathbf{V} \cdot \nabla_{\mathbf{R}} + \tilde{\mathbf{v}} \cdot \nabla_{\tilde{\mathbf{r}}}\right) f_2(\mathbf{R}, \tilde{\mathbf{r}}, \mathbf{P}, \tilde{\mathbf{p}}; t)$$
$$\approx \tilde{\mathbf{v}} \cdot \nabla_{\tilde{\mathbf{r}}} f_2(\mathbf{R}, \tilde{\mathbf{r}}, \mathbf{P}, \tilde{\mathbf{p}}; t), \qquad (3.191)$$

where the second line is the result of the expectation that the center of mass motion is much slower than that of the relative coordinates. Substitute (3.191) in (3.189) to obtain

$$\left(\frac{\partial f_1}{\partial t}\right)_{\text{coll}} = \int d^3 \mathbf{p}_1 \int\limits_{|\mathbf{r}_1| < |\mathbf{r}_0|} \tilde{\mathbf{v}} \cdot \nabla_{\tilde{\mathbf{r}}} f_2 \, d^3 \mathbf{r}_1. \qquad (3.192)$$

To evaluate the integral above, we work in the coordinate system in which the direction $\tilde{\mathbf{v}}$ is the z-axis (see Fig. 3.2) so that (3.192) assumes the form.

Fig. 3.2 Scattering of two particles in the rest frame of one at rest at O. The region inside the sphere of radius $|r_0|$ centered at O is the interaction region

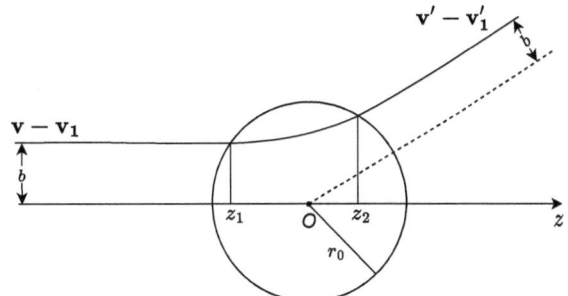

$$\left(\frac{\partial f_1}{\partial t}\right)_{\text{coll}} = \int d^3 p_1 \int\limits_{|\mathbf{r}_1| < |\mathbf{r}_0|} |\mathbf{v} - \mathbf{v}_1| \frac{\partial f_2}{\partial z} d^3 \mathbf{r}_1, \tag{3.193}$$

The plane perpendicular to the direction of propagation is characterized by the impact parameter b and the angle ϕ. The volume element in the said coordinate system is $d^3 \mathbf{r}_1 = b \, db \, d\phi \, dz$. The particle at rest is shown at the center of the sphere S_0 of radius $|r_0|$. The region inside S_0 is the interaction region. The incoming particle meets the sphere S_0 at z_1, enters the interaction region and leaves it from the point z_2 on the sphere. Since integration of (3.193) is restricted to the interior of the sphere S_0, the integration over z can be performed to get

$$\left(\frac{\partial f_1}{\partial t}\right)_{\text{coll}} = \int d^3 p_1 |\mathbf{v} - \mathbf{v}_1| \int \left(f_2(z_2) - f_2(z_1) \right) b \, db \, d\phi, \tag{3.194}$$

where $f_2(z_1)$ is the function of the momenta of the particles before scattering and $f_2(z_2)$ that after scattering. Invoking the expression (3.113) for $d\sigma$, (3.194) assumes the form

$$\left(\frac{\partial f_1}{\partial t}\right)_{\text{coll}}$$
$$= \frac{1}{m} \int |\mathbf{p} - \mathbf{p}_1| \left(f_2(z_2) - f_2(z_1) \right) \frac{d\sigma}{d\Omega} d\Omega d^3 \mathbf{p}_1. \tag{3.195}$$

The points z_1, z_2 are the ones at which the particle enters and leaves the interaction region outside which the assumption of molecular chaos contained in (3.108) is expected to hold:

$$f_2(z_1) = f_1(\mathbf{r}, \mathbf{p}; t) f_1(\mathbf{r}, \mathbf{p}_1; t), \quad f_2(z_2) = f_1(\mathbf{r}, \mathbf{p}'; t) f_1(\mathbf{r}, \mathbf{p}_1'; t). \tag{3.196}$$

In writing the equations above we have replaced $\mathbf{r}_1'$ and $\mathbf{r}_2'$ by $\mathbf{r}$ assuming that f_1 does not vary on the scale r_0. On using (3.196), (3.195) reduces to the collision integral in the Boltzmann (3.111). We have thus shown that, under appropriate assumptions, the equation (3.170) for $f_1(\mathbf{r}, \mathbf{p}; t)$ can be decoupled from the rest of the chain resulting in the Boltzmann equation derived before by working in one-particle phase space.

Next, we derive the equilibrium solution of the Boltzmann equation.

3.8 The H-Theorem

The BBGKY equations determine time evolution of the distribution functions. The system would approach an equilibrium state as $t \to \infty$. That state is described by the asymptotic solution of those equations. In particular, as $t \to \infty$ the one-particle distribution function $f_1(\mathbf{r}, \mathbf{p}; t)$ approaches the solution of the equation

$$\frac{\partial f_1(\mathbf{r}, \mathbf{p}; t)}{\partial t} = 0. \tag{3.197}$$

Assuming no external force, $\partial f_1(\mathbf{r}, \mathbf{p}; t)/\partial t$ satisfies (3.111) with $U(\mathbf{r}) = 0$. We see that, for (3.197) to hold in this case, it is sufficient that f_1 be a function of $\mathbf{p}$ alone, say $f_0(\mathbf{p})$, such that

$$f_0(\mathbf{p}') f_0(\mathbf{p}_1') - f_0(\mathbf{p}) f_0(\mathbf{p}_1) = 0. \tag{3.198}$$

However, the questions that may be asked are: Is (3.198) also a necessary condition? Does the solution $f_1(\mathbf{r}, \mathbf{p}; t)$ of the Boltzmann equation approach $f_0(\mathbf{p})$ as $t \to \infty$? What is the solution of (3.197) when $U(\mathbf{r}) \neq 0$?

The answer to first two questions above turns out to be in the affirmative. The proof is based on the examination of the behavior of the following functional under time evolution,

$$H = \int f_1(\mathbf{r}, \mathbf{p}; t) \ln(f_1(\mathbf{r}, \mathbf{p}; t)) \, \mathrm{d}^3 r \mathrm{d}^3 p, \tag{3.199}$$

called the *Boltzmann's H-Function*, or simply the *H-function*. The answers to the said two questions are the consequences of the *H*-theorem which states that, if $f_1(\mathbf{r}, \mathbf{p}; t)$ obeys the Boltzmann equation (3.111) (or equivalently (3.110)) then

$$\frac{\mathrm{d}H}{\mathrm{d}t} \leq 0. \tag{3.200}$$

Proof: The time-derivative of (3.199) reads

$$\frac{dH}{dt} = \int \frac{\partial f_1(\mathbf{r}, \mathbf{p}; t)}{\partial t} \left(1 + \ln(f_1(\mathbf{r}, \mathbf{p}; t))\right) d^3r d^3p$$

$$= \int \frac{\partial f_1(\mathbf{r}, \mathbf{p}; t)}{\partial t} \ln(f_1(\mathbf{r}, \mathbf{p}; t)) d^3r d^3p, \tag{3.201}$$

where the last line is due to the fact that

$$\int \frac{\partial f_1(\mathbf{r}, \mathbf{p}; t)}{\partial t} d^3r d^3p = \frac{d}{dt} \int f_1(\mathbf{r}, \mathbf{p}; t) d^3r d^3p = 0. \tag{3.202}$$

The second equation above is due to the normalization condition (3.65). On invoking (3.110), (3.201) assumes the form

$$\frac{dH}{dt} = \int \ln(f_1(\mathbf{p})) W(\mathbf{p}', \mathbf{p}_1' | \mathbf{p}, \mathbf{p}_1) \delta^{(4)} \left(P' - P\right)$$

$$\times \left(f_1(\mathbf{p}') f_1(\mathbf{p}_1') - f_1(\mathbf{p}) f_1(\mathbf{p}_1)\right) d^3r d^3p d^3p' d^3p_1 d^3p_1', \tag{3.203}$$

where we have dropped the $\mathbf{r}, t$ from the argument of f_1 and used the fact that, by using the identity (3.135), the contribution from the Poisson bracket term vanishes.

Interchange $\mathbf{p}_1 \leftrightarrow \mathbf{p}$ in (3.203). The only term that changes due to this is the logarithmic term. Add the resulting equation and (3.203) to get

$$\frac{dH}{dt} = \frac{1}{2} \int \ln\left(f_1(\mathbf{p}) f_1(\mathbf{p}_1)\right) W(\mathbf{p}', \mathbf{p}_1' | \mathbf{p}, \mathbf{p}_1) \delta^{(4)} \left(P' - P\right)$$

$$\times \left[\left(f_1(\mathbf{p}') f_1(\mathbf{p}_1') - f_1(\mathbf{p}) f_1(\mathbf{p}_1)\right)\right] d^3r d^3p d^3p' d^3p_1 d^3p_1'. \tag{3.204}$$

Interchange $\mathbf{p}' \leftrightarrow \mathbf{p}$ and $\mathbf{p}_1' \leftrightarrow \mathbf{p}_1$. Make use of the relation $W(\mathbf{p}, \mathbf{p}_1 | \mathbf{p}', \mathbf{p}_1') = W(\mathbf{p}', \mathbf{p}_1' | \mathbf{p}, \mathbf{p}_1)$, add the resulting equation and (3.204) to get

$$\frac{dH}{dt} = -\frac{1}{4} \int W(\mathbf{p}', \mathbf{p}_1' | \mathbf{p}, \mathbf{p}_1) \delta^{(4)} \left(P' - P\right)$$

$$\times \left(\ln(x) - \ln(y)\right)(x - y) d^3r d^3p d^3p' d^3p_1 d^3p_1', \tag{3.205}$$

where

$$x = f_1(\mathbf{p}') f_1(\mathbf{p}_1'), \qquad y = f_1(\mathbf{p}) f_1(\mathbf{p}_1). \tag{3.206}$$

Clearly, the integrand is always positive. Hence follows the desired result (3.200). This shows that, as the system evolves, H decreases monotonically and will approach the constant value for such f_1 for which $dH/dt = 0$.

The H-theorem thus asserts that the one-particle phase space distribution evolves monotonically, i.e. irreversibly to a steady-state form thereby breaking time-reversal symmetry. This is in contrast with time-reversal symmetry of the N-particle distribution function which obeys the Liouville equation exhibiting that symmetry. How is the time-reversal symmetry broken in going from time-reversal symmetric Liouville's equation to Boltzmann's one-particle equation? Is it because of operation of phase space integration involved in obtaining one-particle distribution function from the N-particles one? It does not appear to be so because the said operation does not break time-reversal symmetry. The reason for time acquiring a preferred direction appears to be rooted in the assumption of molecular chaos. There are however subtle issues involved in identifying the source of irreversibility which we do not address (see for example [Huang], [Kardar]).

The equation (3.205) shows that the necessary and sufficient condition for $dH/dt = 0$ to hold is $x - y = 0$. On restoring $\mathbf{r}$ in the argument of f_1 and denoting by $f_0(\mathbf{r}, \mathbf{p})$ the function for which $dH/dt = 0$, the condition $x = y$, with x, y given by (3.206), reads

$$f_0(\mathbf{r}, \mathbf{p}') f_0(\mathbf{r}, \mathbf{p}_1') = f_0(\mathbf{r}, \mathbf{p}) f_0(\mathbf{r}, \mathbf{p}_1). \tag{3.207}$$

Next we find the function solving (3.207) .

3.9 Equilibrium Distribution

Take the logarithm of (3.207) to get

$$\ln\big(f_0(\mathbf{r}, \mathbf{p}')\big) + \ln\big(f_0(\mathbf{r}, \mathbf{p}_1')\big) = \ln\big(f_0(\mathbf{r}, \mathbf{p})\big) + \ln\big(f_0(\mathbf{r}, \mathbf{p}_1)\big). \tag{3.208}$$

On one side of the equation above we have the quantities before collision and on the other side are the same quantities after collision. Hence (3.208) is in the form of a conservation law. This implies that $\ln\big(f_0(\mathbf{r}, \mathbf{p})\big)$ is a combination of independent constants of motion, say, $C_1(\mathbf{r}, \mathbf{p})$, $C_2(\mathbf{r}, \mathbf{p})$, ..., $C_n(\mathbf{r}, \mathbf{p})$:

$$\ln\big(f_0(\mathbf{r}, \mathbf{p})\big) = \sum_{k=1}^{n} C_k(\mathbf{r}, \mathbf{p}). \tag{3.209}$$

Let us consider first the case of the gas which is not acted upon by any external force. The distribution function f_0 in that case will be independent of $\mathbf{r}$ reducing (3.208) thereby to (3.198).

Consider a monatomic gas. The constants of motion in this case are the kinetic energy and three components of momentum. The right side of (3.209) is then an arbitrary sum of the said two constants whereby it reads

$$\ln\big(f_0(\mathbf{p})\big) = -D\,(\mathbf{p}-\mathbf{p}_0)^2 + \ln(B), \tag{3.210}$$

where B, D, $\mathbf{p}_0$ are constants. This implies

$$f_0(\mathbf{p}) = B\exp\left(-D\,(\mathbf{p}-\mathbf{p}_0)^2\right). \tag{3.211}$$

The normalization condition (3.65) demands

$$\int B\exp\left(-D\,(\mathbf{p}-\mathbf{p}_0)^2\right) \mathrm{d}^3 r\, \mathrm{d}^3 p = N. \tag{3.212}$$

The average momentum of a molecule is then given by

$$\langle\mathbf{p}\rangle = \frac{B}{N}\int \mathbf{p}\exp\left(-D\,(\mathbf{p}-\mathbf{p}_0)^2\right)\mathrm{d}^3 r\, \mathrm{d}^3 p = \mathbf{p}_0. \tag{3.213}$$

Assuming the gas has no translational motion as a whole, $\mathbf{p}_0 = 0$, so that

$$f_0(\mathbf{p}) = B\exp\left(-D\mathbf{p}^2\right). \tag{3.214}$$

On integrating this over the volume occupied by the gas we obtain the distribution function for the momentum, expressed in terms of the velocity:

$$F(\mathbf{v}) = C\exp\left(-A\mathbf{v}^2\right). \tag{3.215}$$

This is same as the expression (3.7) for Maxwell distribution of velocities. Its relationship with thermodynamics has been discussed in Sect. 3.2.

The equilibrium distribution function (3.211) is for a gas not subject to any external force. Let us consider a gas in the external potential $U(\mathbf{r})$. In that case, we know that $C - p^2/2m + U(\mathbf{r})$ is the only constant of motion. The equilibrium distribution function (3.209) is then reads

$$f_0(\mathbf{r}, \mathbf{p}) = B\exp\left\{-D\left(p^2/2m + U(\mathbf{r})\right)\right\}. \tag{3.216}$$

That the time-independent distribution function $f_0(\mathbf{r}, \mathbf{p})$ is indeed the equilibrium solution of the Boltzmann equation (3.111) can be ascertained by noting that, since $f_0(\mathbf{r}, \mathbf{p})$ solves (3.207) it reduces the collision term in (3.111) to zero. It is straightforward to check that, with $f_0(\mathbf{r}, \mathbf{p})$ given by (3.216), the streaming term in (3.111) is also zero:

$$\frac{\mathbf{p}}{m}\cdot\nabla_r f_0 - \nabla_r U(\mathbf{r})\cdot\nabla\mathbf{p}\, f_0(\mathbf{r}, \mathbf{p}) = 0, \tag{3.217}$$

Hence $f_0(\mathbf{r}, \mathbf{p})$ in (3.216) is indeed the equilibrium solution of (3.111).

We will see that the equilibrium solutions derived above in the framework of the kinetic theory are same as those arrived at in Chap. 7 by means of the methods of statistical mechanics.

Exercises

Ex. 3.7. Show that an arbitrary sum of $p^2/2m$ and the components p_x, p_y, p_z of momentum can be expressed in the form appearing on the right side of (3.210).

References

1. C. Truesdel, *Archives for History of Exact Sciences 30.XII.1975*, vol. 15 (1975), pp. 1–66. https://www.jstor.org/stable/41133440
2. J.C. Maxwell, Philos. Mag. **19**, 19–32, **20**, 21–37 (1860)

Chapter 4
Boltzmann Entropy

As outlined in Chap. 3, the kinetic theory makes explicit use of the laws of mechanics requiring thereby knowledge of inter-particle interaction. Under several conditions on the inter-particle interaction and the assumption of molecular chaos, we arrived at non-linear Boltzmann's single-particle integro-differential equation. We saw that the state of thermal equilibrium of the gas in that case corresponds to that single-particle distribution function for which Boltzmann's H-function attains its minimum. Subsequently, in [1], Boltzmann showed that, rather than solving a kinetic equation ...*it is possible to calculate the state of the equilibrium of heat by finding the probability of the different possible states of the system. The initial state in most cases is bound to be highly improbable and from it the system will always rapidly approach a more probable state until it finally reaches the most probable state, i.e. that of the heat equilibrium.* His said approach, outlined in this chapter, led to the concept of what is now called Boltzmann entropy which formed the basis for determining the state of thermal equilibrium without the use of the knowledge of the inter-particle potential and without the use of the mechanical equations of motion.

We begin with Boltzmann's said paper [1] which laid the foundation for erecting the edifice of modern statistical mechanics and follow it up by demonstrating some of its important applications.

4.1 Discrete Energy Levels

Boltzmann considered a gas of N molecules assuming that the kinetic energy of each molecule is capable of having discretely spaced values $n\epsilon$ ($n = 0, 1, 2, \ldots, P$) with the remark that this assumption does not correspond to any realistic mechanical model, but it is easier to handle mathematically and the actual problem is solved by letting $\epsilon \to 0$, $P \to \infty$.

The molecules are identified by numbering them from 1 to N. Let E_k denote the energy of the kth molecule where E_k can be any of the possible values $n\epsilon$

R. R. Puri, *Modern Thermodynamics and Statistical Mechanics*, Undergraduate Lecture Notes in Physics, https://doi.org/10.1007/978-3-031-54310-4_4

$(n = 0, 1, 2, \ldots, P)$. The set $(E_1, E_2, \ldots, E_N)$ denoting energy of the molecules numbered 1 to N is called a *configuration* (called *complexion* by Boltzmann). Distinct permutations of the E_k's denote distinct configurations. All distinct configurations in which there are n_k molecules having energy ϵ_k irrespective of the identity of the molecules are said to constitute a *state*. Thus a state is characterized by ordered set of numbers $(n_0, n_1, \ldots, n_P)$ in which n_k is the number of molecules in the level of energy $k\epsilon$ $(k = 0, 1, \ldots, P)$ irrespective of the identity of the molecules such that

$$\sum_{k=0}^{P} n_k = N. \tag{4.1}$$

The number n_k is the *occupation number* of the energy level $k\epsilon$. To determine the number of states and configurations, we can look upon the molecules, assumed to be distinguishable, as distinguishable boxes. A molecule in the state of energy $k\epsilon$ can be looked upon as a box having k identical balls in it. Since $P\epsilon$ is the highest level of energy of a molecule, the number P would correspond to the maximum number of balls that can be placed in a box. In this way, the problem of determining the number of states and configurations of the molecules reduces to the combinatorial problem of distribution of identical balls in distinguishable boxes addressed in Appendix D.

The number of configurations $D(N, P, \{n_k\}_P)$ corresponding to the state $(n_0, n_1, \ldots, n_P)$ is given by (D.4):

$$D(N, P, \{n_k\}_P) = \frac{N!}{n_0! n_1! \ldots, n_P!}, \qquad \sum_{k=0}^{P} n_k = N, \tag{4.2}$$

whereas total number of configuration, given by (D.9), is

$$\text{Total number of configurations} = (P + 1)^N. \tag{4.3}$$

As an example, consider three atoms each having two energy levels $(0, \epsilon)$. With $n_0 + n_1 = 3$, the possible states are $(n_0, n_1) : (3, 0), (2, 1), (1, 2), (0, 3)$. The distinct configurations corresponding to each of the said states are listed in Table 4.1. It may be verified that the number of configurations corresponding to each state and total number of configurations are in accordance with the formulas (4.2) and (4.3), respectively, with $P = 1$, $N = 3$.

In the discussion so far, only restriction placed on identifying states has been total number of atoms. A macroscopic state of a gas is characterized not only by the number of atoms, but also by energy U which in the present case is evidently

$$U = \epsilon \sum_{k=0}^{P} k n_k \equiv \epsilon L, \qquad L = \sum_{k=0}^{P} k n_k. \tag{4.4}$$

Table 4.1 States and corresponding configurations for three atoms having two levels of energy $0, \epsilon$. The n_k is the number of atoms in the level of energy $k\epsilon$ and E_p is the energy of the pth atom

State	Configurations
(n_0, n_1)	(E_1, E_2, E_3)
$(3, 0)$	$(0, 0, 0)$
$(2, 1)$	$(\epsilon, 0, 0), (0, \epsilon, 0), (0, 0, \epsilon)$
$(1, 2)$	$(\epsilon, \epsilon, 0), (\epsilon, 0, \epsilon), (0, \epsilon, \epsilon)$
$(0, 3)$	$(\epsilon, \epsilon, \epsilon)$

We can therefore say that a microscopic state of the gas of N molecules, each having $P + 1$ energy levels $k\epsilon$ $(k = 0, 1, \ldots P)$ and total energy $L\epsilon$, is characterized by the sets of the $n'_k s$ satisfying

$$\sum_{k=0}^{P} n_k = N, \qquad L = \sum_{k=0}^{P} k n_k. \tag{4.5}$$

Though in the example tabulated in Table 4.1 different states have different total energy, in general, as the example discussed next shows, there would be several different states corresponding to same energy but the number of configurations for different states of same energy may or may not be same (see also Ex. 4.1).

Boltzmann postulated that the state of thermal equilibrium of a gas is characterized by that set of occupation numbers $n'_k s$ which satisfy (4.5) and correspond to largest number of configurations.

For illustration, we take the example discussed by Boltzmann [1]. He considered a system of seven atoms, each of which can have energy ranging in unit steps from 0 to 7ϵ such that total energy of the system is 7ϵ. In the notation above, $N = P = L = 7$. Table 4.2 lists possible states and the number $D(N, P, \{n_k\}_P)$ of configurations corresponding to each of them, calculated using (4.2).

The example considered is a special case of (4.5) corresponding to $L = P$. In that case, as shown in Appendix D, following results hold:

1. For a given number N of atoms each having $P + 1$ energy levels $k\epsilon$ $(k = 0, 1, \ldots, P)$ and total energy $L\epsilon$ such that $L \leq P$, total number of configuration $D(N, P, L \leq P)$ is given by (D.23):

$$\mathcal{P}(N, L) \equiv D(N, P, L \leq P) = \frac{(N + L - 1)!}{L!(N - 1)!}, \qquad L \leq P. \tag{4.6}$$

It is independent of P. This result is applicable to the example under consideration as in the present case $L = P = 7$. The expression above yields $\mathcal{P}(7, 7) = 1716$. This is same as the sum of the number of configurations corresponding to different states tabulated in Table 4.2.

Table 4.2 States and configurations for seven atoms which can have energies $k\epsilon$ ($k = 0, 1, \ldots, 7$), n_k is number of atoms having energy $k\epsilon$, total energy being 7ϵ

State	Number of configurations
$(n_0, n_1, n_2, n_3, n_4, n_5, n_6, n_7)$	$D(7, 7, \{n_k\}_P)$
$(6, 0, 0, 0, 0, 0, 0, 1)$	7
$(5, 1, 0, 0, 0, 0, 1, 0)$	42
$(5, 0, 1, 0, 0, 1, 0, 0)$	42
$(5, 0, 0, 1, 1, 0, 0, 0)$	42
$(4, 2, 0, 0, 0, 1, 0, 0)$	105
$(4, 1, 1, 0, 1, 0, 0, 0)$	210
$(4, 0, 2, 1, 0, 0, 0, 0)$	105
$(4, 1, 0, 2, 0, 0, 0, 0)$	105
$(3, 3, 0, 0, 1, 0, 0, 0)$	140
$(3, 2, 1, 1, 0, 0, 0, 0)$	420
$(3, 1, 3, 0, 0, 0, 0, 0)$	140
$(2, 4, 0, 1, 0, 0, 0, 0)$	105
$(2, 3, 2, 0, 0, 0, 0, 0)$	210
$(1, 5, 1, 0, 0, 0, 0, 0)$	42
$(0, 7, 0, 0, 0, 0, 0, 0)$	1

2. Number of configurations in which n_0 is the number of atoms having zero energy, denoted by $C(N, L, n_0)$ ($n_0 = 0, 1, \ldots, N - 1$), is

$$C(N, L; n_0) = \frac{(L-1)!}{(L - N + n_0)!(N - n_0 - 1)!}\, ^N C_{n_0}. \tag{4.7}$$

In the present case, with $N = L = 7$, we see that

(a) When $n_0 = 6$, $C(7, 7, 6) = 7$. This is same as the number of configurations listed in Table 4.2 corresponding to the state having six molecules in the state of zero energy.

(b) When $n_0 = 5$, $C(7, 7, 5) = 126$. This is same as the the number of configurations listed in Table 4.2 corresponding to three states having five atoms in the state of zero energy.

(c) When $n_0 = 4$, $C(7, 7, 4) = 525$. This is same as the number of configurations listed in Table 4.2 corresponding to four states having four atoms in the state of zero energy.

(d) When $n_0 = 3$, $C(7, 7, 3) = 700$. This is same as the number of configurations listed in Table 4.2 corresponding to three states having three atoms in the state of zero energy.

(e) When $n_0 = 2$, $C(7, 7, 2) = 315$. This is same as the number of configurations listed in Table 4.2 corresponding to two states having two atoms in the state of zero energy.
(f) When $n_0 = 1$, $C(7, 7, 1) = 42$. This is same as the number of configurations listed in Table 4.2 corresponding to the state having one atom in the state of zero energy.
(g) When $n_0 = 0$, $C(7, 7, 0) = 1$. This is same as the number listed in Table 4.2 corresponding to the state having no atom in the state of zero energy.

Since the state of thermodynamic equilibrium corresponds to the one having maximum number of configurations, the interest is in determining the state corresponding to maximum number of configurations. We see that the maximum number of configurations in the example under consideration correspond to the state

$$(n_0, n_1, n_2, n_3, n_4, n_5, n_6, n_7) = (3, 2, 1, 1, 0, 0, 0, 0). \tag{4.8}$$

It is, however, not possible to determine the state having maximum number of configurations in the direct way used in the simple example above as real systems are very large. Exploiting the largeness of the system, Boltzmann devised the following method to determine the state of thermal equilibrium.

Boltzmann determined the set of values of the $n'_k s$ which maximizes $D(N, P, \{n_k\}_P)$ in (4.2) subject to the conditions in (4.5) by maximizing logarithm of $D(N, P, \{n_k\}_P)$,

$$\ln(D(N, P, \{n_k\}_P)) = \ln(N!) - \sum_{k=0}^{P} \ln(n_k!), \tag{4.9}$$

using Lagrange's method of undetermined multipliers (see Sect. 5.2) to incorporate the constraints in (4.5) so that the desired set of $n'_k s$ is one which maximizes

$$F(\{n_k\}) = \ln(D(N, P, \{n_k\}_P)) - \sum_{k=0}^{P} (\alpha + k\gamma) n_k, \tag{4.10}$$

where α, γ are the Lagrange multipliers. The α term is due to the normalization condition and the γ term is due to the condition on total energy. Since, in a large system, the numbers N, n_k are large, the logarithms of their factorials may be approximated by Stirling's formula (H.6) so that

$$\ln(D(N, P, \{n_k\}_P)) = N\ln(N) - N - \sum_{k=0}^{P} \{n_k\ln(n_k) - n_k\}, \tag{4.11}$$

which on substitution in (4.10) reduces $F(\{n_k\})$ to the form

$$F(\{n_k\}) = N\ln(N) - N - \sum_{k=0}^{P} (\ln(n_k) - 1 + \alpha + k\gamma)\, n_k. \qquad (4.12)$$

The extremization condition $\partial F/\partial n_k = 0$ yields

$$\ln(n_k) + \alpha + k\gamma = 0. \qquad (4.13)$$

This determines the values of the $n_k's$ which extremize $F(\{n_k\})$:

$$n_k = Ax^k, \qquad A = \exp(-\alpha), \qquad x = \exp(-\gamma). \qquad (4.14)$$

The Lagrange multipliers α, γ are determined in terms of the known quantities N, L by substituting (4.14) in the conditions (4.5) on the n_k's.

The normalization condition in (4.5) yields

$$A = \frac{N(1-x)}{1-x^{P+1}}, \qquad (4.15)$$

whereas the expression for L therein reads

$$L = \sum_{k=0}^{P} k n_k = A \sum_{k=0}^{P} k x^k = Ax \frac{d}{dx}\left[\frac{1-x^{P+1}}{1-x}\right]. \qquad (4.16)$$

We leave it as an exercise to show that, with A given by (4.15), (4.16) leads to the polynomial equation of degree $P+2$ in x ((4.36) in Ex. 4.2). Since x by its definition in (4.14) is positive, only positive roots of (4.36) are acceptable. Boltzmann showed that (4.36) has only one positive real root. Noting also that the maximum kinetic energy $P\epsilon$ of an atom is much larger than its mean kinetic energy $L\epsilon/N$ so that $P >> L/N$, he showed that the said positive root of (4.36) lies in the interval $(0, 1)$. Thus the unique positive root of (4.36) determines x which is one of the two unknowns. The other unknown A is obtained using its expression (4.15) in terms of x. The n_k's in (4.14) then turn out to be given by

$$n_k = \frac{N(1-x)}{1-x^{P+1}} x^k, \qquad (4.17)$$

where x is real positive solution of (4.36) which, as mentioned before, has been shown to be unique by Boltzmann.

Since, as discussed above, the situations of interest correspond to $P >> 1$ in which case (4.15) and (4.16) reduce, respectively, to

$$A = N(1-x), \qquad (4.18)$$

and

$$L = \frac{Nx}{1 - x} \implies x = \frac{L}{L + N}. \tag{4.19}$$

The equation above is a simple expression for x in terms of the known quantities, N, L.

With A given by (4.18) and x by (4.19), the expression (4.17) for the n_k's which maximize $D(N, P, \{n_k\}_P)$ for $P \gg 1$ reads

$$n_k = N(1 - x)x^k. \tag{4.20}$$

This expresses the n_k's in terms of the known quantities N, L.

The maximum number of configurations, denoted by $\mathcal{P}_{max}$, is obtained by substituting the $n'_k s$ given by (4.20) in (4.11):

$$\ln(\mathcal{P}_{max}) = -N\left(\ln(1 - x) + \frac{x}{1 - x}\ln(x)\right). \tag{4.21}$$

Use (4.19) to write x in the equation above in terms of L, N to get

$$\ln(\mathcal{P}_{max}) = (N + L)\ln(N + L) - L\ln(L) - N\ln(N). \tag{4.22}$$

We have thus at hand the formula for $\mathcal{P}_{max}$ in terms of the known quantities.

Boltzmann noted that the expression (4.22) for $\mathcal{P}_{max}$ is same as the expression for the number of configurations of the system of N molecules of total energy $L\epsilon$. For, recall that the number of configurations corresponding to energy $L\epsilon$ of N atoms is given by $\mathcal{P}(N, L)$ in (4.6) provided $L \le P$ where $P\epsilon$ is the highest energy of an atom. Since P is assumed to be large in the derivation of (4.22), formula (4.6) for $\mathcal{P}(N, L)$ is applicable in the present case. Furthermore, (4.22) has been arrived at by applying Stirling's approximation assuming N, L to be large. It is straightforward to see that $\ln(\mathcal{P}(N, L))$ with $\mathcal{P}(N, L)$ as in (4.6) is same as $\ln(\mathcal{P}_{max})$ in (4.22) under Stirling's approximation. Hence, under Stirling's approximation on $\mathcal{P}(N, L))$,

$$\ln(\mathcal{P}_{max}) = \ln(\mathcal{P}(N, L)), \tag{4.23}$$

where $\mathcal{P}(N, L))$ is the number of configurations corresponding to total energy $L\epsilon$ of N atoms.

Boltzmann analyzed his example presented above in the light of his results derived by maximizing the number of configurations of a specified value of energy. In his example, $N = L = P = 7$ and the state with maximum number of configurations is given by (4.8). He solved (4.36) numerically for $N = L = P = 7$ to obtain $x = 0.5078125$. Note that this value is close to the value $x = 0.5$ that will be obtained using x in (4.19) for $P \gg 1$ even though in the present case $P = 7$, not at all a large number. Boltzmann evaluated the n_k's using (4.17). We reproduce below his results displaying in bracket the corresponding values of the n_k's obtained using the expressions in (4.20) for n_k and x applicable for large P:

$$n_0 = 3.4535(3.5), \qquad n_1 = 1.7574(1.75), \qquad n_2 = 0.8943(0.875),$$
$$n_3 = 0.4551(0.4375), \quad n_4 = 0.2316(0.2187), \quad n_5 = 0.1178(0.10937),$$
$$n_6 = 0.0599(0.05468), \quad n_7 = 0.0304(0.02734). \tag{4.24}$$

Boltzmann observed that, since N, P in the example under consideration are small, the agreement of the values of the n_k's in (4.24), based on the assumption of large numbers, with the corresponding exact values in (4.8) is hardly expected. Nevertheless, on comparing the two by taking the nearest integral value for each n_k in (4.24) except for n_3 whose value though is less than 0.5 but is approximated as 1, we get

$$n_0 = 3, \; n_1 = 2, \; n_2 = n_3 = 1, \; n_4 = n_5 = n_6 = n_7 = 0. \tag{4.25}$$

These values agree with the exact result in (4.8).

Boltzmann considered next the example taking $N = 7, L = 19, P = \infty$ and used (4.20) to evaluate the $n_k's$. He found that, except for approximating $n_5 = 1$ though its value is 0.389, the approximation of the n_k' by the nearest integers yields correct values of the $n_k's$ which maximize $\mathcal{P}$.

Following the examples outlined above, Boltzmann states that the formula (4.20) determines the $n_k's$ within one or two units of the true values even for small values of P, N and that, since in the mechanical theory of heat we deal with extremely large number of molecules, these small differences disappear and the approximate formula provides an exact solution of the problem. He determined also the form of the formula (4.20) when average molecular energy is much higher than the separation ϵ between the consecutive energy levels as follows.

When separation ϵ between consecutive energies is very small compared with the average kinetic energy $u \equiv L\epsilon/N$, the expression (4.20) for the number of atoms having energy $k\epsilon$ can be approximated by

$$n_k = \frac{N\epsilon}{u + \epsilon} \left(1 + \frac{\epsilon}{u}\right)^{-k} \approx \frac{N\epsilon}{u} \exp(-k\epsilon/u). \tag{4.26}$$

Concluding at this point the discussion of discrete energy model, Boltzmann remarks: *to achieve a mechanical theory of heat, these formulas must be developed further, particularly through the introduction of differentials and some additional considerations.*

Before outlining in Sect. 4.2 Boltzmann's treatment of continuous molecular motion in terms of number of configurations for continuous variables which led to the notion of entropy named after him, we carry further the discussion of the discrete model to establish its link with thermodynamics.

4.1.1   Connection of Discrete Model with Thermodynamics

Boltzmann did not pursue the discrete model any further and went on to consider the continuum limit $\epsilon \to 0$. In Sect. 4.3, we will see how he used his discrete model results to construct the position and velocity distribution of molecules and established connection with thermodynamics by relating the function, he called the permutability measure, with thermodynamic entropy. For the present, we investigate the thermo-dynamic properties of the collection of molecules having discrete energy levels by adopting the definition (4.77) of Boltzmann's entropy for the discrete case in terms of the number $\mathcal{P}$ of configurations subject to the constraints (4.5). Its connection with thermodynamics is established by that entropy given by (4.78) corresponding to the maximum value $\mathcal{P}_{max}$ of $\mathcal{P}$, through the equations (4.79):

$$S_{B,max} = k_B \ln(\mathcal{P}_{max}), \qquad \frac{1}{T} = \frac{\partial S_{B,max}}{\partial U}. \qquad (4.27)$$

Due to quantum mechanics, discrete models are no more fictitious as they appeared at Boltzmann's time.

1. In terms of energy per atom given by

$$u = \frac{L\epsilon}{N}, \qquad (4.28)$$

the relation (4.19) for x reads

$$x = \frac{u}{u + \epsilon}. \qquad (4.29)$$

The energy per atom can therefore be written in terms of x as

$$u = \frac{x\epsilon}{1 - x}. \qquad (4.30)$$

2. Using (4.29), the expression (4.20) for n_k may be rewritten as

$$n_k = \frac{N\epsilon}{u + \epsilon} \left(\frac{u}{u + \epsilon} \right)^k. \qquad (4.31)$$

3. Recalling (4.21) for $\ln(\mathcal{P}_{max})$ and on writing x therein in terms of u as in (4.29), the expression (4.27) for entropy reads

$$S_{B,max} = Nk_B \left[\left(1 + \frac{u}{\epsilon} \right) \ln \left(1 + \frac{u}{\epsilon} \right) - \frac{u}{\epsilon} \ln \left(\frac{u}{\epsilon} \right) \right]. \qquad (4.32)$$

This would reduce to Planck's expression for entropy of the oscillators of frequency ν in the walls of a blackbody cavity if $\epsilon = h\nu$ where h is Planck's constant (see Sect. 4.5).

4. With $S_{\mathrm{B,max}}$ given by (4.32), the relation (4.27) between $S_{\mathrm{B,max}}$ and temperature yields following expression for u in terms of T:

$$u = \frac{\epsilon}{\exp(\epsilon/k_{\mathrm{B}}T) - 1}. \tag{4.33}$$

This expresses average molecular energy in terms of temperature. It is same as Planck's blackbody law (4.85) if $\epsilon = h\nu$.

5. Substitute (4.33) in (4.29) to obtain x in terms of T:

$$x = \exp(-\epsilon/k_{\mathrm{B}}T). \tag{4.34}$$

On substituting this in (4.20) (or (4.33) in (4.31)), the number of molecules having energy $k\epsilon$ turns out to be given by

$$n_k = N(1 - \exp(-\epsilon/k_{\mathrm{B}}T)) \exp(-k\epsilon/k_{\mathrm{B}}T). \tag{4.35}$$

We thus have the expression for a microscopic quantity, the number of molecules in an energy level, in terms of the thermodynamic one, namely, temperature when the molecular gas is in thermal equilibrium, establishing thereby the desired link between statistical mechanical and thermodynamic descriptions.

Ex. 4.1. Consider a system of three atoms each having energy levels $k\epsilon$ ($k = 0, 1, 2$). Find its states and corresponding configurations for total energy $U = 2\epsilon$. Hint: The states corresponding to total energy 2ϵ are $(n_0, n_1, n_2) \equiv (2, 0, 1), (1, 2, 0)$. There are thus two states corresponding to same energy. The configurations comprising the state $(2, 0, 1)$ are $(2\epsilon, 0, 0), (0, 2\epsilon, 0), (0, 0, 2\epsilon)$, and those corresponding to $(1, 2, 0)$ are $(0, \epsilon, \epsilon), (\epsilon, 0, \epsilon), (\epsilon, \epsilon, 0)$. Note that the number of configurations found for each set of (n_0, n_1, n_2) are consistent with the formula (4.2) and total number of configurations is consistent with the formula (4.6).

Ex. 4.2. Show that (4.16) leads to the equation

$$(Np - L)x^{p+2} - (Np + N - L)x^{p+1} + (L + N) = 0. \tag{4.36}$$

Ex. 4.3. Show that maximum Boltzmann entropy of the discrete model at temperature T is given by

$$S_{\mathrm{B,max}} = N\left[\frac{u}{T} - k_{\mathrm{B}}\ln\{1 - \exp(-\epsilon/k_{\mathrm{B}}T)\}\right]. \tag{4.37}$$

4.2 Continuous Distribution of Energy

Boltzmann used the results derived for energy varying in equally spaced discrete steps to continuous energy values as follows.

Let energy E assume continuously varying values from 0 to ∞. Divide energy in small intervals of length ϵ each. Let $f(E)$ denote the number of atoms per unit energy interval in the vicinity of E. The number n_k of atoms in the interval $(k\epsilon, (k+1)\epsilon)$ is then given by

$$n_k = \epsilon f(k\epsilon). \tag{4.38}$$

Using this, the normalization condition (4.1) reads

$$\epsilon \sum_{k=0}^{\infty} f(k\epsilon) = N. \tag{4.39}$$

Taking limit $\epsilon \to 0$ and invoking the definition of Riemann integral, the equation above reduces to

$$\int_0^\infty f(E)\mathrm{d}E = N. \tag{4.40}$$

Similarly, with $U = Nu$, the expression (4.4) reads

$$\epsilon \sum_{k=0}^{\infty} k\epsilon f(k\epsilon) = Nu. \tag{4.41}$$

In the continuum limit $\epsilon \to 0$ this evidently reduces to

$$\int_0^\infty Ef(E)\mathrm{d}E = Nu. \tag{4.42}$$

Next, we find the continuum limit representation of the number of configurations for a given set of values of the $n'_k s$.

The expression (4.11) for the logarithm of the number of configurations $D(N, P, \{n_k\}_P) \to \mathcal{P}$, with n_k therein given by (4.38), assumes the form

$$\ln(\mathcal{P}) = N\ln(N) - N - \epsilon \sum_{k=0}^{\infty} \left\{ f(k\epsilon)\ln(\epsilon f(k\epsilon)) - f(k\epsilon) \right\}$$

$$= N\ln(N) - N - \epsilon \sum_{k=0}^{\infty} \left\{ f(k\epsilon)\ln(f(k\epsilon)) - f(k\epsilon) \right\}$$

$$-\epsilon \sum_{k=0}^{\infty} f(k\epsilon)\ln(\epsilon)$$

$$= N\ln(N) - \epsilon \sum_{k=0}^{\infty} f(k\epsilon)\ln(f(k\epsilon)) - N\ln(\epsilon), \qquad (4.43)$$

where last line is due to (4.39). In the continuum limit $\epsilon \to 0$, (4.43) reads

$$\ln(\mathcal{P}) = N\ln(N) - \int_{0}^{\infty} f(E)\,\ln(f(E))\mathrm{d}E - N\mathrm{Lt}_{\epsilon \to 0}\ln(\epsilon). \qquad (4.44)$$

This shows that $\ln(\mathcal{P})$ blows up as $\epsilon \to 0$. However, the equilibrium distribution corresponds to that $f(E)$ which maximizes $\ln(\mathcal{P})$ subject to the conditions (4.39) and (4.41). Since its extremization is not changed by addition of constants, we can ignore the additive constants in the expression (4.44) for $\ln(\mathcal{P})$, including the ϵ term which though blows up but is a constant, and extremize instead the quantity

$$\Omega = -\int_{0}^{\infty} f(E)\,\ln(f(E))\mathrm{d}E, \qquad (4.45)$$

with respect to $f(E)$ subject to the conditions (4.40), and (4.42):

$$\int_{0}^{\infty} f(E)\mathrm{d}E = N, \quad \int_{0}^{\infty} Ef(E)\mathrm{d}E = Nu. \qquad (4.46)$$

The Ω defined in (4.45) extends to energy distribution the notion of permutability measure of Boltzmann introduced for velocity distribution (see (4.58)). The said extremization can be carried using the method of Lagrange multipliers (see Sect. 5.2). Presently, it involves extremizing

$$F = \int_{0}^{\infty} \{f(E)\,\ln(f(E)) + \alpha f(E) + \gamma Ef(E)\}\,\mathrm{d}E, \qquad (4.47)$$

where α, γ are Lagrange multipliers. The extremum is attained by varying $f(E)$ such that $\delta F = 0$ leading to the equation

$$1 + \ln(f(E)) + \alpha + \gamma E = 0. \qquad (4.48)$$

This is solved by

$$f_{\max}(E) = A \exp(-\gamma E). \tag{4.49}$$

Using the conditions in (4.46), A and γ are evaluated so that

$$f_{\max}(E) = \frac{N}{u} \exp(-E/u). \tag{4.50}$$

We have thus at hand the equilibrium distribution of continuously varying energy, obtained by extremizing the quantity Ω defined in (4.45). The extremized value $\Omega_{\max}$ of Ω is the value of Ω corresponding to the $f(E)$ that extremizes it:

$$\Omega_{\max} = -\int_0^\infty f_{\max}(E) \ln(f_{\max}(E)) dE, \tag{4.51}$$

where $f_{\max}(E)$ is as in (4.50). Evaluation of the integral above yields

$$\Omega_{\max} = N(1 + \ln(u) - \ln(N)), \tag{4.52}$$

which on use of (4.44) by ignoring the blowing up term therein shows that

$$\ln(\mathcal{P}_{\max}) = N(1 + \ln(u)). \tag{4.53}$$

Recalling the definition of entropy $S_{B,\max}$ in (4.27) we get

$$S_{B,\max} = k_B \ln(\mathcal{P}_{\max}) = N k_B (1 + \ln(u)). \tag{4.54}$$

Invoking the relation between temperature and entropy in (4.27), we obtain $u = k_B T$. We can verify that (4.54) is same as the expression (4.32) for $S_{B,\max}$ arrived at by assuming small separation ϵ between energy in the limit $\epsilon \to 0$. For, in the limit $\epsilon \to 0$, (4.32) reduces to

$$S_{B,\max} = N k_B (1 + \ln(u)) - N k_B \mathrm{Lt}_{\epsilon \to 0} \ln(\epsilon). \tag{4.55}$$

This is same as (4.54) if the blowing up last term in it is ignored which is in accordance with the fact that that term has been ignored even while arriving at (4.54).

4.3 Distribution of Velocities

Boltzmann considered next the problem of determining distribution of molecular velocities. He began by assuming each component of velocity varying discretely: the x, y, z components v_x, v_y, v_z of the velocity take values $v_x = k_x \epsilon_x$, $v_y = k_y \epsilon_y$,

$v_z = k_z \epsilon_z$ $(-p_\mu \le k_\mu \le p_\mu, \mu = x, y, z)$. The number of configurations $\mathcal{P}$ in which there are $n_{k_x}, n_{k_y}, n_{k_z}$ number of molecules having the x, y, z components of velocity given, respectively, by $v_x = k_x \epsilon_x, v_y = k_y \epsilon_y, v_z = k_z \epsilon_x$ is

$$\mathcal{P} = N! \left(\prod_{k_x=-p_x}^{p_x} n_{k_x}! \prod_{k_y=-p_y}^{p_y} n_{k_y}! \prod_{k_z=-p_z}^{p_z} n_{k_z}! \right)^{-1}. \tag{4.56}$$

Proceeding in the same manner as described in Sect. 4.2, he goes over to the continuum limit by defining the function $f(\mathbf{v}) \equiv f(v_x, v_y, v_z)$ which is the number of molecules per unit velocity interval in the vicinity of $\mathbf{v}$ so that

$$n_{k_x,k_y,k_z} = \epsilon_x \epsilon_y \epsilon_z f(k_x v_x, k_y v_y, k_z v_z) \tag{4.57}$$

is the number of molecules having components of velocity in the intervals $(k_x \epsilon_x, (k_x + 1)\epsilon_x), (k_y \epsilon_y, (k_y + 1)\epsilon_y), (k_z \epsilon_z, (k_z + 1)\epsilon_z)$, respectively. As done in Sect. 4.2, it can be shown that, apart from additive constants, $\ln(\mathcal{P})$ is same as Ω where

$$\Omega = -\int f(\mathbf{v})\ln\{f(\mathbf{v})\}\mathrm{d}^3 v. \tag{4.58}$$

Boltzmann calls Ω the *permutability measure*. We call the related quantity

$$S_B = k_B \Omega, \tag{4.59}$$

the *Boltzmann entropy*.

The state of thermodynamic equilibrium corresponds to such $f(\mathbf{v})$ which maximizes (4.58) subject to the normalization constraint

$$\int f(\mathbf{v})\mathrm{d}^3 v = N, \tag{4.60}$$

and the constraint on total kinetic energy

$$\frac{m}{2} \int v^2 f(\mathbf{v})\mathrm{d}^3 v = U \equiv Nu. \tag{4.61}$$

Invoking the method of Lagrange multipliers the equilibrium distribution may be seen to be given by

$$f_{\max}(\mathbf{v}) = A \exp(-mv^2 \gamma), \tag{4.62}$$

where the constants A, γ are determined by the conditions (4.60) and (4.61). The velocity distribution function $f(\mathbf{v})$ in (4.62) is same as the Maxwell distribution derived differently. It is straightforward to verify that

$$\gamma = \frac{3}{4u}, \quad A = N \left(\frac{3m}{4\pi u}\right)^{3/2}, \tag{4.63}$$

so that the equilibrium distribution assumes the form

$$f_{\max}(\mathbf{v}) = N \left(\frac{3m}{4\pi u}\right)^{3/2} \exp(-3mv^2/4u). \tag{4.64}$$

This is Boltzmann's velocity distribution. He then states that complete description of gas is provided not by velocity distribution alone but it must also include the distribution of the positions of molecules. We outline next his method of obtaining the said complete distribution and the way he uses it to establish connection of his theory with thermodynamics.

4.4 Relation Between Thermodynamic and Boltzmann Entropies

Boltzmann introduced the function $f(\mathbf{r}, \mathbf{v}) \equiv f(x, y, z; v_x, v_y, v_z)$ which is such that $f(\mathbf{r}, \mathbf{v})\mathrm{d}^3 r\mathrm{d}^3 v$ gives the number of atoms positioned in the vicinity of $\mathbf{r}$ having velocity in the vicinity of $\mathbf{v}$. The equilibrium distribution is obtained by maximizing the permutability measure Ω defined by generalizing its definition (4.58) for velocity distribution:

$$\Omega = -\int f(\mathbf{r}, \mathbf{v})\ln\{f(\mathbf{r}, \mathbf{v})\}\mathrm{d}^3 r\mathrm{d}^3 v. \tag{4.65}$$

subject to the normalization condition

$$\int f(\mathbf{r}, \mathbf{v})\mathrm{d}^3 r\mathrm{d}^3 v = N, \tag{4.66}$$

and the condition on total kinetic energy

$$\frac{m}{2}\int v^2\, f(\mathbf{r}, \mathbf{v})\mathrm{d}^3 r\mathrm{d}^3 v = U. \tag{4.67}$$

By following the method of Lagrange's multipliers, it is straightforward to show that the said $f(\mathbf{r}, \mathbf{v})$ is given by

$$f_{\max}(\mathbf{r}, \mathbf{v}) = \frac{N}{V}\left(\frac{3m}{4\pi u}\right)^{3/2} \exp(-3mv^2/4u). \tag{4.68}$$

Verify that the kinetic energy per molecule is correctly given by

$$\frac{m}{2}\langle v^2 \rangle = u. \tag{4.69}$$

The permutability measure Ω in (4.65) corresponding to its maximizing distribution (4.68) can be seen to be given by

$$\Omega_{\text{max}} = N\left(\frac{3}{2}\ln(u) + \ln(V)\right) - N\ln(N) + \frac{3N}{2}\left(\ln(4\pi/3m) + 1\right). \tag{4.70}$$

To connect his formalism with thermodynamics, Boltzmann recalls the thermodynamic relation

$$dQ = Ndu + PdV, \tag{4.71}$$

and the ideal gas law $PV = 2Nu/3$ to obtain

$$\int \frac{dQ}{u} = \frac{2N}{3}\left(\frac{3}{2}\ln(u) + \ln(V)\right) + C, \tag{4.72}$$

where C is a constant. Noting that the bracketed terms in (4.70) and (4.72) are same and the term outside the bracket in (4.70) is a constant, with a proper choice of C in (4.72), one can write

$$\int \frac{dQ}{u} = \frac{2}{3}\Omega_{\text{max}}. \tag{4.73}$$

Boltzmann identifies the left side of (4.73) as thermodynamic entropy and hence arrives at the conclusion that the permutability measure Ω_{max} is 3/2 times the thermodynamic entropy at equilibrium. Invoking the relation $u = (3/2)k_B T$ for an ideal gas, (4.73) may be written in terms of the standard definition $dS_{\text{th}} = dQ/T$ of thermodynamic entropy as

$$S_{\text{th}} = k_B \Omega_{\text{max}}. \tag{4.74}$$

Thus the Boltzmann entropy defined in (4.59) corresponding to maximum Ω,

$$S_{\text{B,max}} = k_B \Omega_{\text{max}}, \tag{4.75}$$

provides link with the thermodynamic entropy. Using (4.70) the equation above reads

$$S_{\text{B,max}} = Nk_B\left\{\frac{3}{2}\ln\left(\frac{U}{N}\right) + \ln\left(\frac{V}{N}\right) + \frac{3}{2}\ln\left(\frac{4\pi}{3m}\right) + \frac{3}{2}\right\}. \tag{4.76}$$

This shows that the Boltzmann entropy is extensive. The thermodynamic entropy (4.72), on the other hand, is not extensive unless the constant C therein is chosen appropriately. Since it is extensive, it does not suffer from the Gibbs paradox discussed in Sect. 7.6.

Since Ω is the continuous limit form of the logarithm of discrete number $\mathcal{P}$ of configurations, the expression (4.59) of Boltzmann's entropy can be generalized to systems having discrete energy levels by defining Boltzmann entropy as

$$S_B = k_B \ln(\mathcal{P}). \tag{4.77}$$

The S_B corresponding to $\mathcal{P}_{max}$, denoted by $S_{B,max}$:

$$S_{B,max} = k_B \ln(\mathcal{P}_{max}), \tag{4.78}$$

provides link with the thermodynamic entropy:

$$S_{th} = S_{B,max} \quad \Longrightarrow \quad \frac{1}{T} = \frac{\partial S_{B,max}}{\partial U}. \tag{4.79}$$

We have used (4.78) and (4.79) in Sect. 4.1.1 to investigate connection between Boltzmann's discrete energy level model of molecules and thermodynamics. In the following, we outline some important applications of the said relations.

It may be mentioned that Boltzmann entropy is widely written as $S = k \ln(W)$. This is how it is inscribed even on his tombstone. However, though Boltzmann did introduce the symbol W in his paper, he never used it. Besides, W in his paper is the probability of occupation of a state and not the number of configurations. The said form got stuck in popular memory after it was used first by Planck in his paper on blackbody radiation [2].

4.5 Planck's Distribution

In his theory of blackbody radiation [2], Planck assumed that the atoms in the walls of the cavity behave as oscillators which absorb and emit radiation having energy in integral multiples of some basic unit ϵ and that the energy density $\mathcal{U}_\nu(T)$ of radiation at frequency ν in the cavity at temperature T is related with the average energy u_ν of an oscillator at the same frequency by the relation, with $\rho_\nu d\nu$ standing for the number of radiation modes per unit volume in the interval $(\nu, \nu + d\nu)$,

$$\mathcal{U}_\nu(T) = \rho_\nu u_\nu(T), \qquad \rho_\nu = \frac{8\pi\nu^2}{c^3}. \tag{4.80}$$

This relation is based on the classical theory of interaction of radiation with an atom. Planck evaluated $u_\nu(T)$ in terms of entropy of the system of oscillators as follows.

If N is the number of oscillators, n_k the number of oscillators having energy $k\epsilon$, and $P\epsilon$ total energy of the oscillators, then maximum energy that each oscillator can have is $P\epsilon$ and hence

$$\sum_{k=0}^{P} n_k = N, \quad \sum_{k=0}^{P} kn_k = P. \qquad (4.81)$$

Planck defined entropy as

$$S_P = k_B \ln(W) + \text{constant}, \qquad (4.82)$$

where W is the probability of the configuration in which N atoms have energy $P\epsilon$. Now, $W = \mathcal{R}/\mathcal{J}$ where $\mathcal{J}$ is total number of configurations which is a constant and $\mathcal{R}$ is the number of configuration of N atoms having energy $P\epsilon$. Since entropy is defined up to the addition of a constant, the expression (4.82) may be rewritten as

$$S_P = k_B \ln(\mathcal{R}). \qquad (4.83)$$

The number $\mathcal{R}$ is same as $\mathcal{P}(N, L)$ given by (4.6), with $L = P$ therein.

The $\mathcal{R}$ in Planck's definition of entropy is the number of all configurations corresponding to energy $P\epsilon$ whereas in Boltzmann's definition (4.78) of entropy, the number of configurations considered are only those which correspond to the state from among all the states of energy $P\epsilon$ for which the number of configurations is maximum. As noted in (4.23), both give same result under Stirling's approximation. Planck's entropy is therefore same as Boltzmann entropy in (4.32) in which u is given in terms of temperature by (4.33). Using known empirical laws of blackbody radiation, Planck showed that entropy of the oscillator must be a function of u/ν:

$$S_P = f\left(\frac{u}{\nu}\right). \qquad (4.84)$$

On comparing (4.32) and (4.84) it follows that ϵ must be proportional to ν. Hence, with $\epsilon = h\nu$, where h is a constant known now as Planck's constant. Consequently, the expression (4.33) for energy reads

$$u_\nu(T) = \frac{h\nu}{\exp(h\nu/k_B T) - 1}. \qquad (4.85)$$

Substitution of (4.85) in (4.80) yields Planck's law of blackbody radiation.

See [3] for discussion of relationship between Planck's and Boltzmann entropies.

4.6 Bose Statistics

Bose re-derived in [4] Planck's formula in an entirely different way. In contrast with the approach of Planck which is based on determining the radiation energy in terms of the energy of oscillators in the walls of the cavity, approach of Bose as outlined below does not require the intermediary of said oscillators and therefore neither of the relation (4.80) between energy of the oscillators and the electromagnetic field energy in the cavity. In fact, Bose considered this as the triumph of his theory stating *All existing derivations make use of the relation between the radiation density and mean energy of an oscillator* (relation (4.80) in the present case) *and they make assumptions concerning the number of degrees of freedom of ether as exemplified in the above equation* (the factor ρ_ν in (4.80)). *This factor, however, could be deduced only from the classical theory. This is the unsatisfactory point of all derivations and it is not surprising that again and again efforts are made which try to give a derivation free of this logical deficiency.* The unsatisfactory feature of Planck's and other derivations Bose mentioned in the statement quoted above from his paper were highlighted by him also in the covering letter he mailed to Einstein along with the manuscript of his paper stating: *You will see that I have tried to deduce the coefficient $8\pi\nu^2/c^3$ in Planck's Law independent of the classical electrodynamics.* He requested Einstein to arrange for its publication if he finds worth it.

Following Einstein's 1905 hypothesis to explain the photoelectric effect, Bose considered radiation inside a blackbody as a gas of quanta (named photons by G.N. Lewis in 1926) in which energy of a quantum of frequency ν is $E_\nu = h\nu$. In accordance with the electromagnetic theory, the magnitude of momentum of the quantum of energy $h\nu$ is $h\nu/c$ in the direction of its propagation. He assumed that the state of a quantum is described by its phase space coordinates constituted by its position having components (x, y, z) and momentum having components (p_x, p_y, p_z) where

$$p_x^2 + p_x^2 + p_x^2 = \left(\frac{h\nu}{c}\right)^2. \tag{4.86}$$

He considered the phase space of the quanta in frequency interval $(\nu, \nu + d\nu)$. The volume of the said phase space is evidently $d\omega_\nu = V dV_\nu$ where V is the spatial volume of the region in which the quanta move and dV_ν is the volume of the shell between the spheres of radii $h\nu/c$ and $h(\nu + d\nu)/c$ given by

$$dV_\nu = 4\pi \left(\frac{h\nu}{c}\right)^2 \left(\frac{h}{c}\right) d\nu. \tag{4.87}$$

Hence

$$d\omega_\nu = 4\pi V \nu^2 \left(\frac{h}{c}\right)^3 d\nu. \tag{4.88}$$

Bose divided $d\omega_\nu$ into cells of volume h^3 each.[1] Taking into account two directions of polarization, the number P_ν of the phase space cells available to the quanta in the frequency interval $(\nu, \nu + d\nu)$ is $2d\omega_\nu / h^3$:

$$P_\nu = \frac{8\pi V}{c^3} \nu^2 d\nu. \tag{4.89}$$

Bose evaluated entropy of the system of quanta by following Boltzmann's approach by counting the number of ways of distributing the quanta in the cells. To that end, each cell is considered as a box into which quanta are distributed. There is no restriction on the number of quanta that a box can contain. Hence, if $n_{k\nu}$ is the number of boxes that contain k quanta ($k = 0, 1, \ldots, \infty$) then the number of configurations is given, invoking (4.2), by

$$\mathcal{P}_\nu = \frac{P_\nu!}{n_{0\nu}! n_{1\nu}! \cdots}, \qquad P_\nu = \sum_{k=0}^{\infty} n_{k\nu}. \tag{4.90}$$

The energy U_ν of the gas of quanta of frequency ν is

$$U_\nu = h\nu \sum_{k=0}^{\infty} k n_{k\nu}. \tag{4.91}$$

Total number of ways of distributing the quanta of all frequencies is

$$\mathcal{P} = \prod_{\nu=0}^{\infty} \mathcal{P}_\nu. \tag{4.92}$$

With $\mathcal{P}_\nu$ given by (4.90), the equation above in Stirling's approximation yields

$$\ln(\mathcal{P}) = \sum_{\nu} \left[P_\nu \ln(P_\nu) - \sum_{k} n_{k\nu} \ln(n_{k\nu}) \right]. \tag{4.93}$$

The $\ln(\mathcal{P})$ above is maximized under the constraints

$$U = \sum_{\nu} U_\nu = h \sum_{\nu} \nu \sum_{k=0}^{\infty} k n_{k\nu}, \qquad P_\nu = \sum_{k=0}^{\infty} n_{k\nu}, \tag{4.94}$$

---

[1] It may be mentioned that the idea of dividing phase space into cells was invoked earlier by Sackur and Tetrode to derive formula for entropy of the ideal gas which helped them determine the undetermined constant in the thermodynamic expression for entropy (see Sect. 7.1.2).

where U is total energy of the gas. Using Lagrange's method of undetermined multipliers, it is straightforward to see that the n_{kv}'s which maximize (4.92) subject to the constraint (4.94) are given by

$$n_{kv} = A_v \exp(-k\beta h v). \tag{4.95}$$

Due to second constraint in (4.94),

$$A_v = P_v\{1 - \exp(-\beta h v)\}, \tag{4.96}$$

so that

$$n_{kv} = P_v(1 - \exp(-\beta h v)) \exp(-k\beta h v). \tag{4.97}$$

The $\ln(\mathcal{P}_{\max})$ is obtained by substituting (4.97) in (4.93) so that, due to (4.27),

$$S_{B,\max} = \sum_v S_{vB,\max}, \tag{4.98}$$

where $S_{vB,max}$ is entropy of the quanta of frequency v given by

$$S_{vB} = k_B \left[\beta U_v - P_v \ln(1 - \exp(-\beta h v))\right]. \tag{4.99}$$

Due to second equation in (4.27) it is straightforward to see that $\beta = 1/k_B T$.

With n_{kv} as in (4.97) and $\beta = 1/k_B T$, it may be verified that

1. The number of quanta of frequency v is

$$N_v = \sum_{k=0}^{\infty} k n_{kv} = \frac{P_v}{\exp(h v/k_B T) - 1}. \tag{4.100}$$

2. The energy U_v of the quanta in the frequency interval $(v, v + dv)$ is

$$U_v = h v N_v = \frac{h v P_v}{\exp(h v/k_B T) - 1}. \tag{4.101}$$

In his paper, Bose mentions number of quanta as a constraint but, without specifying any reason, does not use it. We know that the reason for the absence of constraint on the number of quanta is due to the fact that their number is not conserved because the quanta are continuously created and annihilated. However, as we will see, while applying Bose's formalism to material particles, Einstein uses the said constraint.

As stated in the beginning of this section, Bose mailed to Einstein the manuscript of his paper, written in English, with a request to arrange for its translation and publication, if found worth it, in Zeitschrift für Physik. Einstein, impressed by the idea in the paper, himself translated it and sent it for publication in Bose's name with the comment: *In my opinion Bose's derivation of Planck's formula signifies an important advance. The method used also yields the quantum theory of the ideal gas, as I will work out in detail elsewhere.*

Einstein submitted his paper [5] within 2 weeks after receiving Bose's paper. The paper of Bose and that of Einstein drew widespread criticism for using the distribution law of indistinguishable objects in distinguishable boxes thereby treating, without stating, the light quanta in the case of Bose and material particles in Einstein's case as indistinguishable particles (see [6]). The concept of indistinguishable particles was non-existent at the time. Einstein acknowledged this lapse in his second paper [7] submitted about 6 months later stating therein:

> Mr. Ehrenfest and other colleagues have faulted Bose's theory of radiation and my analogous one for ideal gases for not treating quanta, or molecules, as statistically mutually independent structures, without specifically pointing out this circumstance in our papers. This is entirely correct. If one treats the quanta as statistically independent of one another in their localizations, one arrives at Wien's radiation law; if one treats gas molecules analogously, one arrives at the classical equation of state for ideal gases. Even if one otherwise proceeds exactly as Bose and I have done. Here I will compare both considerations for gases in order to clearly bring out differences and to be able to compare our results easily with those of the theory of independent molecules.

That the indistinguishability is a puzzling aspect was recognized by Einstein while stating later in the same paper by referring to the formula (4.106) for number of configurations or complexions treating particles as indistinguishable:

> ... the formula indirectly expresses a certain hypothesis about an initially completely puzzling mutual influence of the molecules....

We outline next the theory mentioned by Einstein in his cited comments.

4.7   Einstein's Quantum Theory of Ideal Gas

Einstein developed the quantum theory of ideal gas by using Bose's method of dividing phase space volume into cells as follows.

Consider the gas of molecules in a container of volume V. Consider molecules having energy in the interval $(E, E + \Delta E)$. Assume the motion to be non-relativistic so that the relation between energy and the magnitude p of momentum of a molecule is $E = p^2/2m$. Let the magnitude of corresponding momenta be in the interval $(p, p + \Delta p)$. The volume of the phase space occupied by these molecules is the product of the spatial volume V and the volume $4\pi p^2 \Delta p$ of the spherical shell between the spheres of radii p and $p + \Delta p$ so that the volume of the phase space traversed by these molecules is

$$\Delta \omega = (2\pi V)(2m)^{3/2} \sqrt{E} \Delta E. \qquad (4.102)$$

Divide the phase space into cells of volume h^3 each. The number of cells in the phase space volume $\Delta\omega$ then is

$$P_E = \frac{2\pi V}{h^3}(2m)^{3/2}\sqrt{E}\,\Delta E. \qquad (4.103)$$

Let N_E be the number of molecules in the interval $(E, E + \Delta E)$. Let $\mathcal{P}_E$ denote the number of configurations for distribution of N_E molecules in P_E cells. The number $\mathcal{P}_E$ will depend on whether the molecules are treated as indistinguishable or distinguishable. In any case, total number of configurations for all energies is

$$\mathcal{P} = \prod_E \mathcal{P}_E, \qquad (4.104)$$

and the entropy is $S = k_B \ln(\mathcal{P})$. The equilibrium state is obtained by maximizing S with respect to variation in N_E subject to the conditions

$$N = \sum_E N_E, \quad U = \sum_E E N_E, \qquad (4.105)$$

where N is total number of molecules.

In his first paper [5] Einstein adopted Bose's method to determine the number of configurations, namely, by grouping the cells by the number of molecules they contain and finding the number of boxes in each group that maximize $\mathcal{P}$.

In his second paper, Einstein adopted a different approach to evaluate $\mathcal{P}$ which could be applied to the system of indistinguishable as well as that of distinguishable particles. We outline below the said approach for the cases of (1) indistinguishable molecules and (2) distinguishable molecules, and (3) revisit Bose's system of light quanta and (4) extend it to the gas of distinguishable particles obeying exclusion principle.

4.7.1 Indistinguishable Molecules

The number of configurations resulting from distribution of N_E indistinguishable objects in P_E distinguishable boxes is given, recalling (D.23), by

$$\mathcal{P}_E = \frac{(N_E + P_E - 1)!}{(P_E - 1)!N_E!}. \qquad (4.106)$$

Using Stirling's approximation for the logarithm of the factorials in (4.106), the expression for entropy reads

$$S = k_{\mathrm{B}} \sum_E [(N_E + P_E)\ln(N_E + P_E) - P_E\ln(P_E) - N_E\ln(N_E)]. \quad (4.107)$$

Invoking Lagrange's method of multipliers, it is straightforward to see that the value of N_E that maximizes S, subject to the constraints (4.105), is

$$N_E = \frac{P_E}{\exp(\alpha + \beta E) - 1}. \quad (4.108)$$

With N_E given by (4.108), (4.107) for entropy reduces to

$$S = k_{\mathrm{B}}(\alpha N + \beta U) - k_{\mathrm{B}} \sum_E P_E\ln\{1 - \exp(-\alpha - \beta E)\}. \quad (4.109)$$

Using $1/T = (\partial S/\partial U)_{V,N}$ follows the relation $\beta = 1/k_{\mathrm{B}}T$. Hence, with P_E given by (4.103), the number of molecules per unit energy interval in the vicinity of E ($n_E = N_E/\Delta E$) turns out to be given by

$$n_E = \frac{2\pi V (2m)^{3/2} h^{-3} \sqrt{E}}{\exp(\alpha + \beta E) - 1}, \qquad \beta = 1/k_{\mathrm{B}}T. \quad (4.110)$$

This is Einstein's formula for an ideal quantum gas of indistinguishable molecules. This is also called Bose–Einstein distribution. We will study the thermodynamics of the gas described by it in Chap. 10.

4.7.2   Distinguishable Molecules

In case the molecules are distinguishable, the number of ways of distributing N_E molecules in P_E cells may be determined as follows. Note that there are P_E ways of distributing a molecule among P_E cells. Since the molecules are independent and distinguishable, total number of ways of distributing N_E molecules would be the product of the individual number of ways which is $(P_E)^{N_E}$ (see 8.2.1 for alternative derivation). Hence the number of distributions for all energy intervals is

$$\mathcal{P}_1 = \prod_E P_E^{N_E}. \quad (4.111)$$

On multiplying this by the number of ways of choosing N_E molecules from among N follows the expression for the number of configurations:

$$\mathcal{P} = \prod_E P_E^{N_E} \left(\frac{N!}{\prod_E N_E!} \right). \quad (4.112)$$

The expression $S = k_B \ln(\mathcal{P})$ for entropy in Stirling's approximation reads

$$S = k_B \sum_E [N_E \ln(P_E) - N_E \ln(N_E)] + k_B N \ln(N). \tag{4.113}$$

The expression for N_E which maximizes (4.113) subject to the conditions in (4.105) reads

$$N_E = P_E \exp(-\alpha - \beta E). \tag{4.114}$$

Substitution of this in first condition in (4.105), with P_E as in (4.103), yields

$$N = \sum_E P_E \exp(-\alpha - \beta E) \equiv \exp(-\alpha) V Z, \tag{4.115}$$

where

$$Z = \frac{2\pi}{h^3}(2m)^{3/2} \sum_E \sqrt{E} \exp(-\beta E) \Delta E$$
$$= \frac{2\pi}{h^3}(2m)^{3/2} \int_0^\infty \sqrt{E} \exp(-\beta E)\, dE = (2\pi m / h^2 \beta)^{3/2}. \tag{4.116}$$

Substitute (4.114) in (4.113) for entropy to get

$$S = k_B\{N \ln(V) + N \ln(Z) + U\beta\}, \tag{4.117}$$

Using $1/T = (\partial S/\partial U)_{V,N}$ we get $\beta = 1/k_B T$, and using $P/T = (\partial S/\partial V)_{U,N}$ we obtain

$$PV = Nk_B T, \tag{4.118}$$

which is the equation of state of an ideal classical gas.

4.7.3 Distribution of Light Quanta

Planck's law of blackbody radiation was derived in Sect. 4.6 using Bose's method of counting the number of different ways of distributing identical quanta in phase space cells. We re-derive that law now applying Einstein's method of counting the distributions.

To that end, recall from Sect. 4.6 that the number P_ν of cells in the phase space of the light quanta in the frequency interval $(\nu, \nu + d\nu)$ is given by (4.89) and the number of quanta in those cells is denoted by N_ν. In Einstein's formulation, the number $\mathcal{P}_\nu$ of ways of distributing those N_ν indistinguishable quanta in P_ν cells is given by (4.106) with $P_E \rightarrow P_\nu$, $N_E \rightarrow N_\nu$ therein:

$$P_\nu = \frac{(N_\nu + P_\nu - 1)!}{(P_\nu - 1)!N_\nu!}, \tag{4.119}$$

whereas the entropy in Stirling's approximation is

$$S = k_B \sum_\nu [(N_\nu + P_\nu)\ln(N_\nu + P_\nu) - P_\nu\ln(P_\nu) - N_\nu\ln(N_\nu)]. \tag{4.120}$$

The equilibrium distribution is found by maximizing S subject to:

$$U = h \sum_{\nu=0}^{\infty} \nu N_\nu. \tag{4.121}$$

In the present case, there is no condition on total number of quanta because they are continuously created and absorbed in the wall. The values of the N_ν's for which S is maximum with respect to variation of N_ν, subject to the condition (4.121) is

$$N_\nu = \frac{P_\nu}{\exp(\beta h\nu) - 1}. \tag{4.122}$$

With $\beta = 1/k_B T$ this is same result as has been derived by Bose's method.

4.8 Distribution of Particles Obeying Exclusion Principle

We can apply Bose's concept of dividing the phase space into cells of volume h^3 to determine distribution of indistinguishable particles obeying exclusion principle, i.e. particles which are such that only one of them can occupy a state. That being the case, the number of configurations of N_E such particles in P_E cells in the interval $(E, E + \Delta E)$ of energy would be the number of ways one can choose N_E cells from among the P_E to accommodate N_E molecules. It is given by

$$P_E = \frac{P_E!}{(P_E - N_E)!N_E!}. \tag{4.123}$$

The entropy in Stirling's approximation in this case is

$$S = k_B \sum_E [P_E\ln(P_E) - (P_E - N_E)\ln(P_E - N_E) - N_E\ln(N_E)]. \tag{4.124}$$

Invoking Lagrange's method of multipliers, the N_E which maximizes S above subject to the conditions (4.105) turns out to be given by

$$N_E = \frac{P_E}{\exp(\alpha + \beta E) + 1}.$$ \hfill (4.125)

Substitution of this in (4.124) results in the following expression for entropy:

$$S = k_{\mathrm{B}}(\alpha N + \beta U) + k_{\mathrm{B}} \sum_E P_E \ln\{1 + \exp(-\alpha - \beta E)\}.$$ \hfill (4.126)

Applying relation $1/T = \partial S/\partial U$ we see that $\beta = 1/(k_{\mathrm{B}}T)$. The resulting distribution is called Fermi–Dirac distribution. We will study the thermodynamics of the gas described by it in Chap. 9.

References

1. L. Boltzmann, K. Akademie der Wissenschaften Mathematisch-Natyrwissen Classe. Abt. II, LXXVI, **1877**, 373–435 (Wien. Ber. **1877** 76, 373–435). Reprinted in Wiss. Abhandlungen, Vol. II, reprint 42, pp. 164–233, Barth, Lepzig, 1909). English translation with commentary by K. Sharp, F. Matschinsky, *Entropy*, **17**, 1971–2009 (2015)
2. M. Planck, Ann. der Phys. **4** 553 (1901)
3. M. Nauenberg, Am. J. Phys. **84** 709 (2016)
4. S.N. Bose, Zeitschrift für Physik **27**, 384 (1924). English translation with commentary by O. Theimer, B. Ram, Am. J. Phys. **44**, 1057 (1976)
5. A. Einstein, Zeitschrift für Physik **22**, 261 (1924). English translation in *Cpllected Papers of Albert Einstein*
6. M. Delbruck, J. Chem. Edu. **57**, 467 (1980)
7. A. Einstein, Zeitschrift für Physik **22**, 261 (1924). English translation in *Cpllected Papers of Albert Einstein*

Chapter 5
Shannon and Statistical Entropies

In Chap. 2, we saw that the thermodynamic variables are expressible in terms of various partial derivatives of entropy. The link between mechanical and thermodynamic descriptions may therefore be established by identifying suitably defined mechanical entropy, i.e. entropy defined in terms of mechanical variables with thermodynamic entropy. In Chap. 4, we saw that identification of Boltzmann entropy with thermodynamic entropy provides the desired link when particles are non-interacting. We will see that entropy for systems of N interacting particles, identifiable with thermodynamic entropy, is Gibbs entropy [1] defined in terms of N-particle phase space distribution function. The Boltzmann entropy for the gas of N molecules on the other hand is in terms of single-particle distribution function. It does not depend on inter-particle interaction even when inter-particle interaction is present and does not correspond to thermodynamic entropy in the presence of inter-particle interaction [2].

Since it is constructed in terms of phase space distribution function, Gibbs entropy cannot describe quantum systems as there is no concept of phase space in quantum mechanics. The quantum von Neumann entropy, introduced in Chap. 13, serves the desired purpose. A simpler approach, adequate for describing widely encountered thermodynamic systems, is Shannon entropy [3] (see also [Balian] and [4]). Its origin is in the information theory and is applicable to any situation described statistically. In this chapter, we introduce the concept of Shannon entropy for discrete as well as continuous probabilities using which we define entropy for quantum and classical mechanical systems called the statistical entropy. We will see that the entropy so defined in terms of continuous probabilities is same as Gibbs entropy.

5.1 Shannon Entropy

In this section, we introduce the concept of Shannon entropy for discrete as well as for continuous probabilities.

© The Author(s), under exclusive license to Springer Nature Switzerland AG 2024   167
R. R. Puri, *Modern Thermodynamics and Statistical Mechanics*, Undergraduate Lecture Notes in Physics, https://doi.org/10.1007/978-3-031-54310-4_5

5.1.1 Discrete Probabilities

Let x be a random variable capable of assuming n discretely spaced values $\{x_1, x_2, \ldots, x_n\}$. Let $p(x_k)$ be the probability that the value realized in some event is x_k. The probabilities are such that $0 \leq p(x_k) \leq 1$ and

$$\sum_{k=1}^{n} p_k = 1, \qquad p_k \equiv p(x_k). \tag{5.1}$$

If $p_i < p_j$ then the occurrence of x_i in repeated experiments is rarer than that of x_j. In other words, occurrence of x_i is more uncertain than that of x_j. Hence the occurrence of x_i may be viewed as removing greater uncertainty or contributing more to the information about x than the occurrence of x_j. Accordingly, it is postulated that the gain in information on observing the realization of the value x_i of the random variable x is a function of the probability of the occurrence of that value. It is denoted by $I(p_i)$. From the discussion above we see that $I(p_i) > I(p_j)$ if $p_i < p_j$.

Now, consider two random variables, x and y where x is as above and y takes m values $\{y_1, y_2, \ldots, y_m\}$ with probabilities $\{q_1, q_2, \ldots, q_m\}$ where $q_k \equiv q(y_k)$. Let x and y be independent so that the joint probability of observation of the value x_i of x and y_k of y in an event is $p_i q_k$. The information gain due to the said observation is $I(p_i q_k)$. Since they are independent, the information gained by the said joint event must be the sum $I(p_i) + I(q_k)$ of the information gained from the occurrence of those events individually. Hence follows the relation

$$I(p_i q_k) = I(p_i) + I(q_k). \tag{5.2}$$

This equation is solved by $I(p) = k' \ln(p)$ where k' is a constant. Since information is to be a positive quantity and $0 \leq p \leq 1$, k' must be negative. The $I(p)$ is then defined as

$$I(p) = -k \ln(p), \qquad k > 0. \tag{5.3}$$

The constant k is arbitrary and only defines the scale for the measurement of information.

Average gain in information in a large number of realizations of x is clearly

$$S(\{p_i\}_n) \equiv S(p_1, p_2, \ldots, p_n) = -k \sum_{i=1}^{n} p_i \ln(p_i), \tag{5.4}$$

with $0\ln(0) = 0$. The $S(\{p_i\}_n)$ is called *Shannon entropy* associated with the probability distribution $p_1, p_2, \ldots, p_n$.

We list below some properties of $S(\{p_i\}_n)$.

1. It follows immediately from its expression (5.4) that entropy is positive.
2. If $p_k = 1$ for some k, $p_j = 0$ ($j \neq k$) so that $S(\{p_i\}_n) = 0$. This means, since there is no uncertainty about an outcome when the variable can take only one value (x_k in the present example), there is no gain in information when that value is realized.
3. If all the outcomes occur with equal probability then $p_i = 1/n$ and the corresponding entropy is

$$S(\{p_i\}_n) = -\frac{k}{n} \sum_{i=1}^{n} \ln\left(\frac{1}{n}\right) = k\ln(n). \tag{5.5}$$

In Sect. 5.2, we will show that this is the maximum entropy for an n-valued random variable. Hence, for a random variable which can assume n values,

$$0 \leq S(\{p_i\}_n) \leq k\ln(n). \tag{5.6}$$

4. It can be shown that $S(\{p_i\}_n)$ is a concave function in the sense elaborated in Appendix C. It has been shown therein that a function is concave if its second derivative is negative (see C.9)). We prove the concavity of $S(\{p_i\}_n)$ by examining its second derivative.

Consider $F(x)$ defined by

$$F(x) = -x\ln(x), \quad x \geq 0. \tag{5.7}$$

Since $d^2 F(x)/dx^2 = -1/x < 0$ for $x > 0$, it follows that $F(x)$ is concave. Rewrite the expression (5.4) for Shannon entropy as

$$S(\{p_i\}_n) = \sum_{i=1}^{n} F_i(p_i), \quad F_i(p_i) = -kp_i\ln(p_i). \tag{5.8}$$

The $F_i(p_i)/k$ is same as the function $F(x)$ defined in (5.7) with $x \to p_i$. Hence $F_i(p_i)$ is concave. This shows that each term in the summation in (5.8) is concave whereby it follows that $S(\{p_i\}_n)$ is concave which proves our assertion.

Being concave, each $F_i(p_i)$ obeys the inequality (C.5) due to which,

$$F_i(\lambda p_i + (1 - \lambda)q_i) \geq \lambda F(p_i) + (1 - \lambda)F(q_i), \quad 0 \leq \lambda \leq 1. \tag{5.9}$$

On using this in (5.8) follows the inequality

$$S(\{\lambda p_i + (1 - \lambda)q_i\}_n) \geq \lambda S(\{p_i\}_n) + (1 - \lambda)S(\{q_i\}_n). \tag{5.10}$$

To understand the physics meaning of the inequality above, consider two sources, say A and B, emitting same particles in one of the n states. Let the probability with which the particles are emitted in the state labeled i by the source A be p_i and that for the particles emitted by the source B be q_i ($i = 1, 2, \ldots, n$). The entropy of the particles emitted by A is $S(\{p_i\}_n)$ and that of the particles emitted by B is $S(\{q_i\}_n)$. Assume that the sources A and B are emitting particles with rate r_A and r_B, respectively. The fraction of particles coming from the source A in time t will be $\lambda = r_A/(r_A + r_B)$ and that from B in the same time will be $r_B/(r_A + r_B) \equiv 1 - \lambda$. Hence the probability of finding a particle in the state labeled i from the mixture of the emitted particles is evidently $\lambda p_i + (1 - \lambda)q_i$. The left side of the inequality (5.10) is the entropy corresponding to the said probability. On the other hand, if the particles were not mixed then the average entropy of the particles emitted by the two sources would have been $\lambda S(\{p_i\}_n) + (1 - \lambda)S(\{q_i\}_n)$ which is the right side of the inequality (5.10).

The inequality (5.10) thus shows that if two probabilities $\{p_i\}_n$ and $\{q_i\}_n$ are mixed then the entropy of the mixture is higher than the average entropy of individual probabilities. In other words, the uncertainty of finding an object in the mixture increases than that when it is in the group it originally belonged to. This is intuitively understandable as well.

In general, if $\{p_i^{(j)}\}_n$ ($j = 1, 2, \ldots, k$) are k probability distributions all for the same random variable x then

$$S\left(\sum_{j=1}^{k} \lambda_j \{p_i^{(j)}\}_n\right) \geq \sum_{j=1}^{k} \lambda_j S\left(\{p_i^{(j)}\}_n\right), \qquad \sum_{j=1}^{k} \lambda_j = 1. \qquad (5.11)$$

This will prove useful in proving the second law of thermodynamics.

5.1.2   Continuous Probabilities

The concept of Shannon entropy for random variables assuming discrete values can be extended to the ones assuming continuous values as follows (see also [4]).

Let x be a real random variable which assumes values in the domain $[a, b]$. Let $p(x)$ be the probability density such that

$$dw = p(x)dx \qquad (5.12)$$

is the probability that the value of x is in $(x, x + dx)$ with

$$\int_a^b p(x)\,dx = 1. \qquad (5.13)$$

Shannon generalized the expression (5.4) of entropy for discrete variables to continuous ones in straightforward manner by replacing discrete probabilities by the continuous one and the summation by integral to define the entropy associated with the probability density $p(x)$ as

$$S(p(x)) = -k \int_a^b p(x)\ln(p(x)) \, dx. \tag{5.14}$$

However, as we will see, if standard mathematical method is followed to convert a discrete sum to an integral in the limit of the discrete variable assuming continuous values then extra blowing up term appears in the resulting expression so obtained.

   To that end, we relate the problem of continuous valued random variable with that of a discrete variable as follows. Divide the interval $a \le x \le b$ into n bins of length ϵ each by the points $a = x_0, x_1, \ldots, x_n = b$ where $\epsilon = (b-a)/n$ with $n \to \infty$, $\epsilon \to 0$ but $n\epsilon = b - a$ remaining finite. Denote by f_i the probability that x takes value in the bin (x_{i-1}, x_i) $(i = 1, 2, \ldots, n)$. Using (5.12), the probability f_i can be expressed in terms of the probability density $p(x)$ as

$$f_i = \epsilon p(x_i), \quad i = 1, 2, \ldots, n. \tag{5.15}$$

The set $f_1, f_2, \ldots, f_n$ defines probability distribution for a random variable which takes discrete values, each lying in one or the other bin defined above. The Shannon entropy corresponding to the said set is therefore given by

$$S = -k \sum_{i=1}^{n} f_i \ln(f_i) = -k \sum_{i=1}^{n} \epsilon p(x_i) \ln(\epsilon p(x_i))$$

$$= -k \sum_{i=1}^{n} \epsilon p(x_i) \left[\ln(p(x_i)) + \ln(\epsilon) \right]. \tag{5.16}$$

Invoking the definition of Riemann integral,

$$\text{Lt}_{\epsilon \to 0} \left(\epsilon \sum_{i=1}^{n} F(x_i) \right) = \int_a^b F(x) \, dx, \tag{5.17}$$

we have

$$\text{Lt}_{\epsilon \to 0} \left(\epsilon \sum_{i=1}^{n} p(x_i) \ln(p(x_i)) \right) = \int_a^b p(x)\ln(p(x)) \, dx,$$

$$\text{Lt}_{\epsilon \to 0} \left(\epsilon \sum_{i=1}^{n} p(x_i) \right) = \int_a^b p(x) \, dx = 1, \tag{5.18}$$

where last equation is due to the normalization condition (5.13). Consequently, in the limit $\epsilon \to 0$, (5.16) reduces to

$$S(p(x)) = -k \int_a^b p(x)\ln(p(x))\, dx - k\mathrm{Lt}_{\epsilon \to 0}\ln(\epsilon).\qquad(5.19)$$

The first term in the expression above is same as Shannon's expression (5.14) but the second term blows as $\epsilon \to 0$. Since it does not depend on $p(x)$, that term will not matter if interest is in the difference in entropy associated with different probability densities. Else (5.14) can be understood as the expression for entropy associated with the probability density of a continuous valued random variable with the second term in (5.19) as its zero.

Equation (5.14) is taken as the definition of Shannon entropy when the random variable takes continuous values.

It must, however, be pointed out that $S(p(x))$ does not share all the properties of the entropy of a discrete valued random variable. Some such important differing properties are

1. Unlike $S(\{p_i\}_n)$, $S(p(x))$ need not be positive (see Ex. 5.1).
2. $S(\{p_i\}_n)$ does not change under transformation of x. However, $S(p(x))$ may not be invariant under transformation of x. For, let x be transformed to y by the relation $x = x(y)$ so that the distribution $p(x)$ transforms to $q(y) = p(x(y))$ such that

$$p(x)dx = q(y)dy = q(y)\left|\frac{dy}{dx}\right|dx.\qquad(5.20)$$

It is straightforward to see that

$$\begin{aligned}
S(p(x)) &= -k \int_a^b p(x)\ln(p(x))\, dx \\
&= -k \int q(y)\ln(q(y))\, dy - k \int q(y)\ln(|dy/dx|)\, dy \\
&= S(q(y)) - k \int q(y)\ln(|dy/dx|)\, dy.\qquad(5.21)
\end{aligned}$$

This shows that entropy of the distribution obtained by transformation of the random variable need not be same as that for the distribution of the original variable. See Exercises 5.3 and 5.4 for examples.

Exercises

Ex. 5. 1. Show that the entropy for the exponential distribution function

$$p(x) = \lambda \exp(-\lambda x), \quad \lambda > 0, \quad 0 \le x \le \infty\qquad(5.22)$$

is given by $S = k(1 - \ln(\lambda))$. This becomes negative for $\lambda > e$.

Ex. 5. 2. Show that the entropy for the distribution function

$$p(x) = Ax^2 \exp(-\alpha x^2), \quad \alpha > 0 \quad 0 \le x \le \infty, \quad A = 4\sqrt{\frac{\alpha^3}{\pi}} \tag{5.23}$$

is given by

$$S/k = \gamma + \frac{1}{2}\ln\left(\frac{\pi}{\alpha}\right) - \frac{1}{2}, \tag{5.24}$$

where γ is Euler's constant. Hint: You may need to use the formula

$$\int_0^\infty x^2 \ln(x^2) \exp(-\alpha x^2)dx = \frac{1}{4}\sqrt{\frac{\pi}{\alpha^3}}(2 - \gamma - \ln(4\alpha)). \tag{5.25}$$

Ex. 5. 3. Show that $S(p(x)) = S(p(x + a))$, where a is a constant. This shows that entropy is invariant under translation of the random variable.

Ex. 5. 4. Show that the entropy $S(q(y))$ for $q(y)$ where $y = ax$ is related with the entropy for $S(p(x))$ where $q(y) = p(x(y))$ by the relation

$$S(q(y)) = S(p(x)) + k\ln(|a|). \tag{5.26}$$

This shows that entropy is not invariant under the given transformation.

5.2 Maximum Entropy

We will see that a problem of interest in statistical mechanics is to find the probability distribution for which entropy is maximum subject to chosen constraints on the moments of the random variables. In view of that we outline the method of evaluating maximum entropy separately for discrete and continuous probabilities.

5.2.1 Discrete Variables

Consider a random variable x capable of taking n values according to the probabilities $\{p_i\}_n$. The p_i's are subject to the normalization condition (5.1) and possibly additional r conditions:

$$\langle A_m(x) \rangle \equiv \sum_{i=1}^{n} A_m(x_i)p_i = C_m, \quad m = 1, 2, \ldots, r, \tag{5.27}$$

where $A_m(x)$ is a function of the random variable x and C_m a given constant. We find the extremum of $-S(\{p_i\}_n)/k$, subject to the normalization condition (5.1) and the constraints in (5.27), by the method of Lagrange multipliers. It consists of finding, as a function of the p_i's, the extremum of $F(\{p_i\}_n)$ defined by

$$F(\{p_i\}_n) = -\frac{S(\{p_i\}_n)}{k} + \alpha_0 \sum_i p_i + \sum_{m=1}^{r} \alpha_m \langle A_m(x) \rangle, \qquad (5.28)$$

where the α_i's are so-called Lagrange multipliers. The extremum of $F(\{p_i\}_n)$ is determined by the solution of the equations,

$$\frac{\partial F(\{p_i\}_n)}{\partial p_i} = 0, \qquad i = 1, 2, \ldots, n. \qquad (5.29)$$

With $F(\{p_i\}_n)$ as in (5.28) wherein $S(\{p_i\}_n)$ is given by (5.4), (5.29) yields

$$1 + \ln(p_i) + \alpha_0 + \sum_{m=1}^{r} \alpha_m A_m(x_i) = 0. \qquad (5.30)$$

This is solved by

$$p_i - C \exp\left(-\sum_{m=1}^{r} \alpha_m A_m(x_i)\right), \qquad (5.31)$$

where $C = \exp(-1 - \alpha_0)$ is independent of i. Using the normalization condition, C turns out to be given by

$$C^{-1} = \sum_{i=1}^{n} \exp\left(-\sum_{m=1}^{r} \alpha_m A_m(x_i)\right). \qquad (5.32)$$

The α_m's are determined by (5.27) defining the constraints.

Equation (5.31) is the desired result determining the probabilities for which $S(\{p_i\}_n)$ is extremum subject to the normalization condition and the constraints (5.27). Though not shown here, the extremum is in fact a maximum. We will refer to the extremum condition on S as the condition of its maximum.

As an example, assume that the normalization condition is the only constraint. The probabilities which maximize $S(\{p_i\}_n)$ in this case are given by (5.31) with $\alpha_m = 0$ so that $p_i = C$. With C given by (5.32) we get $p_i = 1/n$. Thus, if there is no constraint other than the normalization, the maximum entropy is obtained when all the outcomes occur with equal probability.

5.2.2 *Continuous Variable*

Let x be a random variable taking continuously varying values in the domain $[a, b]$ according to the probability distribution $p(x)$. The $p(x)$ is subject to the normalization condition (5.13) and possibly r other conditions,

$$\langle A_m(x) \rangle \equiv \int_a^b A_m(x) p(x)\, \mathrm{d}x = C'_m \qquad m = 1, 2, \ldots, r. \qquad (5.33)$$

To determine $p(x)$ for which entropy $S(p(x))$, defined in (5.14), is maximum subject to the conditions (5.13) and (5.33), we follow the method of Lagrange multipliers which involves finding the extremum of the function

$$F(p(x)) = -\frac{S}{k} + \alpha_0 \int_a^b p(x)\, \mathrm{d}x + \sum_{m=1}^r \alpha_m \langle A_m(x) \rangle, \qquad (5.34)$$

where α_i's are the Lagrange multipliers. The extremum is determined by the solution of the equation

$$\frac{\partial F(p(x))}{\partial p(x)} = 0. \qquad (5.35)$$

For $F(p(x))$ defined in (5.34), with $S(p(x))$ as in (5.14), (5.35) yields

$$1 + \ln(p(x)) + \alpha_0 + \sum_{m=1}^r \alpha_m A_m(x) = 0. \qquad (5.36)$$

This is solved by

$$p(x) - C \exp\left(-\sum_{m=1}^r \alpha_m A_m(x) \right), \qquad (5.37)$$

where C is independent of x, determined by the normalization condition as

$$C^{-1} = \int_a^b \exp\left(-\sum_{m=1}^r \alpha_m A_m(x) \right) \mathrm{d}x. \qquad (5.38)$$

The α_m's are determined by (5.33) defining the constraints.

As an example, let the probability distribution be subject to no constraint other than the normalization. Maximum entropy is then given by (5.37) with $\alpha_m = 0$ yielding $p(x) = C$, where from (5.38) we get $C = 1/(b - a)$. Thus the distribution giving maximum entropy subject to no constraint other than the normalization is uniform.

Based on the concept of Shannon entropy, we introduce that of statistical entropy of a dynamical system.

Exercises

Ex. 5. 5. Let x be a random variable assuming values in the interval $[0, \infty]$. Show
that among all the distributions having given value a of the average $\langle x \rangle$, the
exponential distribution

$$p(x) = \frac{1}{a} \exp(-x/a) \tag{5.39}$$

has maximum entropy.

Ex. 5. 6. Let x be a random variable assuming values in the interval $(-\infty, \infty)$. Show
that among all the distributions having given value a of the average $\langle x \rangle$ of
x and the variance $\sigma^2 = \langle x^2 \rangle - \langle x \rangle^2$ the Gaussian distribution

$$p(x) = \frac{1}{\sqrt{2\pi\sigma^2}} \exp\{-(x-a)^2/(2\sigma^2)\} \tag{5.40}$$

has maximum entropy.

5.3 Statistical Entropy

As discussed in Sect. 3.3, the observed value of an observable is the average over
its values in the microstates which the system passes through during the time of
observation. However, every microstate may not be visited with same frequency.
Following Gibbs, we represent the set of microstates as a collection or an *ensemble*
in which a microstate is represented as many number of times as it occurs during the
process of time averaging. Different external conditions define different ensembles.
Thermodynamic properties of a system are described in terms of the probabilities with
which its microstates are occupied. Construction of the probabilities characterizing
a system in thermodynamic equilibrium is based on the concept of statistical entropy
introduced next, separately for quantum and the classical systems.

5.3.1 Classical Systems

Consider a system of N particles described, in the notation of Chap. 3, by the Hamiltonian $H(\{\mathbf{r}_i, \mathbf{p}_i\}_N)$. Recall also from that chapter that the mechanical state of such
a system is described by the phase space distribution function $f_N(\{\mathbf{r}_i, \mathbf{p}_i\}_N; t)$ in
terms of which the average value of an observable $O(\mathbf{r}, \mathbf{p})$ is given by

$$\langle O(t) \rangle = \int O(\{\mathbf{r}_i, \mathbf{p}_i\}_N) f_N(\{\mathbf{r}_i, \mathbf{p}_i\}_N; t) \mathrm{d}\tau_N, \tag{5.41}$$

with

$$d\tau_N = \frac{d\tau'_N}{N!h^{3N}}, \qquad d\tau'_N = \prod_{i=1}^{N} d^3\mathbf{r}_i d^3\mathbf{p}_i, \tag{5.42}$$

where h is Planck's constant and the factor $N!$ is included in case N particles constituting the system are identical. Note that, in contrast with $d\tau_N$ above, the averages have been defined in Chap. 3 by taking $d\tau'_N$ as the volume element. The factor $N!$ in the volume element given above was conjectured for resolving Gibbs' paradox, discussed in Sect. 7.6. We will see that, in appropriate limit, the quantum theoretic formalism reduces to that based on the phase space formalism when the phase space volume element is taken as in (5.42). The normalization condition on $f_N(\{\mathbf{r}_i, \mathbf{p}_i\}_N; t)$ accordingly reads

$$\int f_N(\{\mathbf{r}_i, \mathbf{p}_i\}_N)d\tau_N = 1. \tag{5.43}$$

The Shannon entropy corresponding to $f_N(\{\mathbf{r}_i, \mathbf{p}_i\}_N; t)$ is

$$S_N(f_N) = -k_B \int f_N(\{\mathbf{r}_i, \mathbf{p}_i\}_N; t)\ln(f_N(\{\mathbf{r}_i, \mathbf{p}_i\}_N; t)) \, d\tau_N. \tag{5.44}$$

The constant k_B, arbitrary at this stage, is determined by comparing the predictions based on entropy defined above with the corresponding ones of thermodynamics. The k_B will turn out to be same as the Boltzmann constant.

The entropy defined above constitutes the basis for determining thermodynamics of a system in terms of the probability distribution which in turn is determined by the Hamiltonian governing its motion linking thereby statistical mechanical and the thermodynamic descriptions. It is called the *statistical entropy* of classical systems.

5.3.2 Quantum Systems

To link the mechanical and the thermodynamic descriptions, we introduced above the notion of statistical entropy. It is in terms of the phase space distribution function which is a function of the positions and momenta of the particles constituting the system. However, since position and momentum of a particle cannot be assigned precise values simultaneously in the quantum theory, the mechanical state of a particle in quantum theory cannot be characterized in terms of its position and momentum. Hence, the concept of phase space distribution function does not exist in quantum theory. We therefore need a different approach, described below, to identify the quantum analog of the phase space statistical entropy.

To that end, consider a quantum system of N particles described by the Hamiltonian $\hat{H}_N$ (a letter with caret on it represents an operator). Let $\{|E_{iN}\rangle\} \equiv |E_{1N}\rangle, |E_{2N}\rangle, \ldots$ be the eigenstates of $\hat{H}_N$ corresponding to the energies $E_{1N}, E_{2N}, \ldots$. At any instant of time, the system will be in the state of one of the said energies. As it evolves, let p_{mN} denote the probability that it is found in the state of energy E_{mN}. Statistical mechanics describes the thermodynamic properties of the system in terms of the set of probabilities $\{p_{mN}\} \equiv p_{1N}, p_{2N}, \ldots$.

The link between statistical mechanics and thermodynamics is established by the Shannon entropy associated with the probabilities $\{p_{mN}\}$:

$$S_N(\{p_{iN}\}) = -k_B \sum_m p_{mN}\ln(p_{mN}), \qquad (5.45)$$

with

$$\sum_m p_{mN} = 1. \qquad (5.46)$$

The S_N defined in (5.45) is statistical entropy of quantum systems. We will see that the expression (5.45) for entropy is same as the von Neumann entropy introduced in Chap. 13.

The phase space distribution and the quantum probabilities depend on time if the system is not in equilibrium. We addressed the question of time evolution of the phase space distributions in Chap. 3 on the kinetic theory. The question of time evolution of the quantum probabilities is addressed briefly in Chaps. 13 and 14. Our interest is, however, in the equilibrium state. It is described by the probabilities obtained as the solution of the time-dependent equations in the limit $t \to \infty$. That approach to determine the equilibrium state is generally a formidable task. However, as we saw in Chap. 4, Boltzmann showed that the equilibrium state of the non-interacting gas can be obtained without solving any time-dependent equation as the one which maximizes the entropy defined by him. The statistical entropy is for any system including the ones consisting of interacting molecules. The equilibrium state of a macroscopic system in general is postulated to be the one in which its statistical entropy is maximum. In Chap. 6, we construct equilibrium states under different external constraints based on the postulate of maximum statistical entropy.

References

1. J.W. Gibbs, *Elementary Principles in Statistical Mechanics* (Dover Publications, 2014. First published by Yale University Press, 1902)
2. E.T. Jaynes, Am. J. Phys. **33**, 391 (1965)
3. C.E. Shannon, Bell Syst. Tech. J. **27**, 379 (1948)
4. T.M. Cover, J.A. Thomas, *Elements of Information Theory* (John Wiley & Sons, 2006)

Chapter 6
Equilibrium Distributions

In Chap. 5 we introduced statistical entropy defined in terms of probabilities of molecular distribution in the energy levels. The energy levels may be discrete or continuous. In this chapter we find the said probability distributions when the system attains thermodynamic equilibrium subject to constraints on it. The probability distribution in question for a system in the state of equilibrium is determined based on the principle of maximum entropy. The entropy so obtained is identified as the thermodynamic entropy. This enables one to establish relationship between statistical mechanical and thermodynamic descriptions.

6.1 Principle of Maximum Entropy

In Chap. 5 we introduced the concept of statistical entropy in terms of the phase space distribution function for classical systems and the probabilities of occupation of energy levels when the system is quantized. The evolution of the said probabilities is governed by relevant equations of motion and it is expected that the distributions approaches a definite limit, called the equilibrium distribution, as $t \to \infty$. The thermodynamic properties derived from the statistical entropy corresponding to the said equilibrium distribution are expected to match with the thermodynamic predictions in the so-called *thermodynamic limit* $N \to \infty$, $V \to \infty$, N/V remaining finite.

As we noted in the kinetic theory, the approach based on determining the equilibrium distribution as the $t \to \infty$ limit of time-dependent distribution is, however, not simple and in fact may not always be possible to work with. Going by the fact that the state of thermodynamic equilibrium is one in which thermodynamic entropy is maximum, it is envisaged that it would be the state of maximum statistical entropy too. Accordingly, the equilibrium distribution is determined using the postulate of maximum entropy:

R. R. Puri, *Modern Thermodynamics and Statistical Mechanics*, Undergraduate Lecture Notes in Physics, https://doi.org/10.1007/978-3-031-54310-4_6

The state of thermal equilibrium is described by such probability distribution of microstates for which the statistical entropy is maximum subject to the constraints on the system.

The constraints of general interest are the ones pertaining to whether the system exchanges energy and/or particles with its environment. Those constraints define three types of ensembles:

1. *Microcanonical ensemble.* It represents a system which is isolated and exchanges neither energy nor particles with the environment.
2. *Canonical ensemble.* It represents a system which exchanges energy but not particles with the environment.
3. *Grand Canonical ensemble.* It represents a system which exchanges energy as well as particles with the environment.

In the following sections we derive the probabilities for each of the above-mentioned ensembles. We treat separately the systems having fixed number of particles and the ones which exchange particles with the environment.

6.2   Systems Having Fixed Number of Particles

Following Chap. 5,

1. We denote by $|E_{1N}\rangle$, $E_{2N}\rangle$, ... the energy eigenstates of the Hamiltonian $\hat{H}_N$ governing the evolution of the quantum system of N particles and by p_{mN} the probability of occupation of $|E_{mN}\rangle$.
2. A classical system of N particles whose evolution is governed by the Hamiltonian $H(\{\mathbf{r}_i, \mathbf{p}_i\}_N)$ is described by the phase space distribution function $f_N(\{\mathbf{r}_i, \mathbf{p}_i\}_N)$.

Accordingly, the averages are defined by

1. In the quantum formalism, the average for a fixed number of particles of an operator $\hat{O}$ (caret on a symbol denotes operator) commuting with the Hamiltonian is given by

$$\langle \hat{O} \rangle_N = \sum_m p_{mN} \langle E_{mN} | \hat{O} | E_{mN} \rangle, \qquad \text{quantum systems.} \qquad (6.1)$$

The general case of an arbitrary $\hat{O}$ is addressed in Chap. 13 in terms of the density matrix formalism.

2. The average of a phase space function $O(\{\mathbf{r}_i, \mathbf{p}_i\}_N)$ in the classical theory for fixed number of particles is given by

$$\langle O \rangle_N = \int f_N(\{\mathbf{r}_i, \mathbf{p}_i\}_N) O(\{\mathbf{r}_i, \mathbf{p}_i\}_N) \, d\tau_N, \quad \text{classical systems.} \qquad (6.2)$$

In particular the expressions for entropy for the two kinds of systems are:

1. The quantum statistical entropy for fixed number of particles is

$$S_N = -k_B \sum_m p_{mN} \ln(p_{mN}). \tag{6.3}$$

2. The classical statistical entropy for fixed number of particles is

$$S_N = -k_B \int f_N(\{\mathbf{r}_i, \mathbf{p}_i\}_N) \ln(f_N(\{\mathbf{r}_i, \mathbf{p}_i\}_N)) \, d\tau_N. \tag{6.4}$$

We extremize S_N subject to the normalization of the probabilities:

$$\sum_m p_{mN} = 1, \qquad \text{quantum systems,} \tag{6.5}$$

$$\int f_N(\{\mathbf{r}_i, \mathbf{p}_i\}_N) \, d\tau_N = 1, \qquad \text{classical systems.} \tag{6.6}$$

and the constraints

$$\langle \hat{A}^{(k)} \rangle \equiv \sum_m p_{mN} A_{mN}^{(k)} \equiv C_k, \quad k = 1, 2, \ldots, r, \tag{6.7}$$

where $\hat{A}^{(k)}$'s are assumed to commute with the Hamiltonian of the system, the C_k's are constants and

$$A_{mN}^{(k)} = \langle F_{mN} | \hat{A}^{(k)} | F_{mN} \rangle. \tag{6.8}$$

For the classical system, the constraints are given by the equations

$$\langle A^{(k)}(\{\mathbf{r}_i, \mathbf{p}_i\}_N) \rangle = C_k, \quad k = 1, 2, \ldots, r. \tag{6.9}$$

Invoking the results derived in Sect. 5.2, the probability distribution which maximizes the statistical entropy for the quantum systems may be seen to be given by

$$p_{mN} = \frac{1}{Z_N} \exp\left(-\sum_{k=1}^{r} \alpha_k A_{mN}^{(k)} \right). \tag{6.10}$$

The normalization condition (6.5) for quantum systems gives

$$Z_N = \sum_m \exp\left(-\sum_{k=1}^{r} \alpha_k A_{mN}^{(k)} \right). \tag{6.11}$$

The Z_N is called the N-particle quantum partition function.

The phase space distribution function which maximizes the classical statistical entropy reads

$$f_N(\{\mathbf{r}_i, \mathbf{p}_i\}_N) = \frac{1}{Z_N} \exp\left(-\sum_{k=1}^{r} \alpha_k A^{(k)}(\{\mathbf{r}_i, \mathbf{p}_i\}_N)\right). \tag{6.12}$$

The normalization condition (6.6) for classical systems leads to

$$Z_N = \int \exp\left(-\sum_{k=1}^{r} \alpha_k A^{(k)}(\{\mathbf{r}_i, \mathbf{p}_i\}_N)\right) d\tau_N. \tag{6.13}$$

The Z_N is the classical partition function.

In the following we list some consequences following from the form of the equilibrium probabilities. We will work with the quantum formalism as the corresponding classical results follow by replacing the quantum averages by the classical ones.

1. It is readily seen that

$$\langle \hat{A}^{(k)} \rangle = -\frac{\partial \ln(Z_N)}{\partial \alpha_k}, \quad k = 1, 2, \ldots, r. \tag{6.14}$$

The Lagrange multipliers $\{\alpha_k\}$ can be determined in terms of given constant values of $\{\langle \mathbf{A}_k \rangle\}$ by inverting the equations above.

2. We leave it as an exercise to show that the expression (6.3) for entropy assumes the form

$$\frac{S_N}{k_B} = \sum_{k=1}^{r} \alpha_k C_k + \ln(Z_N), \quad C_k \equiv \langle \hat{A}^{(k)} \rangle. \tag{6.15}$$

Due to (6.14), it is readily seen that S/k_B is the Legendre transform of $\ln(Z_N)$ with respect to the α_k's to the corresponding conjugate variables C_k's. This implies

$$\alpha_k = \frac{1}{k_B} \frac{\partial S_N}{\partial C_k}. \tag{6.16}$$

This may alternatively be obtained by partial differentiation of (6.15) with respect to C_k keeping α_k's fixed and noting that Z_N is a function of the α_k's.

3. It is straightforward to see that

$$\langle \hat{A}^{(k)2} \rangle = \frac{1}{Z_N} \frac{\partial^2 Z_N}{\partial \alpha_k^2}, \quad k = 1, 2, \ldots, r. \tag{6.17}$$

We leave it as an exercise to show that the variance in the measurement of $\hat{A}_k$ is given by

$$\sigma_k^2 \equiv \langle \hat{A}^{(k)2} \rangle - \langle \hat{A}^{(k)} \rangle^2 = \frac{\partial^2 \ln(Z_N)}{\partial \alpha_k^2}. \tag{6.18}$$

We use the results derived above to construct ensembles for systems with fixed number of particles, namely, microcanonical and canonical ensembles.

Exercises

Ex. 6.1. Show that for p_{mN} as in (6.10), the expression (6.3) for entropy is given by (6.15).

Ex. 6.2. Show that

$$\langle \hat{A}_j^{(k)} \hat{A}_k^{(k)} \rangle - \langle \hat{A}_j^{(k)} \rangle \langle \hat{A}_k^{(k)} \rangle = \frac{\partial^2 \ln(Z_N)}{\partial \alpha_j \alpha_k}. \tag{6.19}$$

For $j = k$ we recover (6.18).

6.2.1 Microcanonical Ensemble

The microcanonical ensemble describes an isolated system. It is characterized by a constant number of particles whereas its energy E is given to lie in some small interval $(E_0 - \Delta, \; E_0 + \Delta)$ about a fixed value E_0. Since in this case, apart from the normalization condition, there is no other constraint, $\alpha_k = 0$. Consequently, the general equilibrium distributions (6.10) and (6.12) lead to the following results for quantum and classical systems:

1. With $\alpha_k = 0$ the (6.12) for classical systems reduces to

$$f_N(\{\mathbf{r}_i, \mathbf{p}_i\}_N) = \frac{1}{Z_N}, \tag{6.20}$$

where Z_N is given by (6.13) in which the integral is to be carried on the part of the phase space for which energy E lies in the interval $(E_0 - \Delta, \; E_0 + \Delta)$ so that if Γ is the volume of the said part then

$$Z_N = \int_{E_0 - \Delta \leq E \leq E_0 + \Delta} d\tau_N = \Gamma. \tag{6.21}$$

The Z_N above is the classical microcanonical partition function. The equilibrium phase space distribution for the microcanonical ensemble is thus given by

$$f_N(\{\mathbf{r}_i, \mathbf{p}_i\}_N) = \frac{1}{\Gamma}, \qquad E_0 - \Delta \leq E \leq E_0 + \Delta. \tag{6.22}$$

On substituting this in (6.4), the expression for entropy for a classical system described by microcanonical ensemble reads

$$S_N = k_B \ln(\Gamma). \tag{6.23}$$

2. For quantum systems, with $\alpha_k = 0$, (6.10) reduces to

$$p_{mN} = \frac{1}{Z_N}, \tag{6.24}$$

where Z_N, given by (6.11), is a sum over states in the energy interval $(E_0 - \Delta, \ E_0 + \Delta)$. If W is the number of states in the said interval then

$$Z_N = W. \tag{6.25}$$

The Z_N above is the quantum microcanonical partition function. On substituting this in (6.3), the expression for entropy for quantum systems reads

$$S_N = k_B \ln(W). \tag{6.26}$$

The number of states W is analogous to the phase space volume Γ accessible to a classical system in the interval $(E_0 - \Delta, \ E_0 + \Delta)$ of energy.

6.2.2 Canonical Ensemble

The canonical ensemble is defined as the one in which the system exchanges energy but not particles with the surroundings. Hence, apart from the normalization conditions, the constraint under which entropy is to be maximized is the value U of internal energy, given for quantum systems by

$$U = \langle \hat{H}_N \rangle = \sum_m E_{mN} p_{mN}, \tag{6.27}$$

where $\hat{H}_N$ is the system Hamiltonian and the second equation is due to the fact that $|E_{mN}\rangle$ is the eigenstate of $\hat{H}_N$ corresponding to energy E_{mN}:

$$\hat{H}_N |E_{mN}\rangle = E_{mN} |E_{mN}\rangle. \tag{6.28}$$

The corresponding constraint for classical systems is

$$U = \int H_N(\{\mathbf{r}_i, \mathbf{p}_i\}_N) f_N(\{\mathbf{r}_i, \mathbf{p}_i\}_N) \, d\tau_N, \tag{6.29}$$

where $H(\{\mathbf{r}_i, \mathbf{p}_i\}_N)$ is classical Hamiltonian.

The equilibrium distribution for quantum systems is then given by (6.10) with $\hat{A}^{(1)} = \hat{H}_N$, $\alpha_k = 0$ ($k \neq 1$). Also, due to (6.28), the definition (6.8) of $A_{mN}^{(1)}$ gives $A_{mN}^{(1)} = E_{mN}$. Hence, with $\alpha_1 \to \beta$, (6.10) assumes the form

$$p_{mN} = \frac{1}{Z_N} \exp\left(-\beta E_{mN}\right),$$

(6.30)

where, due to (6.11),

$$Z_N = \sum_m \exp\left(-\beta E_{mN}\right).$$

(6.31)

The Z_N above is the quantum canonical partition function.

The equilibrium distribution for classical systems may similarly be shown to be given by

$$f_N(\{\mathbf{r}_i, \mathbf{p}_i\}_N) = \frac{1}{Z_N} \exp\{-\beta H_N(\{\mathbf{r}_i, \mathbf{p}_i\}_N)\},$$

(6.32)

where

$$Z_N = \int \exp\{-\beta H_N(\{\mathbf{r}_i, \mathbf{p}_i\}_N)\}\, d\tau_N.$$

(6.33)

The Z_N above is the classical canonical partition function. In the following we derive some relations expressing physical quantities in terms of Z_N assuming the system to be quantum. Analogous relations hold in the phase space description.

1. It is straightforward to see that

$$U = -\frac{\partial \ln(Z_N)}{\partial \beta},$$

(6.34)

for quantum as well as classical systems. Since U is extensive, the equation above shows that $\ln(Z_N)$ must be extensive. Inversion of (6.34) gives undetermined Lagrange multiplier β in terms of the internal energy U.

2. Using (6.30), the expression (6.3) for entropy reads

$$S_N = k_B\left(\beta U + \ln(Z_N)\right),$$

(6.35)

for quantum as well as classical systems.

3. We derive the expression for pressure in terms of Z_N. To that end, recall the well-known expression for pressure,

$$P = -\frac{\partial E}{\partial V},$$

(6.36)

where E is energy and V the volume. Assume that the energy of the gas at some instance of time is $E_{m,N}$ so that the pressure at that instance is

$$P_{mN} = -\frac{\partial E_{mN}}{\partial V}. \tag{6.37}$$

The average pressure would be

$$P = \sum_m P_{mN} P_{m,N} = -\sum_m P_{m,N} \frac{\partial E_{m,N}}{\partial V}$$

$$= -\frac{1}{Z_N} \sum_m \exp(-\beta E_{m,N}) \frac{\partial E_{m,N}}{\partial V}$$

$$= \frac{1}{\beta Z_N} \frac{\partial}{\partial V} \sum_m \exp(-\beta E_{m,N}) = \frac{1}{\beta Z_N} \frac{\partial Z_N}{\partial V}. \tag{6.38}$$

We thus see that

$$P = \frac{1}{\beta} \frac{\partial \ln(Z_N)}{\partial V}. \tag{6.39}$$

This is the expression for pressure in terms of the canonical partition function. See also Sect. 6.4.2 for an alternative derivation.

4. A measure of fluctuations in the measurement of energy is its variance. Invoking (6.18) we have

$$\Delta E^2 \equiv \langle E^2 \rangle - \langle E \rangle^2 = \frac{\partial^2 \ln(Z_N)}{\partial \beta^2}. \tag{6.40}$$

Some consequences of the relation above are:

(a) Due to (6.34), (6.40) may be rewritten as

$$\Delta E^2 = k_B T^2 \frac{\partial U}{\partial T} = k_B T^2 C_V. \tag{6.41}$$

This relates fluctuations in energy with the heat capacity.

(b) Since $\Delta E^2 \geq 0$ it follows from (6.40) that

$$\frac{\partial^2 \ln(Z_N)}{\partial \beta^2} \geq 0. \tag{6.42}$$

(c) On dividing (6.40) by N^2 we have

$$\frac{\Delta E^2}{N^2} = \frac{1}{N^2} \frac{\partial^2 \ln(Z_N)}{\partial \beta^2}. \tag{6.43}$$

As argued circa (6.34), $\ln(Z_N)$ is extensive and hence is proportional to N. The equation above therefore shows that fluctuation in the measurement of energy per particle, $\Delta E/N$, is proportional to $1/\sqrt{N}$. Hence, for large N, fluctuations in energy may be ignored.

(d) Invoking (6.34), rewrite (6.40) as

$$\Delta E^2 = -\frac{\partial U}{\partial \beta}.$$ (6.44)

Since $\Delta E^2 \geq 0$, it follows that

$$\frac{\partial U}{\partial \beta} \leq 0.$$ (6.45)

This shows that the internal energy is a decreasing function of β. The equality holds when $\Delta E = 0$:

$$\frac{\partial U}{\partial \beta} = 0, \qquad \text{if } \Delta E = 0.$$ (6.46)

Hence U is independent of β in the absence of randomness. We will show that $\beta = 1/k_B T$ where T is the temperature of the system. We therefore see that the concept of temperature is related with randomness. In particular we will see that for a non-interacting free classical gas, $U = 3N/2\beta$. The equation (6.44) in that case leads to

$$\Delta E^2 = \frac{3N}{2} k_B^2 T^2.$$ (6.47)

This equation relates temperature directly with energy fluctuations in an ideal classical gas. Clearly, $T = 0$ if $\Delta E = 0$ though the applicability of ideal gas law is questionable at low temperatures.

Exercises

Ex. 6.3. (a) Given that, besides the normalization condition, the probability distribution is constrained by specified average values U and P of energy and
pressure, show that the probability that maximizes entropy is

$$p_{mN} = \frac{1}{Z_{iN}} \exp\left(-\beta E_{mN} - \gamma P_{mN}\right),$$ (6.48)

$$Z_{iN} = \sum_m \exp\left(-\beta E_{mN} - \gamma P_{mN}\right),$$ (6.49)

where $P_{m,N}$ is as in (6.37). This is called *isobaric-isothermal distribution*.
Hint: Recall from (6.38) that pressure is average of $P_{m,N}$.

(b) Show that the entropy corresponding to the probabilities in (6.49) is

$$S = k_{\mathrm{B}} \left(\beta U + \gamma P + \ln(Z_{iN}) \right). \tag{6.50}$$

6.3   Grand Canonical Ensemble

The grand canonical ensemble describes systems which, in addition to exchanging energy with the environment, exchange also the particles. The number of particles then also becomes a variable. We adopt following notation for describing systems in grand canonical ensemble:

1. In a quantum system, the occupation probability of the state $|E_{m,N}\rangle$ when N is variable will be denoted by $p_m(N)$. The argument N of $p_m(N)$ is to distinguish the probabilities for varying number of particles from those for fixed N which have been symbolized by p_{mN}.
2. In a classical system, N-particle probability density in the vicinity of $\{\mathbf{r}_i, \mathbf{p}_i\}_N$ in case the particle number is varying will be denoted by $f(\{\mathbf{r}_i, \mathbf{p}_i\}_N, N)$. The N in the argument of f is to distinguish the distribution for varying number of particles from the one for fixed N which has been symbolized by $f_N(\{\mathbf{r}_i, \mathbf{p}_i\}_N)$.
3. The average $\langle \hat{O} \rangle$ of an operator $\hat{O}$, commuting with the Hamiltonian, in the present case is given by

$$\langle \hat{O} \rangle = \sum_{N=0}^{\infty} \sum_{m} p_m(N)\langle E_{mN}|\hat{O}|E_{mN}\rangle. \tag{6.51}$$

4. The average of a phase space distribution function $O(\{\mathbf{r}_i, \mathbf{p}_i\}_N)$ in the classical description is given by

$$\langle O(\{\mathbf{r}_i, \mathbf{p}_i\}) \rangle = \sum_{N=0}^{\infty} \int f(\{\mathbf{r}_i, \mathbf{p}_i\}_N, N) O(\{\mathbf{r}_i, \mathbf{p}_i\}) \, d\tau_N. \tag{6.52}$$

5. Statistical entropy for varying N for quantum systems is

$$S = -k_{\mathrm{B}} \sum_{N=0}^{\infty} \sum_{m} p_m(N)\ln(p_m(N)). \tag{6.53}$$

6. Statistical entropy for varying N for classical systems is

$$S = -k_{\mathrm{B}} \sum_{N=0}^{\infty} \int f(\{\mathbf{r}_i, \mathbf{p}_i\}_N, N)\ln\{f(\{\mathbf{r}_i, \mathbf{p}_i\}_N, N)\} \, d\tau_N. \tag{6.54}$$

7. The normalization condition for quantum probabilities reads

$$\sum_{N=0}^{\infty} \sum_{m} p_m(N) = 1. \tag{6.55}$$

8. The normalization condition for classical probability distribution is

$$\sum_{N=0}^{\infty} \int f(\{\mathbf{r}_i, \mathbf{p}_i\}_N, N) \, d\tau_N = 1. \tag{6.56}$$

The equilibrium probability distribution in the grand canonical ensemble is one which maximizes entropy subject to its normalization condition, the average values of internal energy, and the number of molecules which in the quantum formalism are

$$U = \langle \hat{H}_N \rangle = \sum_{N=0}^{\infty} \sum_{m} E_{mN} p_m(N), \tag{6.57}$$

and

$$\bar{N} = \langle \hat{N} \rangle = \sum_{N=0}^{\infty} \sum_{m} N p_m(N). \tag{6.58}$$

Following the general procedure outlined in Sect. 5.2, it is straightforward to see that the $p_m(N)$ which maximizes entropy is given by

$$p_m(N) = \frac{1}{Z_G} \exp\left(-\beta E_{mN} + \alpha N\right), \tag{6.59}$$

where α, β are Lagrange multipliers and Z_G is the quantum grand canonical partition function defined by

$$Z_G = \sum_{N=0}^{\infty} \sum_{m} \exp\left(-\beta E_{mN} + \alpha N\right). \tag{6.60}$$

The equation above is the result of the normalization condition.

The equilibrium distribution function for classical systems is

$$f(\{\mathbf{r}_i, \mathbf{p}_i\}_N; N) = \frac{1}{Z_G} \exp\left\{-\beta H_N(\{\mathbf{r}_i, \mathbf{p}_i\}_N) + \alpha N\right\}, \tag{6.61}$$

where the grand partition function is

$$Z_G = \sum_{N=0}^{\infty} \int \exp\{-\beta H_N(\{\mathbf{r}_i, \mathbf{p}_i\}_N) + \alpha N\} \, d\tau_N. \tag{6.62}$$

The Z_G above is the classical grand canonical partition function. In the following we consider quantum systems. The results for the classical systems follow by replacing the quantum averages by classical ones.

1. It is straightforward to see that the internal energy is given by

$$U = -\frac{\partial \ln(Z_G)}{\partial \beta}. \tag{6.63}$$

2. The average number of molecules is

$$\bar{N} = \frac{\partial \ln(Z_G)}{\partial \alpha}. \tag{6.64}$$

The Lagrange multipliers α, β can be determined as functions of U, $\bar{N}$ by inverting the equations (6.63) and (6.64).
3. A useful relation between Z_N and Z_G is obtained by noting that, with Z_N given by (6.31) and (6.33) respectively for quantum and classical systems, the grand partition function Z_G in (6.60) for quantum systems, as well as that in (6.62) for classical systems can be written as

$$Z_G = \sum_{N=0}^{\infty} \exp(\alpha N) Z_N = \sum_{N=0}^{\infty} z^N Z_N, \tag{6.65}$$

where

$$z = \exp(\alpha) \tag{6.66}$$

is called the *fugacity*. From (6.65) follows the inverse relation

$$Z_N = \frac{1}{N!} \frac{\partial^N Z_G}{\partial z^N}\bigg|_{z=0}. \tag{6.67}$$

This express Z_N in terms of Z_G.
4. The expression for pressure may be derived in the same way as has been done to arrive at the one in (6.39) for the canonical ensemble. The resulting expression for pressure is same as the one in (6.39) with Z_N therein replaced by Z_G:

$$P = \frac{1}{\beta} \frac{\partial \ln(Z_G)}{\partial V}. \tag{6.68}$$

Now, $\ln(Z_N)$ and $\ln(Z_G)$ are extensive quantities but with a difference: Whereas $\ln(Z_N)$ is a function of two extensive variables, namely, number of particles

N and volume V, $\ln(Z_G)$ is a function of only one extensive variable, namely, V. For, as shown in (6.65), Z_G is obtained from Z_N by summation over N. Hence $\ln(Z_G)$ must be of the form $\ln(Z_G) = V f(\beta, \alpha)$ so that $\partial \ln(Z_G)/\partial V = f(\beta, \alpha) = \ln(Z_G)/V$. Hence

$$P = \frac{1}{\beta V} \ln(Z_G). \tag{6.69}$$

5. The expression for variance in the measurement of energy can be derived in the same manner as in its derivation for canonical ensemble in (6.40). The resulting expression is same as that in (6.40) with Z_N therein replaced by Z_G:

$$\Delta E^2 = \frac{\partial^2 \ln(Z_G)}{\partial \beta^2} = -\frac{\partial U}{\partial \beta}. \tag{6.70}$$

The consequences of these relations have been discussed circa (6.40).

6. The expression for variance in the measurement of number of molecules, obtained using (6.17), reads

$$\Delta N^2 \equiv \langle N^2 \rangle - \langle N \rangle^2 = \frac{\partial^2 \ln(Z_G)}{\partial \alpha^2}. \tag{6.71}$$

Since $\Delta N^2 \geq 0$ it follows that

$$\frac{\partial^2 \ln(Z_G)}{\partial \alpha^2} \geq 0. \tag{6.72}$$

On using (6.64), (6.71) may be rewritten as

$$\Delta N^2 - \frac{\partial \bar{N}}{\partial \alpha}. \tag{6.73}$$

The quantity $\Delta N / \bar{N}$ serves as a measure of fluctuations in the number of molecules relative to the mean $\bar{N}$. From (6.71) we have

$$\frac{\Delta N^2}{\bar{N}^2} = \frac{1}{\bar{N}^2} \frac{\partial^2 \ln(Z_G)}{\partial \alpha^2}. \tag{6.74}$$

Since $\ln(Z_G) \sim V$ and $V/\bar{N}$ is a constant in the thermodynamic limit $V \to \infty$, $\bar{N} \to \infty$, it follows that $\Delta N / \bar{N} \sim (\bar{N})^{-1/2}$. Thus fluctuation in the number of molecules is negligible in the thermodynamic limit.

In particular, we will see that, for a non-interacting free classical gas, $\partial \bar{N}/\partial \alpha = \bar{N}$ which on substitution in (6.73) yields

$$\Delta N^2 = \bar{N}. \tag{6.75}$$

This shows that variance in the number distribution of a non-interacting classical gas is same as its mean. This is the defining characteristic of the Poisson distribution. Hence the number distribution in an ideal classical gas is Poissonian.

7. A relation of much importance exists between the number fluctuations and isothermal compressibility. To derive it, divide (6.73) by $\bar{N}^2$ to obtain

$$\frac{\Delta N^2}{\bar{N}^2} = \frac{1}{\bar{N}} \left(\frac{\partial \bar{N}}{\partial \alpha}\right)_{V,T} \left(\frac{\partial \ln(Z_G)}{\partial \alpha}\right)_{V,T}^{-1} , = \frac{1}{\beta \bar{N} V} \left(\frac{\partial \bar{N}}{\partial \alpha}\right)_{V,T} \left(\frac{\partial P}{\partial \alpha}\right)_{V,T}^{-1}$$

$$= \frac{1}{\beta V} \frac{V}{\bar{N}} \left(\frac{\partial \bar{N}/V}{\partial P}\right)_{V,T} , \tag{6.76}$$

where (6.64) has been used to write an $\bar{N}$ in the denominator in first equation, (6.69) has been recalled to write $\ln(Z_G)$ in terms of pressure in the second equation, and (A.10) has been invoked to write the last equation. In terms of the specific volume $v = V/\bar{N}$, (6.76) reads

$$\frac{\Delta N^2}{\bar{N}^2} = -\frac{1}{\beta \bar{N}} \left(\frac{\partial v}{\partial P}\right)_T = \frac{k_B T}{V} \kappa_T , \tag{6.77}$$

where κ_T is isothermal compressibility defined in (2.110). The expression above may be derived alternatively by using Gibbs–Duhem equation (see 6.3).

8. On substituting (6.59) in (6.53) follows the expression for entropy:

$$S = k_B \left(\beta U - \alpha \bar{N} + \ln(Z_G)\right) . \tag{6.78}$$

Invoking (6.69) for $\ln(Z_G)$ in terms of the pressure P, we get

$$S = k_B \left(\beta U - \alpha \bar{N} + \beta P V\right) . \tag{6.79}$$

We will show that k_B is Boltzmann's constant, $\beta = 1/k_B T$ and $\alpha = \beta \mu$ where μ is chemical potential. Consequently, (6.79) may be rewritten as

$$S = \frac{1}{T} U - \frac{\mu}{T} \bar{N} + \frac{P}{T} V . \tag{6.80}$$

This is Euler's thermodynamic equation with N therein replaced by $\bar{N}$.

While arriving at (6.80), we compared the statistical mechanical expression (6.79) for S with Euler's equation and, based on the assumption that S in (6.80) is same as the thermodynamic entropy, we deduced the expressions for the Lagrange multipliers β, α in terms of temperature and the chemical potential. Next we show that the assumed equivalence indeed holds.

Ex. 6.4. Derive (6.77) using Gibbs–Duhem relation. Hint: With $v = V/\bar{N}$, $\alpha = \beta \mu$ rewrite (6.73) as

$$\frac{\Delta N^2}{\bar{N}^2} = \frac{V}{\beta \bar{N}^2} \left(\frac{\partial \bar{N}/V}{\partial \mu}\right)_{V,T} = -\frac{1}{\beta V} \left(\frac{\partial v}{\partial \mu}\right)_T$$

$$= -\frac{1}{\beta V} \left(\frac{\partial v}{\partial P}\right)_T \left(\frac{\partial P}{\partial \mu}\right)_T . \tag{6.81}$$

Using Gibbs–Duhem relation (2.49)

$$s dT - v dP + d\mu = 0, \tag{6.82}$$

we have $(\partial P/\partial \mu)_T = 1/v$ which on substitution in (6.81) would lead to the desired result (6.77).

6.4   Relation with Thermodynamics

In this section we establish the relationship of statistical description with thermodynamics by using the standard distributions introduced above. It may appear that we may need to use one or the other canonical ensemble depending on the system under consideration i.e. depending on whether energy and particle numbers are given as exactly known or as averages. However, as shown above, in the thermodynamic limit, fluctuations in energy and number of particles are negligibly small. Hence, in that limit, whether an observable value is given exactly or as an average would lead to the same result. In other words, the three ensembles would lead to the same results in the thermodynamic limit. We may therefore choose any of the three ensembles per our convenience. In what follows we work with the canonical ensemble. See [Balian] for further details.

6.4.1   Zeroth Law of Thermodynamics

Recall that the zeroth law of thermodynamics states that if two bodies are separately in thermal equilibrium with a third body then they are in thermal equilibrium with one another. This law defines temperature as the quantity that is equalized between bodies in thermal equilibrium with each other.

Using the zeroth law, we will show that β is a decreasing function of temperature.

To that end, let the probabilities of occupation of the energy eigenstates of systems 1 and 2 be given by (we drop the index indicating the number of particles)

$$p_m^{(k)} = \frac{1}{Z_k} \exp\left(-\beta_k E_m^{(k)}\right), \quad Z_k = \sum_m \exp\left(-\beta_k E_m^{(k)}\right), \quad k = 1, 2. \tag{6.83}$$

The combined probability when the systems are not interacting is

$$p_{m,n}^{(c)} = p_m^{(1)} p_n^{(2)}.$$ (6.84)

The systems interact when they are brought together. The energy levels of the combined system are obtained by taking account of the interaction potential between them. Let $\{E_M^{(c)}\}$ be the energy eigenstates of the combined system so that the probability of occupation of the state of energy $E_M^{(c)}$ is given by

$$p_M^{(c)} = \frac{1}{Z} \exp\left(-\beta E_M^{(c)}\right), \quad Z = \sum_M \exp\left(-\beta E_M^{(c)}\right).$$ (6.85)

If the interaction is weak then energy eigenvalues of the interacting system will be only negligibly different from the sum of their energies when not interacting. We therefore let $E_M^{(c)} \approx E_m^{(1)} + E_n^{(2)}$ and rewrite (6.85) as

$$p_{m,n}^{(c)} \approx \frac{1}{Z} \exp\left(-\beta\left(E_m^{(1)} + E_n^{(2)}\right)\right), \quad Z = \sum_{m,n} \exp\left(-\beta\left(E_m^{(1)} + E_n^{(2)}\right)\right).$$ (6.86)

This may be rewritten as

$$p_{m,n}^{(c)} = \left(\frac{1}{Z_1'} \exp\left(-\beta E_m^{(1)}\right)\right)\left(\frac{1}{Z_2'} \exp\left(-\beta E_n^{(2)}\right)\right),$$ (6.87)

where

$$Z_k' = \sum_m \exp\left(-\beta E_m^{(k)}\right).$$ (6.88)

Equations (6.87) show that the two systems attain the same value of the parameter β after attaining equilibrium on interaction with each other. Since the equilibrium is caused only due to exchange of energy and thermodynamic quantity that attains the same value due to the exchange of energy is temperature, the parameter β may be identified with temperature.

The manner in which β is related with temperature may be deduced by noting that, since the two systems together are isolated, their total internal energy does not change on interaction, i.e.

$$U_1 + U_2 = U_1' + U_2'.$$ (6.89)

Hence if the internal energy of one increases then that of the other decreases. Also, from (6.45) we know that $\partial U/\partial \beta \leq 0$ i.e. U is a decreasing function of β. This implies that the β value of that system increases whose energy after the interaction is smaller. Since the temperature of a body goes down when it loses energy, we see that β has reciprocal relationship with temperature.

6.4.2 First Law of Thermodynamics

Recall the first law of thermodynamics stated in (1.29):

$$\delta U = \delta Q - \delta W,$$

where δQ is the amount of heat absorbed and δW the amount of work done by the system. To establish this law using statistical mechanical formalism we start with the expression (6.29) of internal energy and obtain

$$\delta U = \sum_m (p_{mN} \delta E_{mN} + E_{mN} \delta p_{mN}). \tag{6.90}$$

A comparison of last two equations suggests that we may identify the amount of heat received by the system as

$$\delta Q = \sum_m E_{mN} \delta p_{mN}, \tag{6.91}$$

and the amount of work done by it as

$$\delta W = - \sum_m p_{mN} \delta E_{mN}. \tag{6.92}$$

In the following we elaborate the meaning of (6.91) and (6.92).

1. The (6.91) shows that the heat exchange is related with the change in probability distribution between energy levels of the system, i.e. it is the redistribution of population among energy levels which leads to change in the heat content of a body. The relations derived above are independent of any specific form of $p_{m,N}$. If the system is described by the canonical ensemble then it is straightforward to see that, due to change only in the parameter β,

$$\delta Q = \sum_m E_{mN} \frac{\partial p_{mN}}{\partial \beta} \delta \beta = \left(\frac{\partial}{\partial \beta} \sum_m E_{mN} p_{mN} \right) \delta \beta$$
$$= \frac{\partial U}{\partial \beta} \delta \beta, \tag{6.93}$$

where the second equation is due to the fact that $E_{m,N}$ is not a function of β. On recalling (6.44), the equation above reads

$$\delta Q = -\Delta E^2 \delta \beta. \tag{6.94}$$

This shows that the amount of heat exchanged is related with the variance in energy. We will see that $\beta \sim 1/T$. Hence increase in temperature is caused by the absorption of heat.

We have thus been able to express heat in terms of the statistical mechanical theoretic entities.

2. The expression (6.92) for the work done by the system shows that the work is related with the change in the energy levels of the system. That change is caused by the changes in external parameters $\{\xi_k\}$ on which energy depends. For example, one such parameter could be the volume. Similarly change in the applied fields like the electromagnetic, gravitational, etc. would result in the change in the energy levels. Hence, considering energy a function of $\{\xi_k\}$, we have

$$\delta E_{mN}(\{\xi_k\}) = \sum_k F_{mN}^k \delta \xi_k, \tag{6.95}$$

where

$$F_{mN}^k = \frac{\partial E_{mN}(\{\xi_k\})}{\partial \xi_k}. \tag{6.96}$$

On substituting this in (6.92) the expression for work done reads

$$\delta W = -\sum_k G_k \delta \xi_k, \qquad G_k = \sum_m F_{mN}^k p_{mN}. \tag{6.97}$$

For example, if $\xi = V$ where V is the volume then corresponding F is pressure $P = -\langle \partial E_{mN}/\partial V \rangle$ and $\delta W = P \delta V$.

6.4.3   Second Law of Thermodynamics

In the following we arrive at the second law of thermodynamics by starting with the statistical description. The statistical entropy will turn out to be proportional to the thermodynamic entropy. This will also enable us to find functional relation between β and temperature. Recall that, while comparing statistical theory with the zeroth law of thermodynamics, we found that β is related inversely with temperature but it does not determine the function relating them.

The change in statistical entropy in canonical ensemble description is

$$
\begin{aligned}
\delta S &= -k_B \sum_m (1 + \ln(p_{mN}))\, \delta p_{mN} \\
&= -k_B \sum_m (1 - \beta E_{mN} - \ln(Z_N))\, \delta p_{mN} \\
&= k_B \beta \sum_m E_{mN} \delta p_{mN} - k_B (1 - \ln(Z_N)) \sum_m \delta p_{mN} \\
&= k_B \beta \delta Q - k_B (1 - \ln(Z_N)) \sum_m \delta p_{mN}, \tag{6.98}
\end{aligned}
$$

where use has been made of the definition (6.91) of δQ. Due to

$$\sum_m p_{mN} = 1 \implies \sum_m \delta p_{mN} = 0, \qquad (6.99)$$

the (6.98) reduces to

$$\delta S = k_B \beta \delta Q. \qquad (6.100)$$

Recall that the change δS_{th} in the thermodynamic entropy of a system when it receives the amount δQ of heat reversibly at temperature T is given by $\delta S_{th} = \delta Q/T$ which on comparison with (6.100) shows that the statistical and the thermal entropies are proportional to each other. If we demand that the statistical entropy be identical with the thermodynamic entropy then we arrive at the relation $\beta = 1/k_B T$. The constant k_B is fixed by the choice of the temperature scale. If the unit of temperature is chosen to be Kelvin then k_B turns out to be the Boltzmann's constant. Assuming that to be the case we are led to the following relations:

$$k_B = \text{Boltzmann's constant}, \quad \beta = \frac{1}{k_B T}. \qquad (6.101)$$

This determines the Lagrange multiplier β in terms of a physical characteristic, namely, temperature.

Having identified statistical entropy S as the thermodynamic entropy, we can legitimately equate its expression (6.79) with Euler's equation to arrive at the relation

$$\alpha = \beta \mu, \qquad (6.102)$$

where μ is the chemical potential.

Having identified the statistical entropy with the thermodynamic one, we need to show that the entropy of an isolated system never decreases. That assertion has been proved in Sect. 13.6. Using it, the second law (1.53) may be arrived at by statistical mechanical considerations as follows. Since the system and reservoir together are isolated, their combined entropy increases:

$$dS + dS_R \geq 0, \qquad (6.103)$$

where dS stands for the change in entropy of the system and dS_R is that for the reservoir. We know that if dQ is the amount of heat transmitted from the reservoir to the system at temperature T then change in its entropy is

$$dS_R = -\frac{dQ}{T}. \qquad (6.104)$$

Substitution of (6.104) in (6.103) leads to the second law (1.53).

6.4.4   Third Law of Thermodynamics

Recall that the third law defines the scale for measuring entropy by asserting that entropy vanishes at absolute zero.

In order to see how it follows from statistical considerations, rewrite p_{mN} for the canonical ensemble in the form

$$p_{mN} = \frac{\exp(-\beta E_{mN})}{\sum_m \exp(-\beta E_{mN})} = \frac{\exp(-\beta(E_{mN} - E_{0N}))}{\sum_m \exp(-\beta(E_{mN} - E_{0N}))}, \tag{6.105}$$

where E_{0N} is the ground state energy so that $E_{mN} > E_{0N}$ for $m \neq 0$. If the ground state is non-degenerate then, in the limit $T \to 0$, the equation above shows that $p_{mN} = \delta_{m0}$ which on substitution in the (6.3) for entropy yields $S = 0$. If the ground state is degenerate with W as the number of states then $S = \ln(W)$. Consequently, the entropy per unit volume is

$$s = \frac{1}{V} \ln(W) \to 0, \qquad V \to \infty, \tag{6.106}$$

provided W does not grow faster than $\exp(V)$. Thus the statistical description leads to the third law.

6.5   Thermodynamic Potentials in Terms of Partition Functions

In this section we derive relations between the thermodynamic potentials and the partition functions.

1. With β given by (6.101), the expression (6.35) for entropy in the canonical ensemble reads (with $S_N \to S$)

$$TS = U + \beta^{-1}\ln(Z_N). \tag{6.107}$$

Since we have identified the statistical entropy as the thermodynamic entropy, the expression above for statistical entropy should be same as the expression (2.57) for the thermodynamic entropy in terms of the Helmholtz free energy $F(T, V, N)$ leading to the following relation between the canonical partition function Z_N and $F(T, V, N)$:

$$F(T, V, N) = -\beta^{-1}\ln(Z_N). \tag{6.108}$$

The equations in (2.60) now assume the form

$$S = \left(\frac{\partial \beta^{-1}\ln(Z_N)}{\partial T}\right)_{V,N}, \quad P = \beta^{-1}\left(\frac{\partial \ln(Z_N)}{\partial V}\right)_{N,T},$$

$$\mu = -\beta^{-1}\left(\frac{\partial \ln(Z_N)}{\partial N}\right)_{V,T}. \tag{6.109}$$

2. With β, α given by (6.101) and (6.102), rewrite (6.78) as

$$ST = U - \mu N + \beta^{-1}\ln(Z_G). \tag{6.110}$$

On comparing this with the expression for entropy in terms of the grand potential $\Omega(T, V, \mu)$ in (2.75) we see that the entropy therein will be same as that in the equation above if

$$\Omega(T, V, \mu) = -\beta^{-1}\ln(Z_G) = -PV, \tag{6.111}$$

where last equation is due to (6.68). We see that Ω in (6.111) has the form (2.76) as it should.
The (2.78) now read

$$S = \left(\frac{\partial \beta^{-1}\ln(Z_G)}{\partial T}\right)_{V,\mu}, \quad P = \beta^{-1}\left(\frac{\partial \ln(Z_G)}{\partial P}\right)_{T,\mu},$$

$$N = \beta^{-1}\left(\frac{\partial \ln(Z_G)}{\partial N}\right)_{T,V}. \tag{6.112}$$

3. The Gibbs potential defined in (2.65) may be written in terms of the Helmholtz potential as

$$G(T, P, N) = F(T, V, N) + PV. \tag{6.113}$$

Using (6.108) follows the expression of the Gibbs potential in terms of the partition function Z_N:

$$G(T, P, N) = -\beta^{-1}\ln(Z_N) + PV. \tag{6.114}$$

Exercises

Ex. 6.5. Using (2.16) for entropy of an ideal gas in (6.107), show that

$$\ln(Z_N) = N\{\ln(V) - c\ln(\beta) - \ln(N) + B\}, \tag{6.115}$$

where $B = c\ln(c) + A - c$ is a constant. With $c = 3/2$, this is same as Z_N derived in (7.18) using the theory of statistical mechanics if the factor $N!$ therein is approximated using Stirling's approximation and the constant B

in (6.115) is identified with the corresponding term therein. For, as we know, the constant B cannot be determined by thermodynamics.

Ex. 6.6. Compare the expression (6.50) of entropy in terms of Z_{iN}, with the expression (2.65) of $G(T, P, N)$ to show that if the undetermined Lagrange multiplies γ in (6.50) is

$$\gamma = \beta V, \tag{6.116}$$

then

$$G(T, P, N) = -\beta^{-1}\ln(Z_{iN}). \tag{6.117}$$

Consequently, the expression (6.49) for the isobaric-isothermal distribution assumes the form

$$p_{mN} = \frac{1}{Z_i} \exp\left(-\beta E_{mN} - \beta P_{mN} V\right). \tag{6.118}$$

Chapter 7
Non-interacting Classical Gas

In this chapter we study thermodynamics of classical gas of non-interacting molecules, including their internal motion, using the equilibrium phase space distribution derived in Chap. 6. We show that, when internal motion is excluded, the thermodynamic properties of the gas, predicted by statistical mechanics, are same as those of the ideal thermodynamic gas. A useful outcome of the phase space approach is the equipartition theorem which is helpful in finding the internal energy of non-interacting gas for certain types of Hamiltonians. We investigate also thermodynamic properties of non-interacting gas in the gravitational field near the surface of the earth.

7.1 Thermodynamics Using Canonical Ensemble

Consider a gas of N molecules whose evolution in classical description is governed by the Hamiltonian $H_N(\{\mathbf{r}_i, \mathbf{p}_i\}_N)$. Recall from Chap. 6 that its state of thermal equilibrium is described by the canonical phase space distribution function (6.32):

$$f_N(\{\mathbf{r}_i, \mathbf{p}_i\}_N) = \frac{1}{Z_N} \exp\{-\beta H_N(\{\mathbf{r}_i, \mathbf{p}_i\}_N)\}, \qquad \beta = 1/k_{\mathrm{B}}T, \qquad (7.1)$$

with the canonical partition function Z_N given by (6.33):

$$Z_N = \int \exp\{-\beta H_N(\{\mathbf{r}_i, \mathbf{p}_i\}_N)\}\, \mathrm{d}\tau_N. \qquad (7.2)$$

We assume H_N to be of the form

$$H_N(\{\{\mathbf{r}_i, \mathbf{p}_i\}_N\}) = \sum_{i=1}^{N} \frac{p_i^2}{2m} + \mathcal{V}(\{\mathbf{r}_i\}_N), \qquad (7.3)$$

© The Author(s), under exclusive license to Springer Nature Switzerland AG 2024
R. R. Puri, *Modern Thermodynamics and Statistical Mechanics*, Undergraduate Lecture Notes in Physics, https://doi.org/10.1007/978-3-031-54310-4_7

where $\mathcal{V}(\{\mathbf{r}_i\}_N) \equiv \mathcal{V}(\mathbf{r}_1, \mathbf{r}_2 \ldots, \mathbf{r}_N)$ is the potential which includes the interaction potential between the particles and that arising from external forces. Using (7.3), the expression for Z_N reads

$$Z_N = \frac{1}{N!h^{3N}} \prod_{i=1}^{N} \left[\int \exp\left(-\frac{\beta}{2m} p_i^2\right) d^3 \mathbf{p}_i \right] \int \exp\left(-\beta \mathcal{V}(\{\mathbf{r}_i\}_N)\right) d^{3N} \mathbf{r}$$

$$= \frac{1}{N!h^{3N}} \left[\int \exp\left(-\frac{\beta}{2m} p^2\right) d^3 \mathbf{p} \right]^{N} \int \exp\left(-\beta \mathcal{V}(\{\mathbf{r}_i\}_N)\right) d^{3N} \mathbf{r}.$$
(7.4)

The momentum integral can be performed using the identity (3.9) to get

$$Z_N = \frac{1}{N!} \lambda_T^{-3N} \int \exp\left(-\beta \mathcal{V}(\{\mathbf{r}_i\}_N)\right) d^{3N} \mathbf{r},$$
(7.5)

where λ_T defined by

$$\lambda_T = \frac{h}{\sqrt{2\pi m k_B T}}$$
(7.6)

is called the *thermal wavelength*. The significance of λ_T and its name will be brought out in Sect. 7.1.2.

It is generally not possible to evaluate the integral in (7.5) analytically in the presence of interaction between the molecules. It can be evaluated if the particles are non-interacting, subject possibly to the external forces. Assuming that the external forces are described by the potential $v(\mathbf{r})$ and that there is no mutual molecular interaction, the potential $\mathcal{V}(\{\mathbf{r}_i\}_N)$ reduces to

$$\mathcal{V}(\{\mathbf{r}_i\}_N) = \sum_{i=1}^{N} v(\mathbf{r}_i),$$
(7.7)

so that

$$\int \exp\left(-\beta \mathcal{V}(\{\mathbf{r}_i\}_N)\right) d^{3N} \mathbf{r} = \prod_{i=1}^{N} \int \exp\left(-\beta v(\mathbf{r}_i)\right) d^3 r_i$$

$$= \left(\int \exp\left(-\beta v(\mathbf{r})\right) d^3 \mathbf{r} \right)^{N}.$$
(7.8)

The Z_N in (7.5) then assumes the form

$$Z_N = \frac{1}{N!} \lambda_T^{-3N} \left(\int \exp\left(-\beta v(\mathbf{r})\right) d^3 \mathbf{r} \right)^{N}.$$
(7.9)

With $\mathcal{V}(\{\mathbf{r}_i\}_N)$ as in (7.7), the phase space distribution function reads

$$f_N(\{\mathbf{r}_i, \mathbf{p}_i\}_N) = \frac{1}{Z_N} \prod_{i=1}^{N} \exp\left\{-\beta\left(\frac{p_i^2}{2m} + v(\mathbf{r}_i)\right)\right\}. \tag{7.10}$$

We use the phase space distribution of molecules to find the probability distribution of their positions.

7.1.1 Position Distribution Function

Often one is interested in the spatial properties of the gas irrespective of their momenta. Such properties are described by the distribution function of the molecular positions obtained by integrating the phase space distribution over their momenta as follows.

The probability density $\tilde{\rho}(\{\mathbf{r}_i\}_N)$ of finding molecules in the vicinity of $\{\mathbf{r}_i\}$ irrespective of their momenta evidently is

$$\tilde{\rho}(\{\mathbf{r}_i\}_N) = \int f_N(\{\mathbf{r}_i, \mathbf{p}_i\}_N) \, d^{3N}\mathbf{p}. \tag{7.11}$$

With $f(\{\mathbf{r}_i, \mathbf{p}_i\}_N)$ given by (7.10), the integral over the momenta in the integral above can be carried as in (7.4). Using also (7.9) for Z_N we have

$$\tilde{\rho}(\{\mathbf{r}_i\}_N) = \tilde{C} \prod_{i=1}^{N} \exp\left(-\beta v(\mathbf{r}_i)\right), \tag{7.12}$$

where

$$\tilde{C}^{-1} = \int \prod_{i=1}^{N} \exp\left(-\beta v(\mathbf{r}_i)\right) d^{3N}\mathbf{r} = \left(\int \exp\left(-\beta v(\mathbf{r})\right) d^3\mathbf{r}\right)^{N}. \tag{7.13}$$

The expression (7.12) for $\tilde{\rho}(\{\mathbf{r}_i\}_N)$ may be rewritten as

$$\tilde{\rho}(\{\mathbf{r}_i\}_N) = \prod_{i=1}^{N} C \exp\left(-\beta v(\mathbf{r}_i)\right), \tag{7.14}$$

where

$$C^{-1} = \int \exp\left(-\beta v(\mathbf{r})\right) d^3\mathbf{r}. \tag{7.15}$$

The equation (7.14) is the probability density for finding molecules numbered $1, 2, \ldots, N$ in the vicinity of $\mathbf{r}_1, \mathbf{r}_2, \ldots, \mathbf{r}_N$ respectively. It is a product of N functions each depending on the position of a particular molecule. The ith factor in the product in (7.14) then stands for the probability density of finding the ith molecule in the vicinity of $\mathbf{r}_i$. Hence the probability density for finding a particular molecule in the vicinity of $\mathbf{r}$ is

$$\rho(\mathbf{r}) = C \exp\left(-\beta v(\mathbf{r})\right). \tag{7.16}$$

The probability density will be independent of position if $v(\mathbf{r})$ is independent of $\mathbf{r}$ in which case system is homogeneous.

We study first the thermodynamics of the gas when $v(\mathbf{r}) = 0$ and follow it up by discussion of the issues that arise when $v(\mathbf{r}) \neq 0$ by working out the problem of the gas in the gravitational field near the surface of the earth.

7.1.2 Free Non-interacting Particles: Ideal Gas

If the molecules are non-interacting and there is no external force on them then $\mathcal{V}(\mathbf{r}) = 0$, so that the phase space distribution function (7.10) reduces to

$$f(\{\mathbf{r}_i, \mathbf{p}_i\}_N) = \frac{1}{Z_N} \prod_{i=1}^{N} \exp\left(-\beta \frac{p_i^2}{2m}\right) \tag{7.17}$$

with Z_N given by (7.9) with $v(\mathbf{r}) = 0$. The value of the space integral therein in that case is V, the volume of the system, so that

$$Z_N = \frac{Z_1^N}{N!}, \tag{7.18}$$

where, with λ_{T} given by (7.6),

$$Z_1 = V\lambda_{\text{T}}^{-3}. \tag{7.19}$$

The Z_1 is the single particle partition function.

With Z_N as in (7.18), the expressions for various thermodynamic quantities are:

1. Pressure is given by

$$P = \frac{1}{\beta}\frac{\partial \ln(Z_N)}{\partial V} = \frac{N}{V}k_{\text{B}}T. \tag{7.20}$$

This is same as the first equations of state of the ideal gas given in (1.80).

2. The expression for internal energy U reads

$$U = -\frac{\partial \ln(Z_N)}{\partial \beta} = \frac{3N}{2} k_B T. \tag{7.21}$$

This is same as the second equation of state of the ideal gas given in (1.80) provided $c = 3/2$ therein. The particular value of c is the result of considering only the three translational degrees of freedom of the molecules. Its value will change on inclusion of other degrees of freedom. The evaluation of U including the rotational and vibrational motions has been carried in Sect. 7.4.

3. We can now provide interpretation of what has been called the thermal wavelength λ_T. Recall that the De Broglie wavelength of a particle having kinetic energy E is given by $\lambda_{DB} = h/\sqrt{2mE}$. Since average energy of a molecule in the gas at temperature T is $3k_B T/2$, the De Broglie wavelength corresponding to that energy is $\lambda_{DB} = h/\sqrt{3mk_B T}$. Clearly, $\lambda_T \sim \lambda_{DB}$. The thermal wavelength λ_T is thus a measure of the De Broglie wavelength of molecules in the gas at temperature T. We will see that it sets scale for the validity of the ideal gas approximation, as well as for applicability of the classical theory beyond which quantum theory must be used even when the particles are non-interacting. The role played by λ_T in determining the limit of applicability of ideal gas law will be brought out, following the equation (7.23), by showing that ideal gas entropy becomes negative if average separation between the molecules is less than approximately $0.435\lambda_T$ and in Sect. 8.8 by showing that the classical theory needs quantum corrections if λ_T far exceeds average separation between the molecules.

4. From (7.21) it follows that the heat capacity at constant volume is

$$C_V \equiv \left(\frac{\partial U}{\partial T}\right)_V = \frac{3Nk_B}{2}. \tag{7.22}$$

This is same as the C_V in (1.77) with $c = 3/2$ therein.

5. Substitute the expressions (7.18) and (7.21) for Z_N and U in the equation (6.35) for entropy and use Stirling's approximation to obtain

$$S_N = Nk_B \left[\frac{5}{2} + \ln\left(\frac{V}{N}\right) + \frac{3}{2}\ln\left(\frac{U}{N}\right) + \frac{3}{2}\ln\left(\frac{4\pi m}{3h^2}\right)\right]. \tag{7.23}$$

This is known as *Sackur–Tetrode equation* [1, 2]. It is same as the one derived in (2.16) using thermodynamic consideration if we take $c = 3/2$ and set A therein equal to the constant term in (7.23). The reason for taking $c = 3/2$ has been explained circa (7.21). However, as has been mentioned before, thermodynamics cannot determine A.

The S_N in (7.23) need not be always positive. To see the condition under which it becomes negative, use (7.21), and the definition (7.6) of the thermal wavelength λ_T to rewrite it in form

$$S_N = N k_B \left[\frac{5}{2} + \ln \left(\frac{V}{N \lambda_T^3} \right) \right]. \tag{7.24}$$

This shows that $S_N < 0$ if

$$\frac{V}{N} < \lambda_T^3 \exp(-5/2) = 0.082 \lambda_T^3. \tag{7.25}$$

Since V/N is the volume per particle, its cube root is a measure of distance between the particles. Hence (7.25) shows that S_N will become negative if average separation between the molecules is less than about $0.435\lambda_T$. Since the thermal wavelength is a measure of molecular De Broglie wavelength corresponding to average thermal energy, the said condition means that entropy will become negative as the De Broglie waves of molecules start overlapping. That is when the quantum effects start showing up and the classical theory begins to break down. Sackur and Tetrode derived (7.23) independently in 1911–12, well before the advent of the quantum theory in 1925–26. Yet Planck's constant h, a purely quantum theoretic entity, appears in the expression in question. The h was "born", of course, in 1900 in Planck's paper on blackbody radiation, but not the quantum machinery employed today to derive (7.23). Sackur and Tetrode derived it by dividing the phase space in cells of volume h^3 and by following Boltzmann's approach to arrive at the equilibrium distribution (see [3] for details of the derivations and references therein to related papers). As has been outlined in Chap. 4, similar approach was followed in 1924 by Bose to re-derive Planck's law and by Einstein to formulate quantum theory of ideal gas. The difference between the distributions derived by Bose and Einstein and the one derived by Sackur and Tetrode lies in treating the molecules as indistinguishable by Bose and Einstein and as distinguishable by Sackur and Tetrode.

Since it is the difference in entropy that is of practical interest, the constant in S would appear to be unmeasurable and in fact, immaterial. It is however remarkable that Sackur and Tetrode could devise and conduct experiments to determine the absolute value of S to confirm the correctness of the value of the constant in question. For details of the experiments of Sackur and Tetrode see [3]. The meaning of measurement of absolute entropy and other experimental verifications of Sackur–Tetrode equation is discussed in [4].

6. The chemical potential μ, determined using (6.109), is given by

$$\frac{\mu}{T} = -k_B \left[\ln \left(\frac{V}{N} \right) + \frac{3}{2} \ln \left(\frac{U}{N} \right) + \frac{3}{2} \ln \left(\frac{4\pi m}{3h^2} \right) \right]. \tag{7.26}$$

Verify that the constant in the expression (7.23) for S and that in (7.26) for μ obey the relation derived in (2.5) with $c = 3/2$ therein.

We have thus shown that the canonical ensemble of the gas of free particles has all the characteristics of the ideal thermodynamic gas.

Next we treat the classical gas by grand canonical ensemble and show that its predictions for free non-interacting gas are the same as the ones arrived at by treating it in the framework of the canonical ensemble.

7.2 Thermodynamics Using Grand Canonical Ensemble

We evaluate the grand canonical partition function Z_G using its expression (6.65) in terms of Z_N:

$$Z_G = \sum_{N=0}^{\infty} \exp(\alpha N) Z_N. \tag{7.27}$$

Substituting the expression (7.18) for Z_N in the equation above we obtain

$$\ln(Z_G) = \exp(\alpha) Z_1 = \frac{V}{\lambda_T^3} \exp(\alpha). \tag{7.28}$$

The thermodynamic properties predicted by the equation above are:

1. Using (6.64), the average number of particles is given by

$$\bar{N} = \frac{\partial \ln(Z_G)}{\partial \alpha} = \ln(Z_G). \tag{7.29}$$

2. Invoking (6.63) we get

$$U = -\frac{\partial \ln(Z_G)}{\partial \beta} = \frac{3k_B T}{2} \ln(Z_G). \tag{7.30}$$

On using (7.29) in the equation above we obtain

$$U = \frac{3k_B T}{2} \bar{N}. \tag{7.31}$$

We see that the U in the equation above is same as the one derived in (7.21) using canonical distribution function with $N \rightarrow \bar{N}$.

3. Using (6.69), the expression for pressure reads

$$P = \frac{1}{\beta V} \ln(Z_G) = \frac{\bar{N}}{V} k_B T. \tag{7.32}$$

This is same as the ideal gas law (7.20) with N therein replaced by $\bar{N}$.

4. To derive expression for chemical potential, combine (7.28) and (7.29), with $\alpha = \beta\mu$, to get

$$\frac{\mu}{T} = -k_B \left[\ln\left(\frac{V}{\bar{N}}\right) + \frac{3}{2}\ln\left(\frac{U}{\bar{N}}\right) + \frac{3}{2}\ln\left(\frac{4\pi m}{3h^2}\right) \right]. \tag{7.33}$$

The expression for μ above is same as that in (7.26) obtained in the framework of canonical ensemble provided N therein is identified as $\bar{N}$.

5. The expression for entropy may be obtained using the relation (2.53) between entropy and chemical potential of the ideal gas. The resulting equation will turn out to be the same as that in (7.23), which has been derived using canonical ensemble, if N therein is replaced by $\bar{N}$.

We see that the results derived using the grand canonical partition function are the same as the corresponding ones obtained by using the canonical partition function if fixed number N in the results of the canonical ensemble are replaced by the average number $\bar{N}$. In Chap. 6 we have shown that the fluctuations in the average value of an observable about its average varies as $1/\sqrt{N}$. Hence interchange of $\bar{N}$ and N is justified for large N which is anyway the requirement for the applicability of thermodynamics.

7.3 Equipartition Theorem

In several situations, the single particle Hamiltonian is expressible as

$$H_1(\{Q_i\}, \{P_i\}, \{q_i\}, \{p_i\}) = \sum_{j=1}^{l} A_j P_j^2 + \sum_{k=1}^{m} B_k Q_k^2 + \sum_{i=1}^{n} C_i(\{q_i\}) p_i^2, \tag{7.34}$$

where the generalized momenta $\{P_i\}$ are canonically conjugate to the coordinates $\{Q_i\}$ and the momenta $\{p_i\}$ are canonically conjugate to the $\{q_i\}$. The sets of the generalized coordinates $\{Q_i\}$ and $\{q_i\}$ are independent of each other. The coefficients A_i's and the B_i's are constants whereas the C_i's may be functions of the q_i's. Note that the q_i's appear only as part of the coefficients multiplying p_i^2's. We will see that the Hamiltonian of the rotational motion of the molecules is of the form of the last term in (7.34).

The equilibrium state of the system governed by this Hamiltonian and interacting with a thermal reservoir at temperature T is

$$f_N = \frac{1}{Z_N} \exp(-N\beta H_1). \tag{7.35}$$

where Z_N is given by (7.18) with Z_1 therein in the present case being

$$Z_1 = \prod_{i=1}^{n} \int_{a_i}^{b_i} dq_i \int_{-\infty}^{\infty} \exp\{-\beta C_i(\{q_i\}) p_i^2\} dp_i$$

$$\times \prod_{j=1}^{l} \int_{-\infty}^{\infty} \exp(-\beta A_j P_j^2) dP_j \prod_{k=1}^{m} \int_{-\infty}^{\infty} \exp(-\beta B_k Q_k^2) dQ_k. \quad (7.36)$$

Transform to the variables $\tilde{p}_i = \sqrt{\beta} p_i$, $\tilde{P}_j = \sqrt{\beta} P_j$, $\tilde{Q}_k = \sqrt{\beta} Q_k$ so that the expression above reduces to

$$Z_1 = \beta^{-(l+m+n)/2} \prod_{i=1}^{n} \int_{a_i}^{b_i} dq_i \int_{-\infty}^{\infty} \exp\{-C_i(\{q_i\}) \tilde{p}_i^2\} d\tilde{p}_i$$

$$\times \prod_{j=1}^{l} \int_{-\infty}^{\infty} \exp(-A_j \tilde{P}_j^2) d\tilde{P}_j \prod_{k=1}^{m} \int_{-\infty}^{\infty} \exp(-B_k \tilde{Q}_k^2) d\tilde{Q}_k. \quad (7.37)$$

The term multiplying $\beta^{-f/2}$ ($f = (l + m + n)/2$) is independent of β. Hence internal energy per particle is given by ($u = U/N$)

$$u = -\frac{\partial \ln(Z_1)}{\partial \beta} = (l + m + n) \frac{k_B T}{2}. \quad (7.38)$$

This shows that each term in (7.34) contributes $k_B T/2$ to internal energy per particle. This is called the *equipartition theorem*.

7.4 Internal Motion

We have so far considered only the translational motion of the molecules. The molecules, however, have internal structure made of atoms bound together. The atoms vibrate and rotate contributing thereby to the molecular energy. We construct the partition function to include the said motion, called the internal molecular motion.

To that end, consider a molecule consisting of n atoms. The molecular energy consists of three parts: the translational motion of its center of mass and the rotational and vibrational motions of its atoms. As the first approximation, it is assumed that the displacement of atoms from their equilibrium position while vibrating is so small that it can be ignored while considering their rotational motion. We will therefore assume that the relative position of the atoms does not change while rotating. The said three contributions to the energy of the molecule are then independent of each other permitting us to write its Hamiltonian as the sum of the Hamiltonians: H_{trans},

H_{rot}, H_{vib} corresponding respectively to the translational motion of its center of mass, its rotational motion as a rigid body and the vibrational motion:

$$H = H_{\text{trans}} + H_{\text{rot}} + H_{\text{vib}}. \tag{7.39}$$

The partition function then reads

$$Z_N = \frac{1}{N!} Z_1^N, \qquad Z_1 = Z_{\text{trans},1} Z_{\text{rot},1} Z_{\text{vib},1}. \tag{7.40}$$

In writing the contributions to the molecular energy, we have ignored contributions from electronic transitions within atoms. Those transitions, however, cannot be described in terms of the phase space distribution function, requiring instead the quantum theoretic treatment to be taken up in Sect. 8.6.2

The partition function $Z_{\text{trans},1}$ is for free evolution of the molecular center of mass which is same as Z_1 in (7.19).

In the following we evaluate the contributions from the rotational and vibrational motions.

7.4.1   Rotational Motion

As stated above, the molecule is assumed to act as a rigid rotor. It has three axes of rotation except when the molecule is diatomic, or when atoms in a polyatomic molecule are arranged linearly in which case rotation is not possible about the line joining them so that the number of axes of rotation becomes two. We evaluate first the partition function for a diatomic molecule.

Rotational Motion: Diatomic Molecules

The Lagrangian of a system of two atoms of mass m_i at the positions $\mathbf{R}_i$ ($i = 1, 2$) with respect to an arbitrarily chosen origin is

$$\mathcal{L} = \frac{1}{2} \sum_{i=1}^{2} m_i \dot{\mathbf{R}}_i^2. \tag{7.41}$$

It is convenient to work in the coordinate system centered at the center of mass (COM) $\mathbf{R}$ where

$$\mathbf{R} = \frac{1}{M} \sum_{i=1}^{2} m_i \mathbf{R}_i, \qquad M = \sum_{i=1}^{2} m_i, \tag{7.42}$$

Let **r** denote the relative position of the atoms with respect to each other:

$$\mathbf{r} = \mathbf{R}_1 - \mathbf{R}_2. \tag{7.43}$$

The (7.42) and (7.43) yield

$$\mathbf{R}_1 = \mathbf{R} + \frac{m_2}{M}\mathbf{r}, \quad \mathbf{R}_2 = \mathbf{R} - \frac{m_1}{M}\mathbf{r}. \tag{7.44}$$

Since relative position of the atoms is assumed to remain unchanged we have

$$|\mathbf{r}| = a, \tag{7.45}$$

where a is the magnitude of separation between the atoms.

On substituting (7.44) in (7.41), the Lagrangian reads

$$\mathcal{L} = \frac{M}{2}\dot{\mathbf{R}}^2 + \frac{\mu}{2}\dot{\mathbf{r}}^2, \quad \mu = \frac{m_1 m_2}{M}. \tag{7.46}$$

The μ is the reduced mass of the molecule. The Lagrangian (7.46) shows that the center of mass executes free translational motion. Hence its contribution to internal energy per molecule is

$$u_{\text{trans}} = \frac{3}{2}k_B T. \tag{7.47}$$

However, evolution of the relative atomic position vector **r**, described by the Lagrangian

$$\mathcal{L}_{\text{rot}} = \frac{\mu}{2}\dot{\mathbf{r}}^2 \tag{7.48}$$

is constrained by the condition (7.45) of constancy of separation between them. Since the motion under the said constraint is rotational, the corresponding Lagrangian is called the rotational Lagrangian. That constraint can be incorporated in the Lagrangian by working in the spherical polar coordinates. To that end, let θ, ϕ be the polar angles of **r** so that

$$\dot{\mathbf{r}} = \dot{r}\,\mathbf{e}_r + r\dot{\theta}\,\mathbf{e}_\theta + r\sin(\theta)\dot{\phi}\,\mathbf{e}_\phi, \tag{7.49}$$

where $\mathbf{e}_r$, $\mathbf{e}_\theta$, $\mathbf{e}_\phi$ are orthonormal unit vectors along the radial and the angular directions. Since $r = a$ all through the motion, the equation above reduces to

$$\dot{\mathbf{r}} = a\left(\dot{\theta}\,\mathbf{e}_\theta + \sin(\theta)\dot{\phi}\,\mathbf{e}_\phi\right). \tag{7.50}$$

Substitute this in (7.48) to get

$$\mathcal{L}_{\text{rot}} = \frac{a^2 \mu}{2} \left(\dot{\theta}^2 + \sin^2(\theta) \dot{\phi}^2 \right).$$ (7.51)

The momenta p_θ, p_ϕ conjugate respectively to θ and ϕ are

$$p_\theta = \frac{\partial \mathcal{L}}{\partial \dot{\theta}} = I\dot{\theta}. \quad p_\phi = \frac{\partial \mathcal{L}}{\partial \dot{\phi}} = I \sin^2(\theta)\dot{\phi}, \quad I = \mu a^2.$$ (7.52)

The I is the moment of inertia of the molecule. Consequently, the Hamiltonian corresponding to the Lagrangian (7.51) turns out to be given by

$$H_{\text{rot}} = \frac{1}{2I} \left(p_\theta^2 + \frac{1}{\sin^2(\theta)} p_\phi^2 \right),$$ (7.53)

called the rotational Hamiltonian. The Hamiltonian above can be written in terms of the angular momentum,

$$\mathbf{L} = \mathbf{r} \times \mathbf{p},$$ (7.54)

as

$$H_{\text{rot}} = \frac{1}{2I} L^2, \quad L^2 = \mathbf{L} \cdot \mathbf{L}.$$ (7.55)

This form will be found useful in the study of the quantum theory of diatomic rotation. The proof of (7.55) is left as an exercise.

The single molecule partition function corresponding to (7.53) is

$$Z_{\text{rot1}} = \frac{1}{h^2} \int \exp\left\{ -\beta H_{\text{rot}}(\theta, \phi; p_\theta, p_\phi) \right\} \mathrm{d}\theta \mathrm{d}\phi \mathrm{d}p_\theta \mathrm{d}p_\phi,$$ (7.56)

where $0 \leq \phi \leq 2\pi, 0 \leq \theta \leq \pi, -\infty \leq p_\phi \leq \infty, -\infty \leq p_\theta \leq \infty$. The evaluation of the integral above yields

$$Z_{\text{rot1}} = \frac{8I\pi^2}{h^2} k_B T.$$ (7.57)

The internal energy per molecule due to rotational motion is

$$u_{\text{rot}} = -\frac{\partial \ln(Z_{\text{rot},1})}{\partial \beta} = k_B T.$$ (7.58)

The H_{rot} in (7.53) is of the form of the Hamiltonian (7.34) with $l = m = 0, n = 2$. Hence, the result derived above is in accordance with that of the equipartition theorem in (7.38).

Rotational Motion: Polyatomic Molecules

Consider a molecule constituted by n atoms. In case the atoms are arranged on a line, the molecule is called linear. Like a diatomic molecule, such a molecule has two axes of rotation whereby the contribution to internal energy per molecule is the same as that for a diatomic one:

$$u_{\text{rot}} = k_B T, \quad \text{linear polyatomic molecules.} \tag{7.59}$$

If the gas is made of non-linear molecules, we can evaluate the contribution of rotation to internal energy as follows.

We treat a polyatomic molecule as a rigid body, denoting the principal components of its moment of inertia by I_j ($j = 1, 2, 3$) and the Euler angles with respect to the principal moment of inertia axes by (θ, ϕ, ψ). The Hamiltonian of the molecule is then given by

$$H_{\text{rot}} = \frac{1}{2I_1 \sin^2(\phi)} \left\{ (p_\theta - p_\psi \cos(\phi)) \cos(\psi) - p_\phi \sin(\phi) \sin(\psi) \right\}^2$$

$$\frac{1}{2I_2 \sin^2(\phi)} \left\{ (p_\theta - p_\psi \cos(\phi)) \sin(\psi) - p_\phi \sin(\phi) \cos(\psi) \right\}^2 + \frac{1}{2I_3} p_\psi^2. \tag{7.60}$$

We can evaluate Z_1 in (7.37) with $H_1 = H_{\text{rot}}$ by transforming to $\tilde{p}_\mu = \sqrt{\beta} p_\mu$ ($\mu = \theta, \phi, \psi$) and show, like we did in proving the equipartition theorem, that

$$u_{\text{rot}} = \frac{3}{2} k_B T, \quad \text{non-linear polyatomic molecules.} \tag{7.61}$$

Else, we can transform the combination of momenta in the curly brackets, along with p_ψ in the last term in (7.60), to three new momenta to reduce H_{rot} to the form (7.34) with $l = m = 0, n = 3$ and apply the equipartition theorem to arrive at (7.61).

Since there are two axes of rotation in a linear molecule and three in a polyatomic molecule, we can say that rotation around each axis contributes $k_B T/2$ to the thermal energy.

In terms of the components L_k ($k = 1, 2, 3$) of the angular momentum along the principal axes, the rotational Hamiltonian assumes the form

$$H_{\text{rot}} = \frac{1}{2I_1} L_1^2 + \frac{1}{2I_2} L_2^2 + \frac{1}{2I_3} L_3^2. \tag{7.62}$$

This form of the Hamiltonian turns out to be useful in the quantum mechanical treatment of the problem carried in Chap. 8.

Summary

The energy per molecule in the ideal gas from translational and rotational motions is

$$u_{\text{trans+rot}} = \frac{5k_B T}{2}, \quad \text{diatomic and linear polyatomic molecules,}$$

$$u_{\text{trans+rot}} = 3k_B T, \quad \text{polyatomic molecules.} \tag{7.63}$$

Next we determine the contribution of vibrations to energy.

7.4.2 Vibrational Motion

The number of degrees of freedom of a molecule made of n atoms is $3n$. There are three translational degrees of freedom associated with the motion of its center of mass. A linear molecule has two degrees of rotational freedom so that remaining $3n - 5$ degrees of freedom are vibrational. If the molecule is not linear, it has three axes of rotation and hence three rotational degrees of freedom. The remaining $3n - 6$ are the degrees of vibrational motion.

Consider a diatomic molecule. Per the discussion above, it has one vibrational degree of freedom corresponding to the vibrational motion along the line joining the atoms. It is governed by the Hamiltonian:

$$H_{\text{vib}} = \frac{p_\eta^2}{2\mu} + \frac{\mu\omega^2}{2}\eta^2, \tag{7.64}$$

where η is the displacement from the equilibrium position of the atoms and p_η is the corresponding momentum. This is of the form of the Hamiltonian (7.34) with $l = m = 1$, $n = 0$. Hence, invoking the equipartition theorem formula (7.38), the contribution to internal energy per molecule due to the vibrational Hamiltonian (7.64) is

$$u_{\text{vib}} = k_B T, \quad \text{diatomic molecules.} \tag{7.65}$$

Thus each vibrational mode contributes $k_B T$ to the thermal energy. There being $3n - 5$ modes of vibration in a linear molecule of n atoms, the vibrational thermal energy per molecule of the gas constituted by such molecules is

$$u_{\text{vib}} = (3n - 5)k_B T, \quad \text{linear polyatomic molecules of } n \text{ atoms.} \tag{7.66}$$

In case the molecule of n atoms is not linear, it has $3n - 6$ vibrational modes. Hence the vibrational thermal energy per molecule of the gas constituted by such molecules is

$$u_{\text{vib}} = 3(n - 2)k_B T, \quad \text{non-linear molecules of } n \text{ atoms.} \tag{7.67}$$

Putting together all the contributions, the energy per molecule consisting of n atoms constituting an ideal gas is

$$u = \left(3n - \frac{5}{2}\right) k_B T, \quad \text{gas of linear molecules,}$$
$$u = 3(n - 1)k_B T, \quad \text{gas of non-linear molecules.} \tag{7.68}$$

This leads to the following expressions for the heat capacity per molecule at constant volume:

$$c_V = \left(3n - \frac{5}{2}\right) k_B, \quad \text{gas of linear molecules,}$$
$$c_V = 3(n - 1)k_B, \quad \text{gas of non-linear molecules.} \tag{7.69}$$

The experimentally observed value of heat capacity is not only temperature dependent but is also much lower than the value predicted above till the temperature is very high. Answer to the question of the reason of said discrepancy is provided by the quantum theoretical treatment, presented in Sect. 8.6.2.

7.5 Gas in Gravitational Field

Consider a gas of N molecules of mass m, enclosed in a box of base area A and height H placed near the surface of earth. With g denoting the acceleration due to gravity. the Hamiltonian of the system is

$$H(\mathbf{r}, \mathbf{p}) = \frac{1}{2m} \sum_{i=1}^{N} p_i^2 + mg \sum_{i=1}^{N} z_i, \quad 0 \le z_i \le H, \tag{7.70}$$

where z_i denotes the height of the ith molecule above the bottom of the box. We study the thermodynamics of the gas in canonical ensemble formalisms assuming that there is no variation in temperature with height. The effect of temperature gradient is discussed in Sect. 7.5.1.

The canonical partition function corresponding to the Hamiltonian (7.70) is given by (7.9) with

$$v(\mathbf{r}) = mgz, \quad 0 \le z \le H, \tag{7.71}$$

so that, with λ_T as in (7.6),

$$Z_N = \frac{Z_1^N}{N!}, \qquad Z_1 = \frac{AH}{\xi}\lambda_T^{-3}(1 - \exp(-\xi)), \tag{7.72}$$

where

$$\xi = mgH\beta. \tag{7.73}$$

Various thermodynamic quantities resulting from Z_N in (7.72) are:

1. Invoking (6.34), energy of the gas is found to be given by

$$U = Nk_BT\left[\frac{5}{2} - \frac{\xi}{\exp(\xi) - 1}\right]. \tag{7.74}$$

   This shows that (i) when $\xi \ll 1$ then $U \to 3Nk_BT/2$. This will be the case when T is high or mgH is low. (ii) In the opposite limit, $\xi \gg 1$, $U \to 5Nk_BT/2$.
2. Using (6.35) it can be shown that the entropy of the gas is

$$S_N = Nk_B\left[\frac{7}{2} - \frac{\xi}{\exp(\xi) - 1} + \ln\left(1 - \exp(-\xi)\right) - \frac{5}{2}\ln(\xi)\right.$$
$$\left. + \ln\left(\frac{V}{N}\right) + \frac{3}{2}\ln(K)\right], \tag{7.75}$$

   where $V = AH$ is the volume of the box and

$$K = \frac{2\pi m^2 g H}{h^2}. \tag{7.76}$$

   The free gas limit $g \to 0$ of S_N can be evaluated conveniently by rewriting (7.75) in the form

$$S_N = Nk_B\left[\frac{7}{2} - \frac{\xi\exp(\xi)}{\exp(\xi) - 1} + \ln\left(\frac{\exp(\xi) - 1}{\xi}\right)\right.$$
$$\left. + \ln\left(\frac{V}{N}\right) + \frac{3}{2}\ln\left(\frac{2\pi m}{h^2\beta}\right)\right]. \tag{7.77}$$

   It is straightforward to show that, in the limit $g \to 0$, i.e. $\xi \to 0$, (7.77) reduces to the free gas entropy formula (7.23). The entropy becomes negative for the values of ξ less than the value ξ_m which is determined by the solution of $S_N(\xi_m) = 0$. As discussed circa (7.24), the negativity of entropy is attributed to the failure of phase space formalism at low temperatures and/or high densities.
3. The probability density of finding a molecule in the vicinity of $\mathbf{r}$ is given by (7.16) with $v(\mathbf{r})$ therein in the present case is given by (7.71) so that

$$\rho(\mathbf{r}) = C \exp\left(-mg\beta z\right) = \frac{\xi \exp\left(-mg\beta z\right)}{V(1 - \exp(-\xi))}, \quad V = AH. \quad (7.78)$$

Some quantities related with $\rho(\mathbf{r})$ are:

a) Since $\rho(\mathbf{r})dxdydz$ is the probability of finding a molecule in the vicinity of $\mathbf{r}$,

$$p(z)dz = \left[\int \rho(\mathbf{r})dxdy\right] dz \quad (7.79)$$

is the probability of finding a molecule in $(z, z + dz)$. Using (7.78) for $\rho(\mathbf{r})$, (7.79) gives

$$p(z) = \frac{\xi \exp\left(-mg\beta z\right)}{H(1 - \exp(-\xi))}. \quad (7.80)$$

Equivalently, the number of molecules in $(z, z + \delta z)$ is given by

$$\delta \bar{n}(z) = N p(z)\delta z. \quad (7.81)$$

4. Since the gas is inhomogeneous in the z-direction, its pressure varies with z. We therefore consider gas in the cylinder constituted by the region between z and $z + \delta z$, its base area being A. The pressure due to molecules within the cylinder moving in the z-direction is the pressure $P(z)$ at z. It can be evaluated using (3.27) by replacing $\langle v_{+x}^2 \rangle$ therein by $\langle v_{+z}^2 \rangle$, and noting that the number of particles in the present case is $\delta\bar{n}(z)$:

$$P(z) = 2m\frac{\delta\bar{n}(z)}{\delta V}\langle v_{+z}^2 \rangle, \quad \delta V = A\delta z. \quad (7.82)$$

It is straightforward to see that $2m\langle v_{+z}^2 \rangle = k_B T$. Recall also (7.81) to show that (7.82) leads to the following expression for pressure at the height z above the earth's surface:

$$P(z) = P(0) \exp(-mgz/k_B T), \quad P(0) = \frac{Nmg}{A(1 - \exp(-\xi))}, \quad (7.83)$$

$P(0)$ being pressure at $z = 0$. This is called the *barometric formula*. In the limit $g \to 0$, the expression above reduces to $P = Nk_B T/V$ which is independent of z and same as for free non-interacting gas. An alternative approach for determining pressure is presented in Sect. 7.5.1.

5. The average pressure is given by

$$\bar{P} = \frac{1}{H}\int_0^H P(z)\,dz = \frac{N}{V}k_B T. \quad (7.84)$$

We see that the gas in the gravitational field obeys the equation of state of a non-interacting free gas if the pressure in that equation is replaced by the average pressure of the gas in the gravitational field.

See also [5–8] for thermodynamics of a column of ideal gas ignoring temperature gradient.

7.5.1 Temperature Gradient

A useful application of the problem addressed above, namely, that of a gas in gravitational field near the surface of earth is to the study of the earth's atmosphere. However, in our formalism so far we have assumed that there is no temperature gradient in the gas column as a function of the height, an assumption which does not hold in the case of earth's atmosphere because we know that the temperature decreases with height. In the following we show how temperature gradient can be incorporated in the theory. We will see that significant temperature change takes place at high altitudes.

Since the earth's atmosphere is unbounded from above, the formulas derived above for finite height H assuming uniform temperature apply to earth's atmosphere in the limit $H \to \infty$. In particular, the (7.80) for the probability per unit height of finding a molecule in the vicinity of z and the equation (7.83) for pressure assume the form

$$p(z) = \frac{mg}{k_B T} \exp(-mg\beta z), \tag{7.85}$$

and

$$P(z) = P(0) \exp(-mgz/k_B T), \qquad P(0) = \frac{Nmg}{A}. \tag{7.86}$$

Recall that we derived the formulas above based on the molecular distribution predicted by statistical mechanics for the gas at uniform temperature. The statistical mechanics formalism as it stands cannot account for temperature gradient in its distribution function. A different approach, presented below, is therefore required when temperature is varying through the height of the gas column.

To that end, consider the gas in the volume of the cylinder formed by the region between the cross-sectional planes of the cylindrical container between the heights z and $z + \delta z$. The pressure of the gas on the plane at z is the sum of pressure on the plane at $z + \delta z$ and the weight per unit area of the column of gas between the two planes:

$$P(z) = P(z + \delta z) + mg\delta\bar{n}(z)/A, \tag{7.87}$$

where $\delta\bar{n}(z)$ is average number of molecules in the cylinder. For small δz, the gas in the volume of the cylinder in question can be considered to be at uniform temperature $T(z)$ at the height z. Consequently the ideal gas law will apply locally so that

$$P(z)(A\delta z) = \delta\bar{n}(z)k_{\mathrm{B}}T(z). \tag{7.88}$$

On eliminating $\delta\bar{n}(z)$ between (7.87) and (7.88) we get

$$\frac{P(z+\delta z) - P(z)}{\delta z} = -mg\frac{P(z)}{k_{\mathrm{B}}T(z)}. \tag{7.89}$$

This leads to the following differential equation for $P(z)$:

$$\frac{\mathrm{d}P(z)}{\mathrm{d}z} = -mg\frac{P(z)}{k_{\mathrm{B}}T(z)}. \tag{7.90}$$

If $T(z)$ is independent of z the solution of (7.90) would be

$$P(z) = P(0)\exp(-mgz/k_{\mathrm{B}}T). \tag{7.91}$$

The constant $P(0)$, which clearly is the pressure at $z = 0$, can be found by noting that, since the formula above is for uniform temperature, the equation (7.81) for $\delta n(z)$ would apply. Use the said formula with $H \to \infty$, to get

$$\delta\bar{n}(0) = \frac{Nmg}{k_{\mathrm{B}}T}\delta z. \tag{7.92}$$

Substitute this in (7.88) to obtain

$$P(0) = \frac{Nmg}{A}. \tag{7.93}$$

The expression (7.91) derived in a different way for the case of uniform temperature is the same as that in (7.86) derived by statistical mechanics approach.

The solution of (7.90) requires knowledge of the functional form of $T(z)$. It may be derived by observing that, in the state of equilibrium, there is no net transfer of heat through the column of the gas but there is an exchange of molecules between the neighboring layers. The assumption of no net transfer of heat means it is transmitted only from the heated earth surface and that the same amount of heat is received by earth from sun all the time. It ignores the direct absorption and emission of heat by gas molecules, as also the effects due to the presence of moisture. Consequently thermodynamic processes operative at equilibrium may be assumed to be adiabatic. This implies the temperature $T(z)$ and pressure $P(z)$ are such that $T(z)P^{-\nu}(z) = $ constant where $\nu = (\gamma - 1)/\gamma$, $\gamma = C_p/C_V$ (see (1.88)) so that

$$T(z)P^{-\nu}(z) = \text{constant}, \qquad \nu = \frac{\gamma - 1}{\gamma}. \tag{7.94}$$

On differentiating the equation above we obtain

$$\frac{\mathrm{d}T(z)}{\mathrm{d}z} = \frac{\nu T(z)}{P(z)} \frac{\mathrm{d}P(z)}{\mathrm{d}z}. \tag{7.95}$$

Invoking (7.90) this reduces to

$$\frac{\mathrm{d}T(z)}{\mathrm{d}z} = -r, \qquad r = \frac{mg\nu}{k_{\mathrm{B}}}. \tag{7.96}$$

This shows that temperature decreases linearly by r K per unit length and

$$T(z) = T(0) - rz. \tag{7.97}$$

The rate r has been evaluated in Exercise 7.3 for some commonly employed values of m and γ. It predicts a reduction of about 0.0097 K per meter above the surface of earth. The observed value of r is ≈ 0.0065 K per meter up to about 11 Km. The discrepancy in the theoretical and observed values of r is evidently attributable to various assumptions which have been made to arrive at (7.95).

The pressure, obtained by substituting (7.97) in (7.94), is

$$P(z) = P(0) \left(1 - \frac{r}{T_0} z \right)^{1/\nu}. \tag{7.98}$$

This gives pressure at height z above the surface of earth.

For further details see [7–9].

Exercises

Ex. 7. 1. Let $P(0)$ denote the atmospheric pressure at the sea level. Assuming $P(0) = 1$ atmosphere, show that the molecular mass per unit area at sea level is approximately $1\,\mathrm{kg/cm^2}$, given $g = 980\ \mathrm{cm\ s^{-2}}$.

Ex. 7. 2. Show that the average height of a molecule in earth's atmosphere is $k_{\mathrm{B}}T/mg$.

Ex. 7. 3. Calculate r defined in (7.95) assuming the mass m of the molecules in the atmosphere to be 28.96 atomic mass unit and the adiabatic constant $\gamma = 1.4$. Hint: Rewrite r in the form $r = (mc^2)(g/c^2 k_{\mathrm{B}})(\gamma/(\gamma - 1))$ and use the following units conversion relations and the values of the constants: $(\mathrm{amu})c^2 = 931.4$ Mev, $k_{\mathrm{B}} = 8.617 \times 10^{-5}$ eV/K, $g = 980\ \mathrm{cm\ s^{-2}}$, $c = 3 \times 10^{10}\ \mathrm{cm\ s^{-1}}$.

7.6 Gibbs Paradox

Gibbs paradox is concerned with paradoxical conclusions which follow in the theory of mixing of ideal gases if the quantum factor,

$$Q_N = N! h^{3N}, \tag{7.99}$$

appearing in the definition (5.42) of the phase space volume element is ignored. We call it the quantum factor because, as brought out in Sect. 8.4.3, it owes its origin to the quantum theory. However, as we will see below, the crucial factor in Q_N is $N!$ which was conjectured by Gibbs much before the advent of the quantum theory.

To that end note that if the phase space volume element (5.42) is defined without the quantum factor Q_N then the partition function for the ideal gas of N molecules will be

$$\tilde{Z}_N = Q_N Z_N, \tag{7.100}$$

where Z_N is the partition function (7.18) for the same gas including the quantum factor in the definition of phase space volume element. Since the multiplying factor Q_N does not change the internal energy, it follows using the expression (6.35) for entropy that the entropy $\tilde{S}_N$ corresponding to $\tilde{Z}_N$ is related with the entropy S_N corresponding to Z_N by the relation

$$\tilde{S}_N = k_B \left(\beta \tilde{U} + \ln(\tilde{Z}_N) \right) = k_B \ln(Q_N) + S_N. \tag{7.101}$$

Consider two boxes each containing an ideal gas at the same density and temperature. One of the boxes is of volume V_1 and contains N_1 molecules of a gas of mass m_1, whereas the volume of the other box is V_2 and has N_2 molecules of mass m_2. Their entropy corresponding to Z_N is

$$S_{N_k} = N_k k_B \left[\frac{5}{2} + \ln \left(\frac{V_k}{N_k} \right) + \frac{3}{2} \ln \left(\frac{U_k}{N_k} \right) + \frac{3}{2} \ln \left(\frac{4\pi m_k}{3h^2} \right) \right]. \tag{7.102}$$

Let the gases mix and attain the state of equilibrium. The quantity of interest is the change in entropy of the gas after mixing. We compute that change with and without the quantum factor Q_N separately for the case of two gases consisting of the same type of molecules and that for the gases consisting of different types of molecules.

Gases Consisting of Same Type of Molecules

The entropy of each of the two gases before mixing in this case is given by (7.102) with $m_1 = m_2 = m$. After mixing it is ($N = N_1 + N_2$, $V = V_1 + V_2$)

$$S_N = Nk_{\mathrm{B}} \left[\frac{5}{2} + \ln\left(\frac{V}{N}\right) + \frac{3}{2}\ln\left(\frac{U}{N}\right) + \frac{3}{2}\ln\left(\frac{4\pi m}{3h^2}\right) \right]. \qquad (7.103)$$

Since the gases before and after mixing have the same number density, and temperature, we have ($k = 1, 2$)

$$\frac{V_k}{N_k} = \frac{V}{N}, \qquad \frac{U_k}{N_k} = \frac{U}{N}. \qquad (7.104)$$

It is then straightforward to see that change in entropy after mixing is

$$\Delta S = S_N - (S_{N_1} + S_{N_2}) = 0. \qquad (7.105)$$

This shows that change in entropy of a gas, initially in two containers at same number densities and temperature, remains unchanged when mixed if the quantum factor Q_N is included in defining the phase space volume.

Now, using (7.101) the change in entropy when the quantum factor is not included may be seen to be given by

$$\Delta \tilde{S} = \tilde{S}_N - (\tilde{S}_{N_1} + \tilde{S}_{N_2}) = k_{\mathrm{B}}\{\ln(Q_N/(Q_{N_1}Q_{N_2}))\} + \Delta S. \qquad (7.106)$$

Using (7.105) for ΔS and the expression for Q_N in (7.99) under Stirling's approximation, the equation above yields

$$\begin{aligned} \Delta \tilde{S} &= k_{\mathrm{B}} \left\{ N_1 \ln(N/N_1) + N_2 \ln(N/N_2) \right\} \\ &\equiv k_{\mathrm{B}} \left\{ N_1 \ln(V/V_1) + N_2 \ln(V/V_2) \right\} > 0, \end{aligned} \qquad (7.107)$$

where second equation is due to (7.104) whereby $N/N_k = V/V_k$. The equation (7.107) shows that entropy of a gas, initially in two containers at same number density and temperature, increases after mixing if the quantum factor is excluded from the definition of the phase space volume.

The conclusion arrived at following (7.107) is unacceptable. For, consider the gas in the volume V consisting of only one kind of molecules, N in number. We can imagine that the said state has been arrived at by starting with the same gas molecules in two containers of volume V_1 and V_2, each containing N_1 and N_2 molecules at the same density and temperature such that $V = V_1 + V_2$, $N = N_1 + N_2$. The equation (7.107) shows that entropy after mixing will depend on the choice of the numbers N_1, N_2. This is absurd because it means the entropy depends on the history of how the state has been arrived at whereas entropy is a function of state and not of how the state has been arrived at. This is known as *Gibbs paradox*.

We see that the paradox arises if the quantum factor is not included in the definition of the phase space volume. There is no paradox if that factor is included in the definition of the phase space volume as in that case, as shown in (7.105), there is no change in entropy after mixing.

The paradoxical conclusion may be traced to the fact that, as is clear from its expression (7.102), S is extensive. On the other hand, due to the factor Q_N, the $\tilde{S}_N$ in (7.23) is not. Inclusion of the factor Q_N has the effect of generating expression for entropy having the desired property of extensivity.

The crucial part in that factor behind the difference in the results of with and without it is, of course, $N!$. Gibbs conjectured it to resolve the paradox. The other constant factor in Q_N comes from the quantum theory alone.

The quantum factor must therefore be included in the definition of the phase space volume.

We evaluate change in entropy when two dissimilar gases are mixed and show that change in entropy then is the same, irrespective of whether or not the quantum factor is included in the definition of the phase space volume.

Gases Consisting of Dissimilar Molecules

The entropy of two gases before mixing is given by (7.102) when the quantum factor is included in the definition of the phase space volume. After mixing, the gases occupy the combined volume V but are still identifiable differently because they consist of different molecules. The entropy of the mixture is the sum of the entropies of each of them. The entropy S_k of the gas numbered k in the mixture is given by (7.102) with $V_k \to V$ ($k = 1, 2$). The change in entropy after mixing therefore is

$$\Delta S = (S_1(V) + S_2(V)) - (S_1(V_1) + S_2(V_2)). \qquad (7.108)$$

It is straightforward to show that

$$\Delta S = k_B \left[N_1 \ln \left(\frac{V}{V_1} \right) + N_2 \ln \left(\frac{V}{V_2} \right) \right] > 0. \qquad (7.109)$$

On the other hand, when the quantum factor is not included then change in entropy is given, following (7.101), by

$$\Delta \tilde{S} = \tilde{S}_N - (\tilde{S}_{N_1} + \tilde{S}_{N_2}) = k_B \{ \ln(Q'_N/(Q_{N_1} Q_{N_2})) \} + \Delta S, \qquad (7.110)$$

where Q'_N is the quantum factor after mixing. Recall that the $N!$ in Q_N is to account for the symmetry under exchange of particles of the gas. Since the molecules of the two gases are distinguishable even after mixing, the exchange symmetry is only between the molecules of one type so that $Q'_N = Q_{N_1} Q_{N_2}$. This show that $\Delta \tilde{S} = \Delta S$. We thus see that entropy after mixing of two gases consisting of dissimilar molecules increases by the same amount whether or not the quantum factor is included in the definition of the phase space volume.

References

1. H. Tetrode, Ann. der Phys. **38** 434 (1912)
2. O. Sackur, Ann. der Phys. **40** 67 (1913)
3. W. Grimus (2013), arxiV:1112.3748v2 [physics.hist-ph]
4. J.P. Fransisco, E. Pérez, Eur. J. Phys. **36**, 055033 (2015)
5. P.T. Landsberg, J. Dunning-Davies, D. Pollard, Am. J. Phys. **62**, 712 (1994)
6. H.-C. Kim, G. Kang, J. Korean Phys. Soc. **69**, 1597 (2016)
7. M.N. Berberan-Santos, E.N. Bodunov, L. Pogliani, Am. J. Phys. **65**, 404 (1997)
8. G. Lente, K. Ösz, Chem. Texts **6**, 13 (2020). https://doi.org/10.1007/s40828-020-0111-6
9. E.N. Bodunov, G.G. Khokhlov, J. Phhys, Conf. Ser. **2131**, 022053 (2021)

Chapter 8
Ideal Quantum Gases

Based on the principle of maximum entropy, we developed in Chap. 6 the classical phase space, as well as the quantum theoretic formalisms of statistical mechanics. Using the phase space formalism, we studied in Chap. 7 the equilibrium statistical thermodynamics of a non-interacting gas. We investigate now the consequences of the quantum theoretic formalism when applied to free non-interacting gases, called *ideal quantum gases*. The probability distribution in quantum theoretic formalism depends on whether or not the molecules are distinguishable. Accordingly we will derive the expressions for the probability distributions treating molecules as (i) distinguishable, (ii) indistinguishable Fermi particles, and (iii) indistinguishable Bose particles.

The equilibrium distributions for Fermi and the Bose gases will turn out to be the same as the corresponding ones derived in Chap. 4 using the approach based on the Boltzmann entropy. We will see that, in the so-called classical limit, the Fermi as well as the Bose distributions reduce to the classical phase space distribution derived in Chap. 6. We discuss thermodynamics of non-interacting Fermi gas in Chap. 9 and that of the Bose gas in Chap. 10.

In this chapter we study thermodynamics of the gas of distinguishable molecules including their internal motion. We will see that when the molecules are considered distinguishable and their internal motion is excluded, the properties of the gas are the same as those predicted by the phase space method. However, the thermodynamic properties associated with the internal motion when treated quantum mechanically, turn out to be vastly different from the ones predicted by the phase space approach in Chap. 7.

225
R. R. Puri, *Modern Thermodynamics and Statistical Mechanics*, Undergraduate Lecture Notes in Physics, https://doi.org/10.1007/978-3-031-54310-4_8

8.1   Canonical Partition Function

Recall that a system of N particles in the quantum formalism is described by the probability of occupation of its collective energy levels $\{E_{n,N}\}$. Since the particles are non-interacting, their collective energy will be the sum of energies of individual particles. To find possible values of collective energy, let $\epsilon_1, \epsilon_2, \ldots, \epsilon_K$ be K possible single particle energy levels. The number of energy levels can even be infinite. Let $\{n_i\} \equiv n_1, n_2, \ldots, n_K$ be the number of particles occupying levels of energies $\epsilon_1, \epsilon_2, \ldots, \epsilon_K$ respectively. A particular set of values of $\{n_i\}'s$ defines the collective energy $E_{n,N}$ by the relation

$$E_{n,N} \equiv E_{n_1,n_2,\ldots,n_K;N} \equiv E_{\{n_i\};N} = \sum_{i=1}^{K} n_i\epsilon_i, \quad \sum_{i=1}^{K} n_i = N. \tag{8.1}$$

Following essentially the terminology introduced in Chap. 6, the n_k's are referred to as the occupation numbers and a particular set $\{n_i\}$ of the values of the $n_i's$ defines a state of the system. If the particles are indistinguishable, there is only one way to choose k particles to occupy the energy level ϵ_k. A particular set $\{n_i\}$ of the values of the $n_i's$ then characterizes uniquely a state of the system. On the other hand, distinguishable particles can be labeled. In that case same number n_k occupying the energy level ϵ_k can be realized by placing differently labeled particles in it thereby characterizing different states of same energy corresponding to same particular set of values of $\{n_i\}$. Following the terminology introduced in Chap. 6, we call those different realizations as different configurations corresponding to the same set $\{n_i\}$. We denote by $g(\{n_k\})$, called the *configurational degeneracy*, the number of configurations corresponding to the state characterized by the given set $\{n_k\}$.

Consider the defining relation (6.31) for the partition function Z_N. The sum over states of different energies therein in the present case is the sum over different sets of values of $\{n_k\}$. In view of the discussion above it assumes the form

$$Z_N = \sum_{\{n_i\}} g(\{n_i\}) \exp\left(-\beta \sum_{i=1}^{K} n_i\epsilon_i\right) \delta\left(\sum_{i=1}^{K} n_i - N\right). \tag{8.2}$$

The $\delta(x)$ ($\delta(0) = 1$, $\delta(x) = 0$ for $x \neq 0$) under the summation ensures that the total number of particles in all the levels equals the given number N. The sum may be carried by choosing first, say, a number n_1 between 0 and N, then the number n_2 which, in order to ensure that $n_1 + n_2 \leq N$, must lie between 0 and $N - n_1$, then the number n_3 which, in order to ensure that $n_1 + n_2 + n_3 \leq N$, must lie between 0 and $N - n_1 - n_2$ and so on till n_{K-1} is chosen from the interval $(0, N - n_1 - n_2 - \cdots - n_{K-2})$. The remaining number n_k is given by $n_k = N - n_1 - n_2 - \cdots n_{K-1}$.

The degeneracy factor $g(\{n_i\})$ for indistinguishable particles is one. We evaluate it next for distinguishable particles.

8.1.1 Configurational Degeneracy

Before evaluating $g(\{n_i\})$ we illustrate the concept of configurational degeneracy by a couple of simple examples.

Assume that the possible values of energy of a particle are ϵ_1 and ϵ_2. Consider a system of two such identical particles labeled A and B. They can be placed in the said two levels in the following ways (see Fig. 8.1):

1. Both the particles can be in the level of energy ϵ_1 (Fig. 8.1(i)). The energy of the system in this case is $2\epsilon_1$ and the corresponding state is characterized by the set of occupation numbers $(n_1, n_2) = (2, 0)$. Since mutual permutation of particles in this case does not lead to distinct configuration, the state of energy $2\epsilon_1$ has no configurational degeneracy.
2. Both the particles can be in the level of energy ϵ_2 (Fig. 8.1(ii)). The energy of the system in this case is $2\epsilon_2$ and the corresponding state is characterized by the set of occupation numbers $(n_1, n_2) = (0, 2)$. Since mutual permutation of the particles does not result in any distinct configuration, the state of energy $2\epsilon_2$ has no configurational degeneracy.
3. One particle can be in the state of energy ϵ_1 and the other in that of energy ϵ_2. The energy of the system then is $\epsilon_1 + \epsilon_2$ characterized by set of occupation numbers $(n_1, n_2) = (1, 1)$. However, in this case the particle A can be in ϵ_1 and B in ϵ_2 (Fig. 8.1(iii)), or the particle B can be in ϵ_1 and A in ϵ_2 (Fig. 8.1(iv)). The said two ways of placing the particles are two distinct configurations corresponding to the same set $(1, 1)$ of the occupation numbers (n_1, n_2). The state of said energy therefore has configurational degeneracy equal to two.

As another example, let three particles labeled A, B, C be distributed in two levels of energies ϵ_1 and ϵ_2. The number of distinct distributions, depicted in Fig. 8.2, are:

1. All particles can be in the level of energy ϵ_1 (Fig. 8.2(i)) so that the energy of the system is $3\epsilon_1$ characterized by the set of occupation numbers $(n_1, n_2) = (3, 0)$. Since it has only one configuration, the said state has no configurational degeneracy.

Fig. 8.1 Distribution of two distinguishable particles in two levels

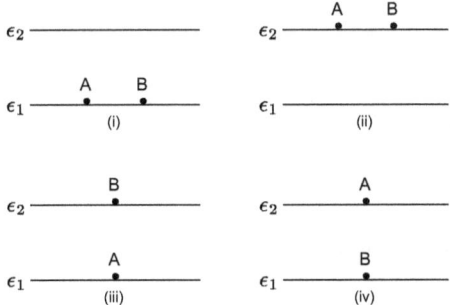

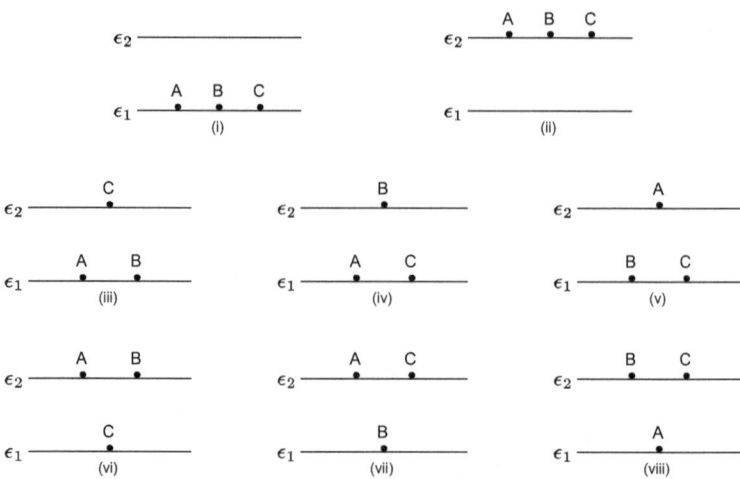

Fig. 8.2 Distribution of three distinguishable particles in two levels

2. All particles can be in the level of energy ϵ_2 (Fig. 8.2(ii)) so that the energy of the system is $3\epsilon_2$ characterized by the set of occupation numbers $(n_1, n_2) = (0, 3)$. Since it has only one configuration, the said state has no configurational degeneracy.
3. One particle (C or B or A) can be in the level of energy ϵ_2 and the remaining two in the level of energy ϵ_1 (Fig. 8.2(iii)–(v)). The energy of the system in this cases is $2\epsilon_1 + \epsilon_2$ corresponding to the occupation numbers $(n_1, n_2) = (2, 1)$. Since it has three configurations, the said state has three-fold configurational degeneracy.
4. One particle (C or B or A) can be in the level of energy ϵ_1 and the remaining two in the level of energy ϵ_2 (Fig. 8.2(vi)–(viii)). The energy of the system in this cases is $\epsilon_1 + 2\epsilon_2$ corresponding to the occupation numbers $(n_1, n_2) = (1, 2)$. Since it has three configurations, the said state has three-fold configurational degeneracy.

To evaluate the degeneracy factor for distinguishable particles in general, consider the set of occupation numbers $(n_1, n_2, \ldots, n_K)$. If the molecules are distinguishable then, following the arguments similar to those given in Appendix D for selecting boxes, the number of ways of choosing the given set of numbers is

$$G(\{n_i\}) = \frac{N!}{n_1! n_2! \cdots n_K!} \delta \left(\sum_{i=1}^{K} n_i - N \right). \tag{8.3}$$

At the end, however, the fact that the particles are identical is invoked to argue that permutation of the particles does not lead to a new state. The desired degeneracy is therefore obtained by dividing the degeneracy factor in (8.3), which has been obtained assuming the particles to be distinguishable, by the number $N!$ of their permutations to get

$$g(\{n_i\}) = \frac{G(\{n_i\})}{N!} = \frac{1}{n_1!n_2!\cdots n_K!}\delta\left(\sum_{i=1}^{K} n_i - N\right). \tag{8.4}$$

We use the expression for $g(\{n_i\})$ derived above to evaluate the partition function of distinguishable particles.

8.2 Distinguishable Particles

On using (8.4) the partition function (8.2) reads

$$Z_{NC} = \frac{1}{N!}\sum_{\{n_i\}} \frac{N!}{n_1!n_2!\cdots n_K!} x_1^{n_1} x_2^{n_2}\cdots x_K^{n_K}\delta\left(\sum_{i=1}^{K} n_i - N\right), \tag{8.5}$$

where

$$x_i = \exp(-\beta\epsilon_i). \tag{8.6}$$

The sum on the right side in (8.5) is the multinomial sum which can be evaluated using the multinomial summation formula (D.8) to obtain

$$Z_{NC} = \frac{1}{N!}\left(\sum_{i=1}^{K} x_i\right)^N. \tag{8.7}$$

This shows that

$$Z_{NC} = \frac{1}{N!}Z_{1C}^N, \tag{8.8}$$

i.e. apart from the multiplying factor of $1/N!$, the N-particle partition function of distinguishable particles is a product of single particle partition functions.

We have thus at hand an analytic expression for the canonical partition function of non-interacting distinguishable particles in terms of single particle energies. In the following we explore its thermodynamic predictions.

8.2.1 Classical Ideal Gas

We will see that the partition function (8.7) predicts same thermodynamic properties as does the partition function corresponding to the phase space distribution function of the classical gas of non-interacting molecules studied in Chap. 7. Since their partition function describes thermodynamic properties of the classical gas, the distinguishable particles are often said to be *classical particles*.

The distinguishability of particles in a non-interacting classical gas can be understood if we recall from the discussion following (7.21) that the classical description is valid if the thermal wavelength of the particles is smaller than their average separation and that the thermal wavelength is a measure of their de Broglie wavelength. The de Broglie wavelength smaller than the separation between particles means their de Broglie waves do not overlap which means they can be distinguished. For further discussion see Sect. 8.4.

With $E_{n,N}$ given by (8.1), the probability (6.30) in the present case reads

$$p_{NC}(\{n_j\}) = \frac{g(\{n_j\})}{Z_{NC}} \prod_{j=1}^{K} x_j^{n_j}. \tag{8.9}$$

This stands for the joint probability of finding n_j particles in the energy level ϵ_j ($j = 1, 2, \ldots, K$). Some consequences of interest of (8.9) are:

1. The probability $p_{NC}(n_i)$ of finding n_i particles in the energy level ϵ_i, irrespective of numbers in other levels, by definition is

$$p_{NC}(n_i) = \sum_{\{n_j \neq n_i\}} p_{NC}(\{n_j\}). \tag{8.10}$$

The notation $\{n_j \neq n_i\}$ stands for the set $(n_1, n_2, \ldots, n_{i-1}, n_{i+1}, \ldots, n_K)$, i.e. for the set of the $n'_j s$ excluding $n_j = n_i$. Recalling (8.9) we have

$$\sum_{\{n_j \neq n_i\}} p_{NC}(\{n_j\}) = \frac{x_i^{n_i}}{n_i! Z_{NC}} \sum_{\{n_j \neq n_i\}} \prod_{j \neq i}^{K} \frac{x_j^{n_j}}{n_j!}, \quad \sum_{j \neq i}^{K} n_j = N - n_i. \tag{8.11}$$

On carrying the multinomial sum in the equation above, the $p_{NC}(n_i)$ in (8.10) turns out to be given by

$$p_{NC}(n_i) = \frac{x_i^{n_i}}{n_i!(N - n_i)! Z_{NC}} \left(\sum_{j \neq i}^{K} x_j\right)^{N-n_i}. \tag{8.12}$$

Recalling the expression (8.7) for Z_{NC}, the equation above reads

$$p_{NC}(n_i) = \frac{x_i^{n_i} N!}{n_i!(N - n_i)!} \left(\sum_{j \neq i}^{K} x_j\right)^{N-n_i} \left(\sum_{k=1}^{K} x_k\right)^{-N}. \tag{8.13}$$

Verify that the sum of $p_{NC}(n_i)$ over n_i is unity, as it should be.

2. Average number of particles in the level of energy ϵ_i is, by definition,

$$\bar{n}_{iC} = \sum_{n_i} n_i \, p_{NC}(n_i). \tag{8.14}$$

We leave it as an exercise to show that

$$\bar{n}_{iC} = \frac{N x_i}{\left(\sum_{i=1}^{K} x_i\right)}. \tag{8.15}$$

Exercises

Ex. 8.1. Using the formula for configurational degeneracy, show that the number of ways of distributing N distinguishable molecules in K distinct energy levels is K^N. This is the same as the result arrived at in Sect. 4.7.2 in a different way. Hint: The desired result is obtained by summing the degeneracy factor (8.3) over the n_k's and using the multinomial formula.

8.3 Quantum Particles

If identical particles are considered indistinguishable even for the purpose of distribution in energy levels, they are called *quantum particles*. That is because the concept of indistinguishability arises in the quantum theory. The configurational degeneracy in the quantum case is $g(\{n_i\}) = 1$. There are, however, restrictions on the number of particles that can occupy the same energy level based on the following classification:

1. **Bosons**: Those are the particles having zero or integral spin quantum number. A level may be occupied by any number of bosons. Hence n_i can take all values between 0 and N so that the partition function Z_{NB} for Bosons is same as Z_N in (8.2) with $g(\{n_i\}) = 1$:

$$Z_{NB} = \sum_{\{n_i\}} \exp\left(-\beta \sum_{i=1}^{K} n_i \epsilon_i\right) \delta\left(\sum_{i=1}^{K} n_i - N\right). \tag{8.16}$$

Unlike the case of classical particles, the sum above cannot be expressed in closed analytic form in general.

2. **Fermions**: Those are the particles having half-odd integral spin quantum number. There cannot be more than one Fermion in any level. Hence possible values of n_i for Fermions are $n_i = 0, 1$. Consequently the partition function Z_{NF} for Fermions is obtained by restricting the values of the n_i's in (8.2) to (0, 1) so that, with $g(\{n_i\}) = 1$,

$$Z_{NF} = \sum_{\{n_i=0,1\}} \exp\left(-\beta \sum_{i=1}^{K} n_i \epsilon_i\right) \delta\left(\sum_{i=1}^{K} n_i - N\right). \qquad (8.17)$$

Unlike the case of classical particles, the sum above cannot be expressed in the closed analytic form in general.

Though, as stated above, we cannot construct Z_N analytically for a quantum gas, as shown below, analytic expression for the grand canonical partition function can be derived for the gas of quantum particles.

Exercises

Ex. 8.2. Derive the canonical partition function for N bosons distributed in levels of energy $\pm\epsilon$. Hint: The expression (8.16) for the canonical partition function for N Bosons in the present case reads

$$Z_{NB} = \sum_{n_2=0}^{N} \sum_{n_1=0}^{N} \exp\{-\beta(-n_1\varepsilon + n_2\varepsilon)\}\delta(n_1 + n_2 - N)$$

$$= \exp(-\beta N\varepsilon) \sum_{n_1=0}^{N} \exp(2\beta n_1\varepsilon)$$

$$= \frac{\exp\{\beta(N+1)\varepsilon\} - \exp\{-\beta(N+1)\varepsilon)\}}{\exp(\beta\varepsilon) - \exp(-\beta\varepsilon)}. \qquad (8.18)$$

Ex. 8.3. Consider the gas of non-interacting particles having single particle energy levels $\epsilon_1, \epsilon_2, \ldots, \epsilon_K$. Determine the canonical partition function for the systems of (i) one particle and (ii) two particles for the case of Bose as well as Fermi particles. The desired results here are derived in Ex. 8.7 by starting from the grand canonical partition function. Hint: The canonical partition function for Bosons is given by (8.16). If there is only one particle, the possible values of $\{n_i\}$ are $n_i = 1$, $n_k = 0$ for $k \neq i$ so that

$$Z_{1B} = \sum_{i=1}^{K} \exp(-\beta\epsilon_i). \qquad (8.19)$$

For two particles, the sum in (8.16) is expressible in the form

$$Z_{2B} = \sum_{i=1}^{K} \exp(-2\beta\epsilon_i) + \sum_{i<j=1}^{K} \exp\{-\beta(\epsilon_i + \epsilon_j)\}. \qquad (8.20)$$

The first term in the equation above corresponds to two particles occupying the same state and the second term is for the particles occupying different states. Since interchange of particles in different states does not change the state, the sum in the second term is restricted to retain only one of the terms symmetric under the exchange of the labels i, j. Equation (8.20) may be rewritten as

$$Z_{2B} = \sum_{i=1}^{K} \exp(-2\beta\epsilon_i) + \frac{1}{2} \left(\sum_{i,j=1}^{K} \exp\{-\beta(\epsilon_i + \epsilon_j)\} - \sum_{i=1}^{K} \exp(-2\beta\epsilon_i) \right). \quad (8.21)$$

Hence

$$Z_{2B} = \frac{1}{2} \left(\sum_{i=1}^{K} \exp(-\beta\epsilon_i) \right)^2 + \frac{1}{2} \sum_{i=1}^{K} \exp(-2\beta\epsilon_i). \quad (8.22)$$

The canonical partition function for Fermions is given by (8.17). If there is only one particle, the possible values of $\{n_i\}$ in it are $n_i = 1$, $n_k = 0$ for $k \neq i$. Hence

$$Z_{1F} = \sum_{i=1}^{K} \exp(-\beta\epsilon_i). \quad (8.23)$$

Since two Fermions cannot occupy the same state, the partition function for two Fermions will have only the terms corresponding to two particles occupying different energy levels so that

$$Z_{2F} = \sum_{i<j=1}^{K} \exp\{-\beta(\epsilon_i + \epsilon_j)\}. \quad (8.24)$$

This may be rewritten as

$$Z_{2F} = \frac{1}{2} \left(\sum_{i=1}^{K} \exp(-\beta\epsilon_i) \right)^2 - \frac{1}{2} \sum_{i=1}^{K} \exp(-2\beta\epsilon_i). \quad (8.25)$$

Note that, unlike the classical gas, apart from the factor $1/N!$, the N particle partition function for Bosons and Fermions is not a product of corresponding single particle partition functions.

Ex. 8.4. Using the partition functions derived in Ex. 8.3 find the energy of systems of one and two Bosons and Fermions.

8.4   Grand Canonical Partition Function

We derive expressions for the grand canonical partition functions of the classical and quantum particles by expressing it in terms of the canonical partition function using (6.65).

8.4.1   Classical Particles

On substituting the expression (8.7) for the classical partition function Z_{NC} in (6.65) the classical grand partition function is found to be given by

$$Z_{GC} = \sum_{N=0}^{\infty} \frac{\exp(\alpha N)}{N!} \left(\sum_{i=1}^{K} \exp(-\beta \epsilon_i) \right)^N$$

$$= \exp \left(\exp(\alpha) \sum_{i=1}^{K} \exp(-\beta \epsilon_i) \right). \tag{8.26}$$

Hence

$$\ln(Z_{GC}) = \exp(\alpha) \sum_{i=1}^{K} \exp(-\beta \epsilon_i). \tag{8.27}$$

It is straightforward to see that the average number of particles is

$$\bar{N} \equiv \frac{\partial \ln(Z_{GC})}{\partial \alpha} = \ln(Z_{GC}). \tag{8.28}$$

We will evaluate (8.27) in Sect. 8.6 for the gas of particles in free space and show that it leads to the thermodynamic properties of an ideal gas.

8.4.2   Quantum Particles

On substituting in (6.65) the expression (8.2) for Z_N with $g(\{n_i\}) = 1$, the grand canonical partition function of quantum particles reads

$$Z_{GQ} = \sum_{N=0}^{\infty} \sum_{\{n_i\}} \exp(\alpha N) \exp \left(-\beta \sum_{i=1}^{K} n_i \epsilon_i \right) \delta \left(\sum_{i=1}^{K} n_i - N \right). \tag{8.29}$$

It is shown in Ex. 8.5 that the equation above is reducible to the form

$$Z_{GQ} = \sum_{\{n_k\}=\{0\}}^{\{\infty\}} \prod_{j=1}^{K} \exp\left\{-n_j(\beta\epsilon_j - \alpha)\right\}.$$ (8.30)

The probability for the system to have N particles, and energy $E_{n,N}$ is given by (6.59). In the present case, with $E_{n,N}$ as in (8.1), the said probability is

$$p_Q(\{n_j\}, N) = \frac{\exp(N\alpha)}{Z_{GQ}} \prod_{j=1}^{K} \exp(-\beta n_j \epsilon_j).$$ (8.31)

This is the joint probability for the system to have N particles of which n_i are in the single particle energy level ϵ_i ($i = 1, 2, \ldots, K$). We apply the results above to systems of Bosons and those of Fermions.

Bosons

1. Since in this case there is no restriction on the values of the n_i's, the (8.30) reads ($\alpha = \mu\beta$),

$$Z_{GB} = \prod_{j=1}^{K} \sum_{n_j=0}^{\infty} \exp\left[-\beta\left\{\epsilon_j - \mu\right\} n_j\right].$$ (8.32)

If $\mu < \epsilon_j$ for all j, which means if $\mu < \epsilon_1$ where ϵ_1 is the single particle ground state energy, then each of the sums in the equation above is convergent and can be carried to yield

$$Z_{GB} = \prod_{j=1}^{K} \left[1 - \exp\left\{-\beta(\epsilon_j - \mu)\right\}\right]^{-1}, \qquad \mu < \epsilon_1. \qquad (8.33)$$

Hence

$$\ln(Z_{GB}) = -\sum_{j=1}^{K} \ln\left[1 - \exp\left\{-\beta\left(\epsilon_j - \mu\right)\right\}\right], \qquad \mu < \epsilon_1. \qquad (8.34)$$

We have thus at hand an analytic expression for the grand partition function for a gas of non-interacting bosons.
2. The occupation probability (8.31) for bosons reads

$$p_B(\{n_j\}, N) = \frac{\exp(N\alpha)}{Z_{GB}} \prod_{j=1}^{K} \exp(-\beta n_j \epsilon_j).$$ (8.35)

3. The probability $p_B(n_i)$ for the system to have n_i bosons in the level of energy ϵ_i, irrespective of the numbers in other levels and the number of bosons is, by definition,

$$p_B(n_i) = \sum_{N=0}^{\infty} \sum_{\{n_j \neq n_i\}} p_B(\{n_j\}, N), \tag{8.36}$$

where the sum over the occupation numbers excludes that over the occupation number n_i of the level of energy ϵ_i whose probability of occupation is sought. We leave it as an exercise to show that

$$p_B(n_i) = \exp\left(-\beta(\epsilon_i - \mu)n_i\right)\left(1 - \exp\{-\beta(\epsilon_i - \mu)\}\right). \tag{8.37}$$

4. Average number of particles in energy level ϵ_i is

$$\bar{n}_{iB} = \sum_{n_i=0}^{\infty} n_i p_B(n_i) = \frac{1}{\exp\{\beta(\epsilon_i - \mu)\} - 1}, \tag{8.38}$$

where the summation has been carried using (8.37) for $p_B(n_i)$.

Fermions

1. For fermions, $n_i = 0, 1$. Hence, with $\alpha = \mu\beta$, (8.30) reads

$$Z_{GF} = \prod_{i=1}^{K} \sum_{n_i=0,1} \exp\{-\beta(\epsilon_i - \mu)n_i\}. \tag{8.39}$$

On carrying each of the summations above we get

$$Z_{GF} = \prod_{i=1}^{K} \left[1 + \exp\{-\beta(\epsilon_i - \mu)\}\right]. \tag{8.40}$$

Hence

$$\ln(Z_{GF}) = \sum_{i=1}^{K} \ln\left[1 + \exp\{-\beta(\epsilon_i - \mu)\}\right]. \tag{8.41}$$

We thus have at hand an analytic expression for the grand partition function for a gas of non-interacting fermions.

2. The occupation probability (8.31) for fermions reads

$$p_F(\{n_i\}, N) = \frac{\exp(N\alpha)}{Z_{GF}} \prod_{j=1}^{K} \exp(-\beta n_j \epsilon_j). \tag{8.42}$$

3. The probability to have n_i ($n_i = 0, 1$) fermions in the level of energy ϵ_i irrespective of numbers in other levels and total number is given by

$$p_F(n_i) = \frac{\exp\{-\beta(\epsilon_i - \mu)n_i\}}{1 + \exp\{-\beta(\epsilon_i - \mu)\}}, \qquad n_i = 0, 1. \tag{8.43}$$

Proof of the equation above is left as an exercise.

4. Average number of fermions in the energy level ϵ_i is given by

$$\bar{n}_{iF} = \sum_{n_i} n_i \, p_F(n_i) = \frac{1}{\exp\{\beta(\epsilon_i - \mu)\} + 1}. \tag{8.44}$$

Summary

The grand partition function, the probability to have n_i particles in the energy level ϵ_i and average number of particles in that level for Bosons and Fermions are given respectively by ($z = \exp(\alpha) = \exp(\mu\beta)$)

$$\ln(Z_{G\eta}) = -\eta \sum_{i=1}^{K} \ln\{1 - \eta z \exp(-\beta\epsilon_i)\}, \tag{8.45}$$

$$p_\eta(n_i) = \exp\{-\beta(\epsilon_i - \mu)n_i\}\left[1 - \eta \exp\{-\beta(\epsilon_i - \mu)\}\right]^\eta, \tag{8.46}$$

$$\bar{n}_{i\eta} = [\exp\{\beta(\epsilon_i - \mu)\} - \eta]^{-1}, \tag{8.47}$$

where $\eta = 1$ gives corresponding quantity for Bosons, and $\eta = -1$ that for Fermions with the understanding that $n_i = 0, 1$ for Fermions whereas n_i is unrestricted for Bosons.

Exercises

Ex. 8.5. Show that

$$\sum_{N=0}^{\infty} \sum_{n_1, n_2, n_3} z^N x_1^{n_1} x_2^{n_2} x_3^{n_3} \delta(N - n_1 - n_2 - n_3)$$

$$= \sum_{n_1=0}^{\infty} \sum_{n_2=0}^{\infty} \sum_{n_3=0}^{\infty} (zx_1)^{n_1} (zx_2)^{n_2} (zx_3)^{n_3}. \tag{8.48}$$

Hint: We have

$$I \equiv \sum_{N=0}^{\infty} \sum_{n_1,n_2,n_3} z^N x_1^{n_1} x_2^{n_2} x_3^{n_3} \delta (N - n_1 - n_2 - n_3)$$

$$= \sum_{N=0}^{\infty} \sum_{n_1=0}^{N} \sum_{n_2=0}^{N-n_1} \sum_{n_3=0}^{N-n_1-n_2} z^N x_1^{n_1} x_2^{n_2} x_3^{n_3} \delta (N - n_1 - n_2 - n_3)$$

$$= \sum_{n_1=0}^{\infty} \sum_{N=n_1}^{\infty} \sum_{n_2=0}^{N-n_1} \sum_{n_3=0}^{N-n_1-n_2} z^N x_1^{n_1} x_2^{n_2} x_3^{n_3} \delta (N - n_1 - n_2 - n_3). \qquad (8.49)$$

Let $N - n_1 \to N$ in the sum over N to rewrite the equation above as

$$I = \sum_{n_1=0}^{\infty} \sum_{N=0}^{\infty} \sum_{n_2=0}^{N} \sum_{n_3=0}^{N-n_2} z^{N+n_1} x_1^{n_1} x_2^{n_2} x_3^{n_3} \delta (N - n_2 - n_3)$$

$$= \sum_{n_1=0}^{\infty} \sum_{n_2=0}^{\infty} \sum_{N=n_2}^{\infty} \sum_{n_3=0}^{N-n_2} z^N (zx_1)^{n_1} x_2^{n_2} x_3^{n_3} \delta (N - n_2 - n_3). \qquad (8.50)$$

Let $N - n_2 \to N$ in the sum over N to rewrite the equation above as

$$I = \sum_{n_1=0}^{\infty} \sum_{n_2=0}^{\infty} \sum_{N=0}^{\infty} \sum_{n_3=0}^{N} z^N (zx_1)^{n_1} (zx_2)^{n_2} x_3^{n_3} \delta (N - n_3)$$

$$= \sum_{n_1=0}^{\infty} \sum_{n_2=0}^{\infty} \sum_{n_3=0}^{\infty} (zx_1)^{n_1} (zx_2)^{n_2} (zx_3)^{n_3}. \qquad (8.51)$$

The procedure outlined above may be generalized to an arbitrary number of n_i's to get the identity (8.30).

Ex. 8.6.   Consider a system of N Bosons distributed in levels of energy $\pm \epsilon$. (a) Using the expression for Z_{NB} derived in Ex. 8.2, construct the corresponding grand canonical function Z_{GB}. (b) Derive Z_{GB} directly using the general formula for the grand partition function for Bosons. Hint: (a) The grand partition function is given by (with $\alpha = \mu\beta$)

$$Z_{GB} = \sum_{N=0}^{\infty} \exp(\mu\beta N) Z_{NB} = \frac{Z}{\exp(\beta\epsilon) - \exp(-\beta\epsilon)}, \qquad (8.52)$$

where

$$Z = \sum_{N=0}^{\infty} \Big[\exp(\beta\epsilon) \exp\{(\epsilon + \mu)N\beta\}$$

$$- \exp(-\beta\epsilon) \exp\{-(\epsilon - \mu)N\beta\} \Big]. \qquad (8.53)$$

First sum above will converge only if $\mu < -\epsilon$, in which case second sum will also converge. This is in accordance with the general result that the chemical potential of non-interacting Boson gas is less than the single particle ground state energy (see (8.33)). We let $\mu \rightarrow -|\mu|$ so that $\mu > \epsilon$ to obtain

$$Z = \frac{\exp(\beta\epsilon) - \exp(-\beta\epsilon\}}{[1 - \exp\{-(|\mu| - \epsilon)\beta\}][1 - \exp\{-(|\mu| + \epsilon)\beta\}]}. \tag{8.54}$$

Substitution of the equation above in (8.52) yields the desired expression:

$$Z_{GB} = \frac{1}{[1 - \exp\{-(|\mu| - \epsilon)\}][1 - \exp\{-(|\mu| + \epsilon)\beta\}]}. \tag{8.55}$$

General formula for Bosonic grand partition function is given by (8.45) for $\eta = 1$. It will yield the same result as in (8.55) with $\epsilon_1 = -\epsilon$, $\epsilon_2 = \epsilon$ and $\mu \rightarrow -|\mu|$.

Ex. 8.7. Consider the gas of non-interacting particles having single particle energy levels $\epsilon_1, \epsilon_2, \ldots, \epsilon_K$. Starting from the grand canonical partition function, determine the canonical partition function for the systems of (i) one particle and (ii) two particles for the case of Bose as well as Fermi particles. The desired results here are derived directly in Ex. 8.3 using the definition of the canonical partition function. Hint: Recall that

$$Z_G(z) = \sum_{N=0}^{\infty} z^N Z_N, \quad Z_N = \frac{1}{N!} \frac{d^N Z_G}{dz^N}\bigg|_{z=0}. \tag{8.56}$$

The grand canonical partition function for free Bosons and Fermions is given by (8.45):

$$\ln(Z_{G\eta}) = -\eta \sum_{i=1}^{K} \ln\{1 - \eta z \exp(-\beta\epsilon_i)\}. \tag{8.57}$$

Note that

$$Z_{G\eta}(0) = 1. \tag{8.58}$$

With $Z_G(z) \rightarrow Z_{G\eta}(z)$, use (8.56) to evaluate the canonical partition function $Z_{N\eta}$ for different values of N.

(i) $N = 1$. In this case

$$Z_{1\eta} = \frac{dZ_{G\eta}(z)}{dz}\bigg|_{z=0} = Z_{G\eta}(z)\frac{d\ln Z_{G\eta}(z)}{dz}\bigg|_{z=0} = \sum_{i=1}^{K} \exp(-\beta\epsilon_i). \tag{8.59}$$

(ii) $N = 2$. In this case

$$
\begin{aligned}
Z_{2\eta} &= \frac{1}{2}\frac{d^2 Z_{G\eta}(z)}{dz^2}\bigg|_{z=0} = \frac{1}{2}\frac{d}{dz}\left[Z_{G\eta}(z)\frac{d\ln Z_{G\eta}(z)}{dz}\right]\bigg|_{z=0} \\
&= \frac{1}{2}\left[Z_{G\eta}(z)\left(\frac{d\ln Z_{G\eta}(z)}{dz}\right)^2 + Z_{\eta}(z)\frac{d^2\ln Z_{G\eta}(z)}{dz^2}\right]\bigg|_{z=0} \\
&= \frac{1}{2}\left[\left(\sum_{i=1}^{K}\exp(-\beta\epsilon_i)\right)^2 + \eta\sum_{i=1}^{K}\exp(-2\beta\epsilon_i)\right].
\end{aligned}
\tag{8.60}
$$

The results derived above are the same as those derived in (Ex. 8.3).

8.4.3 Classical Limit of Quantum Distributions

When $z \ll 1$ then, under the approximation $\ln(1+x) \approx x$ for $x \ll 1$ and noting that $\eta^2 = 1$, (8.45) reduces to

$$
\ln(Z_{G\eta}) = \exp(\alpha)\sum_{i=1}^{K}\exp(-\beta\epsilon_i).
\tag{8.61}
$$

This is same as the grand canonical partition function (8.27) of the classical gas. Hence, both the quantum distributions reduce to the classical one under the condition $z \ll 1$. To understand its meaning in terms of observables, recall the expression (8.28) for the average number of particles in the classical gas to recast the condition $z \ll 1$ in the form

$$
\bar{N} \ll \sum_{i}\exp(-\beta\epsilon_i).
\tag{8.62}
$$

Written in this form, the condition under which quantum gases behave like a classical gas is useful for relating it with observable entities. We will discuss the meaning of (8.62) in Sect. 8.8 for gas in free space.

We thus see that it is not that there are two categories of particles: distinguishable and indistinguishable. The particles are fundamentally indistinguishable. It is only under certain conditions, like the one in (8.62), that their thermodynamic behavior is akin to that of distinguishable particles. As was pointed out in Sect. 8.2.1 and will be brought out again in Sect. 8.8 by examining the condition (8.62) when particles are in free space, the distinguishability shows up when the thermal wavelength of particles, a measure of their de Broglie wavelength, becomes smaller than their average separation.

Next we determine single particle energy levels in free space and find explicit expressions for the partition functions of different classes of particles.

8.5 Single Particle Energy Levels in Free Space

In this section we address the problem of evaluating the sum over energy levels in (8.45) for the gas of non-interacting molecules in a box of large volume. The energy levels of a particle in a box are discrete. However, in the limit of large volume, the separation between successive energy levels turns out to be so small that the discrete sum may be transformed, to good approximation, to an integral by the replacements,

$$\epsilon_i \to \epsilon, \qquad \sum_i f(\epsilon_i) \to \int f(\epsilon) D(\epsilon) d\epsilon, \tag{8.63}$$

where $D(\epsilon)$, called the *density of states*, denotes the number of states in the unit interval of energy. The problem, addressed below, then reduces to finding $D(\epsilon)$ which depends on (i) the functional relationship between energy and momentum and (ii) the dimensionality of the system.

8.5.1 Determining Density of States

Let us assume that energy is related with momentum by the relation

$$\epsilon = \epsilon(p), \qquad p = p(\epsilon), \qquad p \equiv |\mathbf{p}|. \tag{8.64}$$

For example,

1. For a particle of mass m moving in free space with non-relativistic speed,

$$\epsilon(p) = \frac{p^2}{2m}. \tag{8.65}$$

2. For a particle of mass m moving in free space with relativistic speed,

$$\epsilon(p) = \sqrt{p^2 c^2 + m_0^2 c^4}. \tag{8.66}$$

3. In ultra-relativistic limit $m_0 c^2 \ll pc$ which corresponds to the particle energy being much greater than its rest mass energy, (8.66) reduces to

$$\epsilon(p) = pc. \tag{8.67}$$

4. The relation (8.67) applies also to photons as they are massless.

We find density of states for motion in different dimensional spaces.

Density of States in Three-Dimensional Motion

Consider a particle moving in a box whose three sides are of length L_x, L_y, L_z respectively. The wave function $\psi(\mathbf{r})$ of such a particle obeying periodic boundary conditions,

$$\psi(0, y, z) = \psi(L_x, y, z), \quad \psi(x, 0, z) = \psi(x, L_y, z),$$
$$\psi(x, y, 0) = \psi(x, y, L_z) \tag{8.68}$$

is known to be given by

$$\psi(\mathbf{r}) = C \exp(i\mathbf{p} \cdot \mathbf{r}/\hbar), \quad C = \frac{1}{\sqrt{L_x L_y L_z}}, \tag{8.69}$$

where, due to (8.68),

$$p_x = \frac{n_x}{L_x}h, \quad p_y = \frac{n_y}{L_y}h, \quad p_z = \frac{n_z}{L_z}h, \tag{8.70}$$

with $n_x, n_y, n_z = 0, \pm 1, \pm 2 \ldots.$

Consider the problem of evaluating sum over i of some function $f(\epsilon_i)$. The summation index i stands for different states. In the present case, the states are labeled by the numbers (n_x, n_y, n_z) due to which sum over i is that over (n_x, n_y, n_z):

$$I = \sum_i f(\epsilon_i) \equiv \sum_{n_x, n_y, n_z} f(n_x, n_y, n_z). \tag{8.71}$$

For large (n_x, n_y, n_z), the separation between successive levels becomes small enough to permit replacement of sum by integral:

$$\sum_{n_x, n_y, n_z} f(n_x, n_y, n_z) = \int_{-\infty}^{\infty} dn_x \int_{-\infty}^{\infty} dn_y \int_{-\infty}^{\infty} f(n_x, n_y, n_z) dn_z$$

$$= \frac{L_x L_y L_z}{h^3} \int_{-\infty}^{\infty} dp_x \int_{-\infty}^{\infty} dp_y \int_{-\infty}^{\infty} f(p_x, p_y, p_z) dp_z, \tag{8.72}$$

where the second line is due to (8.70). The sum over (n_x, n_y, n_z) is thus transformed to an integral by means of the relation,

$$\sum_{n_x, n_y, n_z} f(n_x, n_y, n_z) = \frac{V}{h^3} \int f(p_x, p_y, p_z) \, \mathrm{d}^3 p, \tag{8.73}$$

where V is the volume of the box.

We are interested in evaluating the sum (8.71) where $f(p_x, p_y, p_z)$ is a function of energy ϵ which, in turn, is a function of $p \equiv |\mathbf{p}|$. The integral over the momentum in (8.73) is then evaluated by transforming it to that over ϵ as follows. In spherical polar coordinates in the momentum space, we have $\mathrm{d}^3 p = p^2 \mathrm{d}p \sin(\theta) \mathrm{d}\theta \mathrm{d}\phi$, $(0 \le \theta < \pi$, $0 \le \phi < 2\pi, 0 \le p \le \infty)$ the angular integration in (8.73) can be performed leading to

$$\frac{V}{h^3} \int f(p_x, p_y, p_z) \, \mathrm{d}^3 p = \frac{4\pi V}{h^3} \int_0^\infty f(\epsilon(p)) \, p^2 \mathrm{d}p = \int_0^\infty f(\epsilon) \, D(\epsilon) \mathrm{d}\epsilon, \tag{8.74}$$

where we have transformed integration over momentum to that over energy with the density of states given by

$$D(\epsilon) = \left(\frac{4\pi V}{h^3}\right) \left(p^2 \frac{\mathrm{d}p}{\mathrm{d}\epsilon}\right). \tag{8.75}$$

We evaluate below the expression above for the density of states for various commonly encountered functional forms of $p(\epsilon)$.

1. For non-relativistic motion of a free particle of mass m, the use of the relation (8.65) between ϵ and p in (8.75) yields

$$D(\epsilon) = A\sqrt{\epsilon}, \quad A = 2\pi V \left(\frac{2m}{h^2}\right)^{3/2}. \tag{8.76}$$

2. For a massless particle, or for particles moving with relativistic speed, with $p(\epsilon)$ given by (8.67), (8.75) yields

$$D(\epsilon) = \frac{4\pi V}{h^3 c^3} \epsilon^2. \tag{8.77}$$

Similar relation is obeyed by phonons in an isotropic solid but with c replaced by the velocity c_s of sound.

Density of States in Two-Dimensional Motion

If a particle is constrained to move in two-dimensional space, say, in the x–y plane, then it is described by the wave function (8.69) but with $\mathbf{r} = x\mathbf{e}_x + y\mathbf{e}_y$, $C = 1/\sqrt{L_x L_y}$ and $\mathbf{p} = p_x\mathbf{e}_x + p_y\mathbf{e}_y$ with (p_x, p_y) given in terms of integers n_x, n_y as in (8.70). The summation over n_x, n_y is then approximated by the integral as follows:

$$\sum_{n_x, n_y} f(n_x, n_y) = \frac{L_x L_y}{h^2} \int_{-\infty}^{\infty} dp_y \int_{-\infty}^{\infty} f(p_x, p_y) dp_x$$

$$= \frac{\sigma}{h^2} \int f(p_x, p_y) \, d^2 p, \qquad (8.78)$$

where $\sigma = L_x L_y$ is the area of the region occupied by the gas. The $f(p_x, p_y)$ in our case shall be a function of $\epsilon(p)$, $p \equiv |\mathbf{p}|$. With $d^2 p = p\,dp\,d\phi$, $(0 \le \phi < 2\pi)$, the angular integration in (8.78) can be performed to obtain

$$\sum_{n_x, n_y} f(\epsilon(p)) = \frac{2\pi\sigma}{h^2} \int_0^{\infty} f(\epsilon(p)) \, p \, dp \equiv \int_0^{\infty} f(\epsilon) \, D(\epsilon) d\epsilon, \qquad (8.79)$$

with the density of states given by

$$D(\epsilon) = \left(\frac{2\pi\sigma}{h^2} \right) \left(p \frac{dp}{d\epsilon} \right). \qquad (8.80)$$

1. For non-relativistic motion of a free particle of mass m, with $p(\epsilon)$ given by (8.65), (8.80) yields

$$D(\epsilon) = \frac{2m\pi\sigma}{h^2}. \qquad (8.81)$$

The density of states in this case is independent of energy.

2. For a massless particle, or for the particles moving with relativistic speed, the $p(\epsilon)$ is given by (8.67) on substituting which in (8.80) we obtain

$$D(\epsilon) = \frac{2\pi\sigma}{c^2 h^2} \epsilon. \qquad (8.82)$$

Density of States in One-Dimensional Motion

If a particle is constrained to move in one dimension, say, along the x-axis $(0 \le x \le L)$, then it is described by the wave function (8.69) but with $\mathbf{r} = x\mathbf{e}_x$, $C = \sqrt{1/L}$, $\mathbf{p} = p\mathbf{e}_x$ with $p = hn/L$. The summation over n may then be approximated by the integral as follows:

$$\sum_n f(n) = \frac{L}{h} \int_{-\infty}^{\infty} f(\epsilon) \, dp = \frac{2L}{h} \int_0^{\infty} f(\epsilon(p)) \, dp$$

$$= \int_0^{\infty} f(\epsilon) \, D(\epsilon) d\epsilon, \qquad (8.83)$$

with the density of states given by

$$D(\epsilon) = \left(\frac{2L}{h}\right)\left(\frac{\mathrm{d}p}{\mathrm{d}\epsilon}\right). \tag{8.84}$$

For non-relativistic motion of a free particle of mass m, the use of the relation (8.65) between ϵ and p in (8.80) yields

$$D(\epsilon) = \sqrt{\frac{2m}{h^2}}\, L\epsilon^{-1/2}. \tag{8.85}$$

Summary

The sum over states of a function of closely spaced discrete energies is converted to an integral by the relation:

$$\sum_i f(\epsilon_i) = \int f(\epsilon)D(\epsilon)\mathrm{d}\epsilon, \tag{8.86}$$

where $D(\epsilon)$ is the density of states given for different spatial dimensions by

$$D(\epsilon) = A_d p^{d-1}\frac{\mathrm{d}p}{\mathrm{d}\epsilon}, \tag{8.87}$$

where d is the dimension of space in which particles are moving with

$$A_3 = \frac{4\pi V}{h^3}, \quad A_2 = \frac{2\pi\sigma}{h^2}, \quad A_1 = \frac{2L}{h}. \tag{8.88}$$

In the expressions above V, σ, L are respectively the volume, area, and length of space in which particles move. Several widely encountered energy-momentum relations are of the form:

$$\epsilon = C_n p^n. \tag{8.89}$$

The expression (8.87) then reads

$$D_{nd}(\epsilon) = A_{nd}\epsilon^{(d-n)/n}, \quad A_{nd} = \frac{A_d}{nC_n^{d/n}}. \tag{8.90}$$

This is the density of states for a d-dimensional system when the energy-momentum relation is of the form (8.89). The expressions for D_{nd} for the values of (n, d) of common interest are tabulated in Table 8.1.

We apply the results derived above to investigate thermodynamic properties of classical and quantum gases in free space.

Table 8.1 Density of states D_{nd} given by (8.90) for different motion types and spatial dimensions

Motion type	Spatial Dimension	$\epsilon = C_n p^n$	D_{nd}
Non-Relativistic	3	$\dfrac{p^2}{2m}$	$2\pi V \left(\dfrac{2m}{h^2}\right)^{3/2} \sqrt{\epsilon}$
Ultra-Relativistic	3	cp	$\dfrac{4\pi V}{h^3 c^3} \epsilon^2$
Non-Relativistic	2	$\dfrac{p^2}{2m}$	$\dfrac{2m\pi\sigma}{h^2}$
Ultra-Relativistic	2	cp	$\dfrac{2\pi\sigma}{h^2 c^2} \epsilon$
Non-Relativistic	1	$\dfrac{p^2}{2m}$	$\sqrt{\dfrac{2m}{h^2}} L \epsilon^{-1/2}$
Ultra-Relativistic	1	cp	$\dfrac{2L}{hc}$

8.6   Thermodynamics of Ideal Classical Gas

We studied the thermodynamics of the gas of free non-interacting molecules in Chap. 7 in phase space formalism. In Sects. 8.1 and 8.4 we derived expressions for the canonical and grand canonical partition functions in the quantum formalism. In this section we investigate thermodynamics of the gas of non-interacting molecules in a box of large volume in the quantum formalism assuming the molecules to be distinguishable. We will see that, if the internal motion of the molecules is ignored, then the quantum formalism is equivalent with the phase space formalism. However, the quantum treatment of the internal degrees of freedom leads to results at variance with those predicted by the phase space approach.

8.6.1   Translational Motion

The partition function of the gas of N free non-interacting distinguishable molecules is given by (8.7). If the molecules are in a large three-dimensional box then the sum can be replaced by an integral using the correspondence (8.86) if the molecular speeds are non-relativistic:

$$\sum_i \exp(-\beta\epsilon_i) = \int_0^\infty \exp(-\beta\epsilon) D(\epsilon)\mathrm{d}\epsilon = \frac{V}{\lambda_T^3}, \qquad (8.91)$$

where in writing the second equation we have used the expression (8.76) for $D(\epsilon)$ and (7.6) of thermal wavelength λ_T. On combining this with (8.7) we see that the resulting partition function is same as (7.18) derived using phase space approach.

Similarly, on using (8.91), in (8.27) we get the same grand partition function as the expression (7.28) for the grand partition function derived using the method of phase space distribution function.

8.6.2 Internal Motion

We have constructed the partition function for the translational motion of molecules. We construct now the partition function taking in to account their internal motion.

To that end, we need to know the molecular energy levels. We assume that they are constituted by two components: the energies $\{\epsilon_i^{\text{trans}}\}$ due to the translational motion of the molecular center of mass and $\{\epsilon_i^{\text{int}}\}$ due to its internal motion. The internal motion could be rotational, vibrational, or even electronic transitions between atomic levels. The single particle energy levels are then specified in terms of two indices, one referring to the kinetic energy, and the other to the internal energy:

$$\epsilon_{ij} = \epsilon_i^{\text{trans}} + \epsilon_j^{\text{int}}, \quad i = 1, 2, \ldots; \quad j = 1, 2, \ldots, M, \tag{8.92}$$

where we have assumed that there are M internal energy levels. We construct the partition function using (8.7) which involves finding the sum over the energy states. Since the energy states are now identified by two indices, the sum therein reads

$$\sum_i \sum_{j=1}^M \exp(-\beta \epsilon_{ij}) = \left(\sum_i \exp(-\beta \epsilon_i^{\text{trans}}) \right) \left(\sum_{j=1}^M \exp(-\beta \epsilon_j^{\text{int}}) \right). \tag{8.93}$$

Consequently the partition function (8.7) assumes the form

$$Z_{NC} = Z_{NC}^{\text{trans}} Z_{NC}^{\text{int}}. \tag{8.94}$$

The translational motion has been investigated in the Sect. 8.6.1 above.

The partition function for the internal motion is

$$Z_{NC}^{\text{int}} = \left(\sum_{j=1}^M \exp(-\beta \epsilon_j^{\text{int}}) \right)^N. \tag{8.95}$$

In the following we evaluate the partition function Z_{NC}^{int} in (8.95) for internal motion of polyatomic molecules.

As discussed in Sect. 7.4, the internal energy of a molecule in its phase space description consists of contributions from its rotational and the vibrational degrees of freedom. Since they do not admit phase space description, the electronic transitions inside the atoms could not be included in the evaluation of the partition function by phase space approach. We now include the energy of electronic transitions within the atoms, calling it the electronic energy. The molecular energy levels are consequently expressed in terms of three indices, one each for rotational, vibrational, and electronic energy levels:

$$\epsilon_{ijk}^{\text{int}} = \epsilon_i^{\text{rot}} + \epsilon_j^{\text{vib}} + \epsilon_k^{\text{elect}}. \tag{8.96}$$

The partition function (8.95) for internal energy consequently reads

$$Z_{NC}^{\text{int}} = (Z_{\text{rot1}} Z_{\text{vib1}} Z_{\text{elect1}})^N, \tag{8.97}$$

where

$$Z_{\text{rot1}} = \sum_i \exp\left(-\beta \epsilon_i^{\text{rot}}\right), \quad Z_{\text{vib1}} = \sum_j \exp\left(-\beta \epsilon_j^{\text{vib}}\right),$$

$$Z_{\text{elect1}} = \sum_k \exp\left(-\beta \epsilon_k^{\text{elect}}\right) \tag{8.98}$$

are single molecule partition functions. In the following we evaluate Z_{rot1} and Z_{vib1} for a diatomic as well as for polyatomic molecules and show that they are very much different from their counterparts evaluated using the phase space approach. We also evaluate Z_{elect1} for a two-level system.

Rotational Motion: Diatomic Molecules

The classical Hamiltonian of a diatomic molecule describing its rotational motion is given by (7.55). The quantum Hamiltonian corresponding to it is

$$\hat{H}_{\text{rot}} = \frac{1}{2I}\hat{L}^2. \tag{8.99}$$

We know that the eigenvalues of the angular momentum operator $\hat{L}^2$ are $\hbar^2 L(L+1)$ where $L = 0, 1, \ldots \infty$. Hence the rotational energy levels of the molecule are

$$\epsilon_L = \frac{1}{2I}\hbar^2 L(L+1). \tag{8.100}$$

The eigenvalue of $\hat{L}^2$ corresponding to L is $2L + 1$-fold degenerate. Hence the single particle partition function is

$$Z_{\text{rot1}} = \sum_{L=0}^{\infty} (2L+1)\exp\{-\alpha L(L+1)\}, \quad \alpha = \frac{\hbar^2 \beta}{2I}. \tag{8.101}$$

It is not possible to perform the sum above analytically. However, we can see that, with the partition function given by (8.101), the internal energy is no longer proportional to T and consequently heat capacity is temperature dependent. We can study temperature dependence of Z_{rotl} by evaluating it approximately in the limit of low and high temperatures in comparison with what is called the characteristic temperature Θ_r of rotation. To define it, note that the separation between nearest levels is

$$\epsilon_{L+1} - \epsilon_L = \frac{\hbar^2(L+1)}{I}. \tag{8.102}$$

The characteristic temperature Θ_r of rotation is defined as half the temperature equivalent of the smallest rotational transition energy:

$$\Theta_r = \frac{1}{2k_B}(\epsilon_1 - \epsilon_0) = \frac{\hbar^2}{2Ik_B}. \tag{8.103}$$

The Θ_r is a measure of minimum thermal energy required to induce a transition between rotational levels. Using (8.103), the rotational partition function (8.101) may be rewritten as

$$Z_{\text{rotl}} = \sum_{L=0}^{\infty}(2L+1)\exp\{-(\Theta_r/T)L(L+1)\}. \tag{8.104}$$

We consider two cases: (1) $T << \Theta_r$ and (2) $T >> \Theta_r$.

1. Consider the case $T << \Theta_r$. In this case only a small number of terms will contribute to the sum in (8.104). Higher the value of Θ_r/T, smaller is the number of terms contributing significantly to the said sum. For example, for Hydrogen molecule, $\Theta_r \approx 85.3$ K. It may be seen that in that case first three terms in the sum would contribute to within 0.1% even at room temperature $T = 300$ K. On the other hand, much lower temperatures are required to get similar accuracy from similar number of terms for molecules like O_2, N_2, CO, NO for which the value of Θ_r is in the range $2 - 3$ K. In case Θ_r/T is so high that contribution from $L \geq 1$ becomes insignificant, $Z_{\text{rotl}} = 1$ which being independent of temperature does not contribute to the heat capacity. In this case average thermal energy $k_B T$ is not enough to excite even the lowest rotational level.
2. In case $T >> \Theta_r$, we can treat L as a continuous variable and replace the sum in (8.104) by an integral so that

$$Z_{\text{rotl}} = \int_0^{\infty}(2x+1)\exp\{-(\Theta_r/T)x(x+1)\}dx = \frac{T}{\Theta_r}. \tag{8.105}$$

The integral above has been carried by changing the variable of integration from x to $y = x(x+1)$. Clearly, the energy per molecule in this case is $k_B T$ and the heat capacity per molecule is k_B, which is same as when the gas is treated classically. In this case average thermal energy $k_B T$ is good enough to excite even the higher

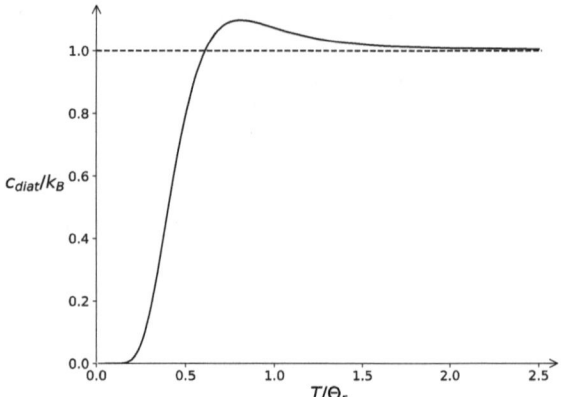

Fig. 8.3 Heat capacity per molecule c_{diat}/k_B of the gas of diatomic molecules as a function of T/Θ_r

rotational levels. We thus see that the classical limit is achieved when temperature is high enough for the thermal energy to excite high energy levels.

Figure 8.3 is a plot of c_{diat}/k_B, where c_{diat} is rotational specific heat per molecule of the gas of diatomic molecules, as a function of normalized temperature T/Θ_r evaluated by numerical computation of the relevant series.

It exhibits the predicted limiting behavior of c_{diat} namely it approaches 0 as $T \to 0$ and approaches k_B as $T \to \infty$. The c_{diat}, however, is not a monotonic function of T; it exhibits a maximum from which it decreases to approach the classical limit. This is called *Schottky anomaly* or *Schottky hump*. We will see similar non-monotonic behavior of the heat capacity in cases where the energy spectrum is non-linear.

Rotational Motion: Polyatomic Molecules

The classical Hamiltonian describing rotation of a molecule is given by (7.62). Its quantum analog is [1]

$$\hat{H}_{\text{rot}} = \frac{1}{2I_1}\hat{L}_1^2 + \frac{1}{2I_1}\hat{L}_2^2 + \frac{1}{2I_3}\hat{L}_3^2. \tag{8.106}$$

We restrict our attention to spherical molecules defined as the ones for which $I_1 = I_2 = I_3 \equiv I$ so that (8.106) reduces to

$$\hat{H}_{\text{rot}} = \frac{1}{2I}\hat{L}^2, \qquad \hat{L}^2 = \hat{L}_1^2 + \hat{L}_2^2 + \hat{L}_3^2. \tag{8.107}$$

This is of the form (8.99) of a diatomic molecule. Its energy levels are given by (8.100). This similarity between diatomic, and the spherically symmetric molecule in the quantum theory makes one wonder: if so, what leads to different heat capacities of the two? For, in classical treatment, we saw that the energy per molecule of a diatomic gas is $k_B T$ whereas it is $3k_B T/2$ for a non-linear polyatomic gas. The difference lies in the fact that the eigenvalue of $\hat{L}^2$ corresponding to L in a spher-

Fig. 8.4 Heat capacity per molecule c_{poly}/k_B of gas of spherical polyatomic molecules as a function of T/Θ_r

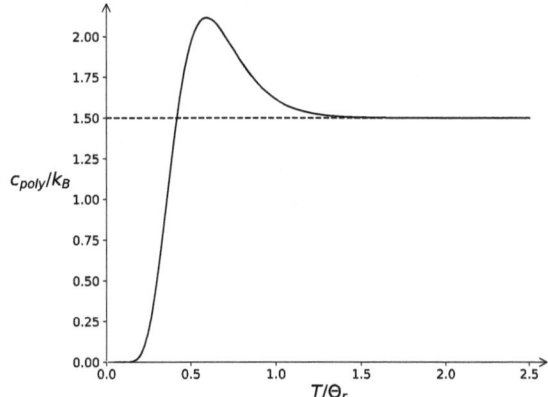

ical molecule is $(2L + 1)^2$-fold degenerate [1]. Recall that the said degeneracy in a diatomic molecule is $2L + 1$-fold. Hence single particle partition function for the gas of spherical molecules is

$$Z_{rot1} = \sum_{L=0}^{\infty} (2L + 1)^2 \exp\{-(\Theta_r/T)L(L + 1)\}. \tag{8.108}$$

We can study the limiting cases of low and high temperatures in the same way as we did for the gas of diatomic molecules. In particular, if temperature is high so that $\Theta_r/T \ll 1$ then the sum in (8.108) can be approximated by an integral so that

$$Z_{rot1} = \int_0^{\infty} (2x + 1)^2 \exp(-\alpha x(x + 1))dx \approx \left(\frac{2I}{\hbar^2 \beta}\right)^{3/2}. \tag{8.109}$$

This shows that energy per molecule is $3k_B T/2$ which is same as that when the polyatomic gas is treated classically. We once again see that the quantum and classical predictions agree in the high-temperature limit. Figure 8.4 is a plot of c_{poly}/k_B, where c_{poly} is rotational specific heat per molecule of the gas of spherical polyatomic molecules, as a function of normalized temperature T/Θ_r evaluated by numerical computation of the relevant series. It exhibits the predicted limiting behavior of c_{poly} namely it approaches 0 as $T \to 0$ and approaches $3k_B/2$ as $T \to \infty$. Like the gas of diatomic molecules the heat capacity in the present case also exhibits Schottky anomaly i.e. c_{poly} is not monotonic as a function of T.

See [2] for detailed study of temperature dependence of heat capacity of methane, which is a spherical top, along with experimental data.

Vibrational Motion

We construct the partition function for the vibrational motion of atoms constituting a molecule assuming the constituent atoms to be non-identical. Such molecules are called heteronuclear. The molecules consisting of identical atoms, called homonu-clear, require different treatment, not undertaken here, because of the quantum the-oretic requirement of definite parity of the wave function under the exchange of identical atoms.

The classical Hamiltonian describing vibration of atoms of mass μ with frequency ω along the line joining them is given by (7.64). The corresponding quantum Hamil-tonian is

$$\hat{H}_{\text{vib}} = \frac{\hat{p}_\eta^2}{2\mu} + \frac{\mu\omega^2}{2}\hat{\eta}^2. \tag{8.110}$$

The energy eigenvalues of the Hamiltonian above are known to be given by

$$\epsilon_n^{\text{vib}} = \hbar\omega\left(n + \frac{1}{2}\right), \quad n = 0, 1, \ldots, \infty. \tag{8.111}$$

Hence the partition function for the said vibrational motion is

$$Z_{\text{vib},\omega} = \sum_{n=0}^{\infty} \exp\left\{-\hbar\omega\beta(n + 1/2)\right\} = \frac{1}{2\sinh(\beta\hbar\omega/2)}. \tag{8.112}$$

The corresponding energy and the heat capacity are

$$U_{\text{vib},\omega} = \frac{\hbar\omega}{2}\coth(\beta\hbar\omega/2), \tag{8.113}$$

and

$$C_{\text{vib},\omega} = k_{\text{B}}\left(\frac{\beta\hbar\omega}{2}\right)^2 \text{cosech}^2(\beta\hbar\omega/2). \tag{8.114}$$

This shows that the heat capacity is temperature dependent. In the limit $\beta\hbar\omega \gg 1$, $C_{\text{vib},\omega} \to k_{\text{B}}$ which is the classical result.

As discussed in Sect. 7.4.2, there is one vibrational mode in a diatomic molecule and $3N - 6$ in a polyatomic molecule consisting of $N \geq 3$ atoms. Total contribu-tion from the vibrational modes will be the sum of contributions from all those modes.

Electronic Transitions

Consider an atom modeled as the one in which an electron can make transitions between only two levels of energies 0 and $\epsilon > 0$. Its partition function is evidently

$$Z_{\text{elect1}} = 1 + \exp(-\beta\epsilon). \tag{8.115}$$

Fig. 8.5 Heat capacity per atom c_{elect}/k_B of a system of two-level atoms as a function of $k_B T/\epsilon$

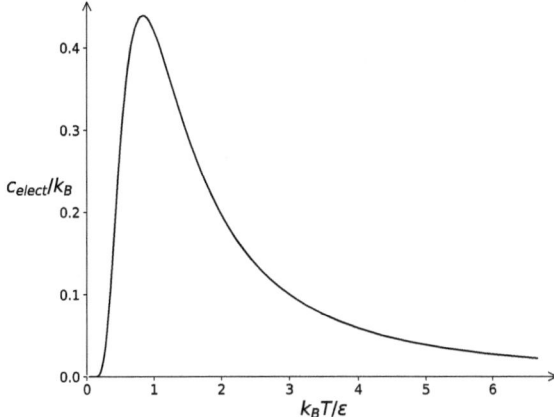

We leave it as an exercise to show that energy and heat capacity per atom of system of such atoms are given by

$$u_{\text{elect}} = \frac{\epsilon \exp(-\beta\epsilon)}{1 + \exp(-\beta\epsilon)}, \tag{8.116}$$

$$\frac{c_{\text{elect}}}{k_B} = (\epsilon/2k_B T)^2 \text{sech}^2(\epsilon/2k_B T). \tag{8.117}$$

Figure 8.5 depicts c_{elect}/k_B as a function of $k_B T/\epsilon$. It exhibits maximum at $k_B T/\epsilon \approx 0.416$ which is approximately the value solving the equation

$$\frac{k_B T}{\epsilon} - (1/2)\tanh(c/2k_B T). \tag{8.118}$$

We once again witness Schottky anomaly in the behavior of the heat capacity as a function of T similar to that observed in the rotational motion.

Exercises

Ex. 8.8. N distinguishable non-interacting particles of mass m are moving in one-dimensional zero potential between infinite potential walls at $x = 0$ and $x = L$ at temperature T. The single particle energy levels are given by

$$E_n = \epsilon_0 n^2, \quad \epsilon_0 = \frac{\pi^2 \hbar^2}{2m L^2} \quad n = 1, 2, \ldots \tag{8.119}$$

Fig. 8.6 Heat capacity per
particle c_{box}/k_B of the gas of
particles in an infinitely deep
potential well as a function
of $k_B T/\epsilon_0$

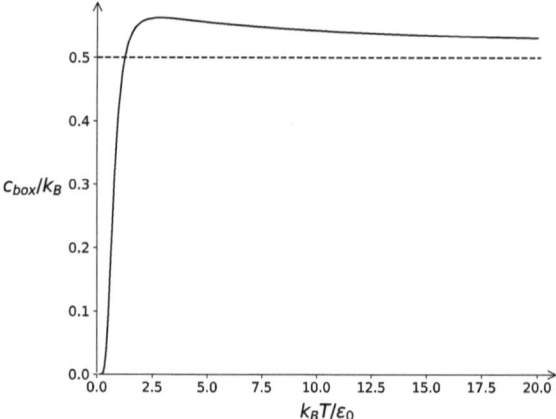

Construct the partition function and evaluate it in the limits of $\epsilon_0 \ll k_B T$
and $\epsilon_0 \gg k_B T$. Hint: The single particle partition function is

$$Z_1 = \sum_{n=1}^{\infty} \exp(-\beta \epsilon_0 n^2). \tag{8.120}$$

This sum cannot be evaluated analytically exactly. In case $\epsilon_0 \ll k_B T$, the
sum can be approximated by an integral:

$$Z_1 \approx \int_0^{\infty} \exp(-\beta \epsilon_0 x^2)\, dx - 1 = \frac{1}{2}\sqrt{\frac{\pi}{\epsilon_0 \beta}}. \tag{8.121}$$

At high temperatures the energy per particle is therefore $k_B T/2$. In the opposite limit $\epsilon_0 \gg k_B T$, dominant contribution comes from $n = 1$ whereby

$$Z_1 = \exp(-\beta \epsilon_0). \tag{8.122}$$

The energy per particle is therefore ϵ_0. This is understandable as the particles
tend to occupy the lowest energy level ϵ_0 as $T \to 0$. Figure 8.6 is a plot of
scaled heat capacity per particle c_{box}/k_B as a function of $k_B T/\epsilon_0$ obtained
by numerical computation of the relevant series. Like other systems we
encountered having non-linear spectrum, the heat capacity in this case too
exhibits Schottky anomaly.

Ex. 8.9. Find the entropy of the system having two levels of energy: 0 and $\epsilon > 0$.

8.7 Thermodynamics of Quantum Gases

We now consider a gas of indistinguishable non-interacting atoms in three-dimensional free space. As mentioned before, we do not have closed-form analytic expression for the canonical partition function of such a gas but have it for its grand canonical partition function, given in (8.45). We therefore investigate thermodynamic properties of such a gas working with its grand canonical partition function.

1. The grand partition function $Z_{G\eta}$ is defined in (8.45) in terms of a sum over single particle energies. We will see that for a Bosonic gas, above some critical value of $N\lambda_T^3/V$ at which $z \equiv \exp(\alpha) = 1$, one needs to separate the ground state term before replacing the rest of the discrete sum by the integral. Bearing in mind that $\eta = 1$ corresponds to the system of Bosons, we rewrite (8.45) for all z as (assuming $\epsilon_1 = 0$)

$$\ln(Z_{G\eta}) = -\eta \sum_{k \neq 1} \ln\{1 - \eta z \exp(-\beta \epsilon_k)\} - \ln(1 - z)\delta_{\eta 1}. \qquad (8.123)$$

Since removing a point from the range of integration does not change its value, we invoke (8.86) to replace the sum by the integral to get

$$\ln(Z_{G\eta}) = -\eta \int_0^\infty D(\epsilon)\{\ln\{1 - \eta z \exp(-\beta\epsilon)\}d\epsilon - \ln(1 - z)\delta_{\eta 1}$$

$$= -\eta \frac{A}{\beta^{3/2}} \int_0^\infty \sqrt{x} \ln\{1 - \eta z \exp(-x)\}dx - \ln(1 - z)\delta_{\eta 1}, \qquad (8.124)$$

where we have recalled (8.76) defining $D(\epsilon)$ and changed the variable of integration from ϵ to $x = \beta\epsilon$. On integrating (8.124) by parts we obtain

$$\ln(Z_{G\eta}) = \frac{2A}{3\beta^{3/2}} \int_0^\infty \frac{x^{3/2} \, dx}{z^{-1} \exp(x) - \eta} - \ln(1 - z)\delta_{\eta 1}. \qquad (8.125)$$

2. Invoking (6.69), the expression for pressure reads

$$P = \frac{2A}{3\beta^{5/2}V} \int_0^\infty \frac{x^{3/2} \, dx}{z^{-1} \exp(x) - \eta} - \frac{1}{\beta V}\ln(1 - z)\delta_{\eta 1}. \qquad (8.126)$$

It will be shown that a significant contribution from the second term comes from $1 - z \approx 1/V$. Hence $\ln(1 - z)/V \sim -\ln(V)/V \to 0$ as $V \to \infty$ leading to the following equation for pressure

$$P = \frac{2A}{3\beta^{5/2}V} \int_0^\infty \frac{x^{3/2} \, dx}{z^{-1} \exp(x) - \eta} . \qquad (8.127)$$

3. Using (6.63), the internal energy may be seen to be given by

$$U = \frac{A}{\beta^{5/2}} \int_0^\infty \frac{x^{3/2}dx}{z^{-1}\exp(x) - \eta}.$$

(8.128)

The fact that it corresponds to the zero-energy state, the second term in (8.125) does not contribute to internal energy.

4. Recalling (6.64) with $\ln(Z_\eta)$ as in (8.124), the average number of particles may be seen to be given by

$$\bar{N} = \frac{A}{\beta^{3/2}} \int_0^\infty \frac{\sqrt{x}dx}{z^{-1}\exp(x) - \eta} + \frac{z}{1-z}\delta_{\eta 1}.$$

(8.129)

5. On comparing (8.127) and (8.128) we get

$$PV = \frac{2}{3}U.$$

(8.130)

This relation holds for all non-interacting gases including Bosonic, Fermionic as well as the classical gas.

6. The average number of particles $\bar{n}_{i\eta}$ in energy level ϵ_i is given by (8.47). In the continuum limit in free space, ϵ_i stands for energy corresponding to momentum $\mathbf{p}$ so that the average number of particles having momentum $\mathbf{p}$ corresponding to energy ϵ would be

$$\bar{n}_\eta(\mathbf{p}) = [\exp\{\beta(\epsilon(p) - \mu)\} - \eta]^{-1}.$$

(8.131)

Hence the number of particles $n_\eta(\epsilon)\Delta\epsilon$ having energy ϵ in energy interval $(\epsilon, \epsilon + d\epsilon)$ would be

$$n_\eta(\epsilon)d\epsilon = n_\eta(\mathbf{p})d^3 p,$$

(8.132)

so that

$$n_\eta(\epsilon) = \frac{D(\epsilon)}{\exp\{\beta(\epsilon - \mu)\} - \eta}.$$

(8.133)

For gas consisting of molecules moving in three-dimensional space with non-relativistic speeds, $D(\epsilon)$ is given by (8.76). In that case, (8.133) reduces to the Bose–Einstein distribution (4.110), for $\eta = 1$ and to the Fermi–Dirac distribution (4.125) for $\eta = -1$.

7. Equation (8.130) is the relation between P, V and U. It is not the equation of state because the equations of state are relations between P, V, $\bar{N}$ and T, and U, V, $\bar{N}$ and T. Those are obtained, in principle, by inverting (8.129) to express z in terms of $\bar{N}$, and T and substituting z so obtained in other thermodynamic quantities

which are functions of z. That is generally a formidable task. We will derive the equations of state in the limiting case of small z.

When $z < 1$, we can rewrite various quantities derived above in terms of the *polylogarithm function* defined by

$$\text{Li}_p(x) = \sum_{k=1}^{\infty} \frac{x^k}{k^p}. \tag{8.134}$$

To that end, recall the expression (7.6) for λ_T and (8.76) for A, to obtain

$$\frac{A}{\beta^{3/2}} = \frac{2V}{\sqrt{\pi}} \lambda_T^{-3}. \tag{8.135}$$

We use also the identity, proved in the exercises,

$$\int_0^{\infty} \frac{x^p \, dx}{z^{-1} \exp(x) - \eta} = \eta \Gamma(p+1) g_{p+1}^{(\eta)}(z), \tag{8.136}$$

where

$$g_p^{(\eta)}(z) = \sum_{k=1}^{\infty} \frac{(\eta z)^k}{k^p} \tag{8.137}$$

is evidently in the form (8.134) of the polylogarithmic function. It is then straightforward to see that:

1. The expression (8.125) for $\ln(Z_{G\eta})$ may be rewritten as

$$\ln(Z_{G\eta}) = \frac{\eta V}{\lambda_T^3} g_{5/2}^{(\eta)}(z) - \ln(1-z)\delta_{\eta 1}. \tag{8.138}$$

2. Equation (8.126) for pressure assumes the form

$$P = \frac{\eta}{\beta \lambda_T^3} g_{5/2}^{(\eta)}(z) - \frac{1}{\beta V} \ln(1-z)\delta_{\eta 1}. \tag{8.139}$$

3. The expression (8.128) for internal energy U reads

$$U = \frac{3\eta V}{2\beta \lambda_T^3} g_{5/2}^{(\eta)}(z). \tag{8.140}$$

4. Equation (8.129) for average number of particles reduces to

$$\bar{N} = \frac{\eta V}{\lambda_T^3} g_{3/2}^{(\eta)}(z) + \frac{z}{1-z}\delta_{\eta 1}. \tag{8.141}$$

In Sect. 8.4.3 we saw that the Bose as well as the Fermi gas behaves as a classical gas when $z \ll 1$ which we called the classical limit. In the following we compute quantum corrections to the classical limit.

See [3] for computing fugacity using Mathematica functions.

Exercises

Ex. 8.10. Prove (8.136). Hint:

$$\int_0^\infty \frac{x^p dx}{z^{-1}\exp(x) - \eta} = \int_0^\infty \frac{x^p z \exp(-x)dx}{1 - z\eta\exp(-x)}.$$
$$= \eta \sum_{k=0}^\infty (\eta z)^{k+1} \int_0^\infty x^p \exp\{-x(k+1)\}dx.$$

$$(8.142)$$

Recalling the integral representation (H.3) of Γ-function it is straightforward to reduce this to the form in (8.136).

Ex. 8.11. Show that

$$\frac{dg_p(z)}{dz} = z^{-1}g_{p-1}(z), \qquad g_p(z) \equiv g_p^{(1)}(z). \qquad (8.143)$$

Ex. 8.12. In a d-dimensional system in which the energy-momentum relation is given by (8.89) and the corresponding density of states by (8.90), show that the equations for $\ln(Z_{G\eta})$, U, $\bar{N}$ and P assume the form

$$\ln(Z_{G\eta}) = \frac{n}{d} \frac{A_{nd}}{\beta^{d/n}} \int_0^\infty \frac{x^{d/n}dx}{z^{-1}\exp(x) - \eta} - \ln(1 - z)\delta_{\eta 1}. \qquad (8.144)$$

$$U = \frac{A_{nd}}{\beta^{(d+n)/n}} \int_0^\infty \frac{x^{d/n}dx}{z^{-1}\exp(x) - \eta}. \qquad (8.145)$$

$$\bar{N} = \frac{A_{nd}}{\beta^{d/n}} \int_0^\infty \frac{x^{(d-n)/n}dx}{z^{-1}\exp(x) - \eta}. \qquad (8.146)$$

$$PV_d = \frac{n}{d}U, \qquad (8.147)$$

where $V_d = V$ in 3-d, $V_d = \sigma$ in $2 - d$ and $V_d = L$ in $1 - d$ systems. In terms of the $g_p^{(\eta)}(z)$ function, the expressions above read

$$\ln(Z_{G\eta}) = \frac{n\eta}{d} \frac{A_{nd}}{\beta^{d/n}} \Gamma((d/n) + 1)g_{(d/n)+1}^{(\eta)}(z) - \ln(1 - z)\delta_{\eta 1}. \qquad (8.148)$$

$$U = \frac{A_{nd}\eta}{\beta^{(d/n)+1}}\Gamma((d/n)+1)g^{(\eta)}_{(d/n)+1}(z).$$
(8.149)

$$\bar{N} = \frac{A_{nd}\eta}{\beta^{d/n}}\Gamma(d/n)g^{(\eta)}_{d/n}(z).$$
(8.150)

Ex. 8.13. Show that the adiabatic equation of state for the ideal quantum gas is

$$PV_d^{(n+d)/d} = \text{constant}.$$
(8.151)

Since the relation above for adiabatic transformations holds for any non-interacting gas in free space, we see that for a given value of n and d it is same for all gases. In particular, for the gas of molecules moving with non-relativistic speeds in three dimensions, $n = 2$, $d = 3$ in which case the relation (8.151) is same as that in (1.87) for classical gas. However the relation between P and T and that between T and V under adiabatic transformation cannot be determined unless the equation of state is known. Furthermore, whereas $5/3$ in the relation $PV^{5/3} = \text{constant}$ for the classical gas is equal to $\gamma \equiv C_P/C_V$, it is not so in general. This has been demonstrated by evaluation of C_P/C_V in (9.63) for the Fermi gas at low temperature and for the Bose gas in (10.48).

Hint: In an adiabatic process, entropy and number of particles are constants. Hence the first law of thermodynamics in this case reduces to $dU + PdV = 0$. Use (8.147) to get the desired result.

Ex. 8.14. Show that fugacity of the $2 - d$ non-relativistic quantum gas is given by

$$z_\eta = \eta\left\{1 - \exp(-\eta\beta\bar{N}\beta/A_{22})\right\}, \quad A_{22} = \frac{2m\pi\sigma}{h^2},$$
(8.152)

where z_{-1} is the fugacity of the Fermi gas and z_1 that of the Bose gas. Note that (8.152) determines z in terms of $\bar{N}$ and T. Its substitution in (8.145) would determine U, which in turn when combined with (8.147) would determine PV_d in terms of $\bar{N}$, and T yielding thereby the equations of state. Hint: For 2-dimensional non-relativistic gas $n = d = 2$. The desired result follows by straightforward integration of (8.146).

8.8 Quantum Corrections to the Classical Limit

In Sect. 8.4.3 we showed that the quantum grand partition function reduces to the classical one under the condition (8.62). That condition is written in terms of sum over energy levels. For molecules in free space, that sum has been evaluated in (8.91) using which the said condition reads

$$\lambda_T << \left(\frac{V}{\bar{N}}\right)^{1/3} \quad \text{or} \quad \frac{\bar{N}}{V}\left(\frac{h^2}{2\pi m k_B T}\right)^{3/2} << 1. \tag{8.153}$$

Now, $V/\bar{N}$ is the volume per particle. Its cube root is a measure of distance between the particles. Hence (8.153) states that the condition under which quantum gases behave as a classical gas is valid if the thermal wavelength of the particles is much smaller than the distance between them, i.e. since thermal wavelength is a measure of their de Broglie wavelength, if there is no significant overlap between De Broglie waves of the particles. Equivalently, the second form of the condition in (8.153) shows that the classical limit will hold if the particle density N/V is low or/and temperature T is high.

In this section we evaluate quantum corrections to the classical limit. To that end, it is useful to write the condition (8.153) as $z_0 << 1$ where

$$z_0 = \frac{\bar{N}}{V}\lambda_T^3, \quad z_0 << 1. \tag{8.154}$$

As stated before, the first step in the study of thermodynamic properties when system is described by grand canonical partition function is to invert the (8.141) to express z in terms of the average number $\bar{N}$ of particles. Presently we invert it assuming z takes values close to those for which classical limit is achieved. Since classical results are obtained in the limit $z \to 0$ and the parameter z_0 defined above is a measure of the closeness of z to the classical limit, we expand z in powers of z_0:

$$z = \sum_{j=1}^{\infty} a_j z_0^j \tag{8.155}$$

and determine the coefficients of expansion by substituting z in the expression (8.141) for $\bar{N}$. Since the second term therein contributes when z is close to unity and our interest at present is in the values of z close to zero, we can ignore the second term in (8.141) to rewrite it as

$$z_0 = \eta g_{3/2}^{(\eta)}(z), \tag{8.156}$$

where the definition (8.154) of z_0 has been used. Substitute the expression (8.155) for z in the series representation (8.137) of $g_p^{(\eta)}(z)$ to get

$$z_0 = \sum_{j=1}^{\infty} a_j z_0^j + \frac{\eta}{\sqrt{8}} \sum_{i,j=1}^{\infty} a_i a_j z_0^{i+j} + \frac{1}{\sqrt{27}} \sum_{i,j,k=1}^{\infty} a_i a_j a_k z_0^{i+j+k} + \cdots$$

$$\tag{8.157}$$

Retaining terms up to z_0^3 we have

$$(a_1 - 1)z_0 + \left(a_2 + \frac{\eta}{\sqrt{8}}a_1^2\right) z_0^2 + \left(a_3 + \frac{\eta}{\sqrt{2}}a_1 a_2 + \frac{a_1^3}{\sqrt{27}}\right) z_0^3 = 0. \quad (8.158)$$

Equating to zero the coefficient of various powers of z_0 in (8.158) we get

$$a_1 = 1, \quad a_2 = -\frac{\eta}{\sqrt{8}}, \quad a_3 = \frac{1}{4} - \frac{1}{3\sqrt{3}}. \quad (8.159)$$

Hence, to order z_0^3,

$$z = z_0 - \frac{\eta}{\sqrt{8}}z_0^2 + \left(\frac{1}{4} - \frac{1}{3\sqrt{3}}\right) z_0^3. \quad (8.160)$$

Substitute this in the (8.140) for U retaining terms up to z^3 to get

$$\begin{aligned}
U &= \frac{3\bar{N}}{2z_0\beta}\left[z + \frac{\eta}{4\sqrt{2}}z^2 + \frac{1}{9\sqrt{3}}z^3\right] \\
&= \frac{3\bar{N}}{2\beta}\left[1 + \left(a_2 + \frac{\eta}{4\sqrt{2}}\right)z_0 + \left(a_3 + \frac{\eta}{2\sqrt{2}}a_2 + \frac{1}{9\sqrt{3}}\right)z_0^2\right]. \quad (8.161)
\end{aligned}$$

Recalling the values of a_2, a_3 derived in (8.159) this reduces to

$$U = \frac{3\bar{N}k_B T}{2}\left[1 - \frac{\eta}{4\sqrt{2}}z_0 + \left(\frac{1}{8} - \frac{2}{9\sqrt{3}}\right)z_0^2\right]. \quad (8.162)$$

The first term in the equation above gives energy of a classical system. The second term is correction to the classical value. Since Fermi gas corresponds to $\eta = -1$, it shows that the energy of the Fermi gas is more than that of the classical gas of same density and at same temperature. On the other hand, since the Bose gas corresponds to $\eta = 1$, we see that energy of the Bose gas is less than that of classical gas of same density and at same temperature. Substitution of (8.162) in the relation (8.130) between PV and U leads to the following equation of state correct up to z_0^2:

$$PV = \bar{N}k_B T\left[1 - \frac{\eta}{4\sqrt{2}}z_0 + \left(\frac{1}{8} - \frac{2}{9\sqrt{3}}\right)z_0^2\right]. \quad (8.163)$$

This shows that, other parameters remaining same, the pressure of Fermi gas is more than that of the classical gas whereas the pressure of Bose gas is less than that of the classical gas.

The fact that the internal energy and the pressure of the Fermi gas is more than that of the classical gas may be understood by recalling that no two Fermi particles can occupy same state whereas there is no restriction on the number of classical particles that can be placed in same state as a consequence of which the number of Fermi particles in higher energy states is more than the number of classical particles.

It may be verified that the (8.162) for U leads to the following expression for the heat capacity at constant volume:

$$\begin{aligned} \frac{C_V}{k_B \bar{N}} &= \frac{3}{2}\left[1 + \frac{\eta}{8\sqrt{2}}z_0 - \left(\frac{1}{4} - \frac{4}{9\sqrt{3}}\right)z_0^2\right] \\ &= \frac{3}{2}\left[1 + 0.0884\eta z_0 + 0.0066 z_0^2\right]. \end{aligned} \qquad (8.164)$$

The same procedure can be extended to obtain higher order quantum corrections to the classical limit.

For a fixed $\bar{N}/V$, the expansion in powers of z_0 employed above is basically an expansion in terms of $1/T$, i.e. it is a high temperature expansion. We found that in that limit Fermi and Bose gases approach the same limit, namely, the classical gas. We will see that the Fermi and the Bose gases behave vastly differently at low temperatures.

References

1. W. Gordy, R.L. Cook, *Microwave Molecular Spectroscopy* (Wiley, 1984)
2. A. Yu Zakharov, A.V. Leont'eva, A. Yu Prokhorov, IOP Conf. Series: Mat. Sci. Eng. **441**, 012060 (2018)
3. B. Cowan, J. Loe, Temp. Phys. **197**, 412 (2019)

Chapter 9
Ideal Fermi Gas

In this chapter we investigate the properties of the gas of free non-interacting Fermi particles at low temperatures.

9.1 Fermi Gas at Zero Temperature

Consider the gas of free fermions having spin quantum number S. They can be distinguished by $2S + 1$ spin projection quantum numbers $m_S = -S, -S + 1, \ldots, S$. Since an energy level can not be occupied by two Fermions having same value of m_s it follows that each free single particle energy level can be occupied by at most $2S + 1$ Fermions, each characterized by one of the $2S + 1$ different values of m_S. Since the density of states $D(\epsilon)$ is evaluated for single particle occupancy, the spin of the particles in the gas containing particles having all possible values of m_S can be accounted for by multiplying $D(\epsilon)$ by $2S + 1$, i.e. by the correspondence

$$D(\epsilon) \rightarrow (2S + 1)D(\epsilon). \tag{9.1}$$

For the sake of simplicity, we assume that all the Fermions in the gas have same value of m_S so that an energy level can be occupied by only one Fermion. Since the occupancy of a level then is same as $2S + 1$ when $S = 0$, such a system is also called a system of "spinless" Fermions. The results for the finite value of spin may be obtained from the ones for the $S = 0$ case by changing the density of states per the correspondence in (9.1).

Let the gas be at $T = 0$. It will then be in its ground state. If N is the number of Fermions in the gas and ϵ_F is the energy of the highest occupied energy level then we must have

$$N = \int_0^{\epsilon_F} D(\epsilon)\mathrm{d}\epsilon. \tag{9.2}$$

© The Author(s), under exclusive license to Springer Nature Switzerland AG 2024
R. R. Puri, *Modern Thermodynamics and Statistical Mechanics*, Undergraduate Lecture Notes in Physics, https://doi.org/10.1007/978-3-031-54310-4_9

If the motion is non-relativistic and in three spatial dimensions, then (9.2) assumes the form

$$N = A \int_0^{\epsilon_F} \sqrt{\epsilon}\, d\epsilon = \frac{2A}{3} \epsilon_F^{3/2},$$ (9.3)

where we have invoked (8.76) for the density of states $D(\epsilon)$ for free non-relativistic particles in three spatial dimensions. The highest possible energy ϵ_F is called the *Fermi energy*. It corresponds to the temperature

$$T_F = \frac{\epsilon_F}{k_B}$$ (9.4)

called *Fermi temperature*. The internal energy is given by

$$U = \int_0^{\epsilon_F} \epsilon D(\epsilon) d\epsilon = \frac{2A}{5} \epsilon_F^{5/2}.$$ (9.5)

The last two equations lead to the relation

$$U = \frac{3}{5} N \epsilon_F.$$ (9.6)

On combining this with the exact expression (8.130) relating PV with U in any non-interacting gas follows the equation of state

$$PV = \frac{2}{5} \epsilon_F N$$ (9.7)

of the gas of free non-interacting Fermions at $T = 0$. We see that the energy and pressure of the Fermi gas are finite at $T = 0$. That is because two fermions in same state cannot occupy same energy level and hence they are forced to occupy distinct higher energy states as their number increases leading to finite energy, and pressure. Its entropy, however, will be shown to be zero at $T = 0$ (see (9.48)), as it should be.

It is instructive to calculate ϵ_F for the gas of electrons in a solid. Since $S = 1/2$ for electrons, with A given by (8.76) and due to the correspondence (9.1), the (9.3) reads

$$\frac{N}{V} = \frac{8\pi (2m)^{3/2}}{3 h^3} \epsilon_F^{3/2}.$$ (9.8)

This shows that

$$\epsilon_F = \left[\left(\frac{N}{V} \right)^{2/3} \frac{1}{2mc^2} \right] \left[\left(\frac{3}{\pi} \right)^{2/3} \frac{(hc)^2}{4} \right].$$ (9.9)

On using $(hc)^2 \approx 1.54 \times 10^{-8}(eV)^2 - cm^2$, $2mc^2 \approx 1MeV$, (9.9) yields

$$\epsilon_F = 3.72 \times 10^{-15} eV - cm^2 \left(\frac{N}{V}\right)^{2/3}, \tag{9.10}$$

where V is in cm^3. In solids N/V is generally in the range $10^{22} - 10^{23}/cm^3$. The value of ϵ_F therefore is few electron volts. Since $k_B \approx 8.6 \times 10^{-5} eV/K$, the Fermi temperature is $\sim 10^4 - 10^5\ K$. The working temperatures, being much lower than the Fermi temperature, may be taken as close to $T = 0$.

The expressions for N, U, and P derived above for $T = 0$ without recourse to statistical mechanical formalism, as shown next, follow also using it.

1. The probability that a fermion in an ideal Fermi gas occupies energy level ϵ is given, recalling (8.43), by

$$p_F(\epsilon) = \frac{1}{\exp\left(\beta(\epsilon - \mu)\right) + 1}. \tag{9.11}$$

In the limit $T \to 0$,

$$p_F(\epsilon) = 1 \quad \text{if } \epsilon < \mu, \qquad p_F(\epsilon) = 0 \quad \text{if } \epsilon > \mu. \tag{9.12}$$

This shows that at $T = 0$ the highest value of energy of the fermions is μ which is therefore the Fermi energy of the system:

$$\mu = \epsilon_F, \qquad T = 0. \tag{9.13}$$

2. Recalling (8.129), the number of particles is

$$N = A \int_0^\infty \frac{\sqrt{\epsilon}\ d\epsilon}{\exp\left(\beta(\epsilon - \mu)\right) + 1} = A \int_0^\mu \sqrt{\epsilon}\ d\epsilon, \quad \beta \to \infty. \tag{9.14}$$

This is same as (9.3).
3. Invoking (8.128), internal energy is

$$U = A \int_0^\infty \frac{\epsilon^{3/2}\ d\epsilon}{\exp\left(\beta(\epsilon - \mu)\right) + 1} = A \int_0^\mu \epsilon^{3/2}\ d\epsilon, \quad \beta \to \infty. \tag{9.15}$$

This is same as (9.5).
4. Recall the expression (8.127) for P. Note that it has been obtained by changing the variable of integration from ϵ to $x = \beta\epsilon$. Carry the integral in the limit $\beta \to \infty$ after transforming x back to ϵ to get

$$P = \frac{4A}{15 V}\epsilon_F^{5/2} = \frac{2N}{5 V}\epsilon_F, \tag{9.16}$$

where we have invoked (9.3) in writing the last equality. The equation above is the same as (9.7).

We have thus at hand the thermodynamic quantities for the ideal Fermi gas at $T = 0$. Next we evaluate finite temperature corrections to its $T = 0$ behavior.

Exercises

Ex. 9. 1. A non-interacting Fermi gas of N particles of mass m is in one-dimensional zero potential between infinite potential walls at $x = 0$ and $x = L$ at $T = 0$. The single particle energy levels are given by

$$E_n = \epsilon_0 n^2, \quad \epsilon_0 = \frac{\pi^2 \hbar^2}{2mL^2} \quad n = 1, 2, \ldots \tag{9.17}$$

(a) Find the maximum occupied energy level. (b) Find the Fermi energy. (c) Find total energy of the system.

Ex. 9. 2. Find the Fermi energy of a three-dimensional ultra-relativistic ($\epsilon = pc$) ideal Fermi gas at $T = 0$.

Ex. 9. 3. The energy levels of the one-dimensional harmonic oscillator of frequency ω are given by $E_n = \hbar\omega(n + 1/2)$ ($n = 0, 1, 2, \ldots$) in which N identical non-interacting Fermions are to be distributed. (i) What is the Fermi energy of the system? (ii) What is total energy of the system?

Ex. 9. 4. The energy levels of an isotropic two-dimensional harmonic oscillator of frequency ω are given by $E_n \equiv E_{n_1,n_2} = \hbar\omega(n_1 + n_2 + 1)$, ($n_1, n_2 = 0, 1, 2, \ldots$) in which N identical non-interacting fermions are to be distributed. (i) How many Fermions can be placed in the levels of energy $E_M = \hbar\omega(M + 1)$? (ii) What is the total number of Fermions if, starting from E_0, all the levels up to the level of energy E_M are occupied? (iii) If Fermions occupy all the levels up to the levels of energy E_M then what is the energy of the system? Hint: The energy is determined by the sum $n_1 + n_2$, and the states obtained by interchanging distinct n_1 and n_2 are different. Hence the degeneracy of the level of energy E_n is given by the number of ways of writing n as the sum of two positive integers. That number can be seen to be $n + 1$

Ex. 9. 5. The energy levels of an isotropic three-dimensional harmonic oscillator of frequency ω are given by $E_n \equiv E_{n_1,n_2,n_3} = \hbar\omega(n_1 + n_2 + n_3 + 3/2)$, ($n_1, n_2, n_3 = 0, 1, 2, \ldots$) in which N identical non-interacting fermions are to be distributed. (i) How many Fermions can be placed in the levels of energy $E_M = \hbar\omega(M + 3/2)$? (ii) What is the total number of Fermions if, starting from E_0, all the levels up to the level of energy E_M are occupied? (iii) If Fermions occupy all the levels up to the levels of energy E_M then what is the energy of the system? Hint: The energy is determined by the sum $n_1 + n_2 + n_3$, and the states obtained by interchanging distinct n_1, n_2, n_3 are different. Hence the degeneracy of the level of energy E_n is given by the number of different ways of writing n as the sum of three positive integers. That number is $(n + 1)(n + 2)/2$.

Ex. 9. 6. Show that the Fermi energy and total energy of a d-dimensional ideal Fermi gas obeying the energy-momentum relation (8.89) and having $\bar{N}$ average number of molecules of mass m is given at $T = 0$ by

$$\epsilon_F = \left(\frac{\bar{N}}{A_{nd}} \frac{d}{n} \right)^{n/d},$$

$$U = A^{-n/d} \frac{n}{d+n} \left(\frac{\bar{N}d}{n} \right)^{(d+n)/d}, \tag{9.18}$$

where A_{nd} is as in (8.90).

9.2 Fermi Gas at Low Temperature

The study of thermodynamic properties of the ideal Fermi gas involves evaluating integrals of the type

$$I = \int_0^\infty \frac{\phi(\epsilon)\, d\epsilon}{1 + \exp(\beta(\epsilon - \mu))}. \tag{9.19}$$

We rewrite the integral above as

$$
\begin{aligned}
I &= \int_0^\mu \frac{\exp(-\beta(\epsilon - \mu))\, \phi(\epsilon)\, d\epsilon}{1 + \exp(-\beta(\epsilon - \mu))} + \int_\mu^\infty \frac{\phi(\epsilon)\, d\epsilon}{1 + \exp(\beta(\epsilon - \mu))} \\
&= \int_0^\mu \phi(\epsilon)\, d\epsilon - \int_0^\mu \frac{\phi(\epsilon)\, d\epsilon}{1 + \exp(-\beta(\epsilon - \mu))} \\
&\quad + \int_\mu^\infty \frac{\phi(\epsilon)\, d\epsilon}{1 + \exp(\beta(\epsilon - \mu))}.
\end{aligned}
\tag{9.20}
$$

Change the variable of integration in the last two integrals to $x = \beta(\epsilon - \mu)$. The limits of second integral then are $x = 0$ corresponding to $\epsilon = \mu$ and $x = -\beta\mu$ corresponding to $\epsilon = 0$. We assume T to be close to zero so that $\beta \gg 1$. Consequently, with $x = -\beta\mu \approx -\infty$, followed by the transformation $x \to -x$, (9.20) assumes the form

$$I = \int_0^\mu \phi(\epsilon)\, d\epsilon + J, \tag{9.21}$$

where

$$J = \frac{1}{\beta} \int_0^\infty \frac{1}{1 + \exp(x)} \left[\phi\left(\mu + \frac{x}{\beta}\right) - \phi\left(\mu - \frac{x}{\beta}\right) \right] dx$$

$$= \frac{2}{\beta} \sum_{n=1}^\infty \frac{1}{(2n-1)!} \frac{d^{2n-1}\phi(\mu)}{d\mu^{2n-1}} \int_0^\infty \frac{dx}{1 + \exp(x)} \left(\frac{x}{\beta}\right)^{2n-1}, \qquad (9.22)$$

$$\frac{d^{2n-1}\phi(\mu)}{d\mu^{2n-1}} \equiv \frac{d^{2n-1}\phi(\epsilon)}{d\epsilon^{2n-1}}\Bigg|_{\epsilon=\mu}. \qquad (9.23)$$

Evaluate (9.22) invoking the identity (H.14) and substitute the resulting expression in (9.21) to obtain

$$I = \int_0^\mu \phi(\epsilon)\, d\epsilon + 2 \sum_{n=1}^\infty \left[(k_B T)^{2n} \frac{d^{2n-1}\phi(\mu)}{d\mu^{2n-1}} \left(1 - 2^{-2n+1}\right) \zeta(2n) \right]. \qquad (9.24)$$

The (9.24) is called the *Sommerfeld expansion*. The first term which is independent of T gives the properties of the Fermi gas at $T = 0$ already studied. We find corrections to the $T = 0$ behavior by expressing (9.24) in powers of z_0:

$$z_0 = k_B T / \epsilon_F. \qquad (9.25)$$

The expansion of thermodynamic quantities in terms of z_0 will determine their behavior for temperatures which are such that $k_B T \ll \epsilon_F$.

The functions $\phi(\epsilon)$ of our interest are of the form

$$\phi(\epsilon) = A\epsilon^{m+1/2}. \qquad (9.26)$$

The expression (9.24), with $I \to I_m$, may then be rewritten as

$$I_m = \frac{2A}{2m+3} \mu^{(2m+3)/2} \left[1 + \sum_{n=1}^\infty a_{mn} z^{2n} \right], \qquad (9.27)$$

where

$$z = k_B T / \mu, \qquad (9.28)$$

and

$$a_{mn} = \frac{2\Gamma(m+5/2)}{\Gamma(m-2n+5/2)} \left(1 - 2^{-2n+1}\right) \zeta(2n). \qquad (9.29)$$

It is convenient to express (9.27) in the form

$$I_m = \frac{2A}{2m+3} \epsilon_F^{(2m+3)/2} \left(\frac{z_0}{z}\right)^{(2m+3)/2} \left[1 + \sum_{n=1}^{\infty} a_{mn} z^{2n}\right]. \tag{9.30}$$

We evaluate U and $\bar{N}$ using (8.128) and (8.129) with $\eta = -1$ and $z = \exp(\beta\mu)$ therein. Clearly, the integral in the expression for U is then same as (9.19) corresponding to $\phi(\epsilon) = A\epsilon^{3/2}$ and that in the expression for $\bar{N}$ is same as (9.19) corresponding to $\phi(\epsilon) = A\epsilon^{1/2}$ with A as in (8.76) so that

$$\bar{N} \equiv I_0 = \frac{2A}{3} \epsilon_F^{3/2} \left(\frac{z_0}{z}\right)^{3/2} \left[1 + \sum_{n=1}^{\infty} a_{0n} z^{2n}\right], \tag{9.31}$$

and

$$U \equiv I_1 = \frac{2A}{5} \epsilon_F^{5/2} \left(\frac{z_0}{z}\right)^{5/2} \left[1 + \sum_{n=1}^{\infty} a_{1n} z^{2n}\right], \tag{9.32}$$

with

$$a_{0n} = \frac{2\Gamma(5/2)}{\Gamma(5/2 - 2n)} \left(1 - 2^{-2n+1}\right) \zeta(2n),$$

$$a_{1n} = \frac{2\Gamma(7/2)}{\Gamma(7/2 - 2n)} \left(1 - 2^{-2n+1}\right) \zeta(2n). \tag{9.33}$$

Use (9.3) to rewrite (9.31) and (9.32) as

$$z \left[1 + \sum_{n=1}^{\infty} a_{0n} z^{2n}\right]^{-2/3} = z_0, \tag{9.34}$$

and

$$U = \frac{3}{5} \bar{N} \epsilon_F \left(\frac{z_0}{z}\right)^{5/2} \left[1 + \sum_{n=1}^{\infty} a_{1n} z^{2n}\right]. \tag{9.35}$$

Assuming $z_0 \ll 1$, we express z in powers of z_0, solve (9.34) up to the desired power of z_0 and substitute the resulting expression of z in (9.35) to determine energy. The (9.34) shows that z/z_0 has only even powers of z. Hence the expansion of z in terms of z_0 must be of the form

$$z = z_0 \left(1 + \sum_{k=1}^{\infty} c_{2k+1} z_0^{2k}\right). \tag{9.36}$$

The c_{2k+1} are the unknowns to be determined by substituting (9.36) in (9.34) and equating like powers of z_0.

Recalling the definitions (9.25), (9.28) of z_0, z the expression for chemical potential reads

$$\mu = \epsilon_F \left(1 + \sum_{k=1}^{\infty} c_{2k+1} z_0^{2k} \right)^{-1}. \tag{9.37}$$

We derive corrections to the $T = 0$ behavior of the Fermi gas retaining terms up to second order in z_0^2 in the series in the expression (9.36) for z.

First-Order Corrections

The first-order correction to the $T = 0$ behavior is obtained by keeping the lowest order term in the series in (9.36):

$$z = z_0(1 + c_3 z_0^2). \tag{9.38}$$

Substitute this in (9.34) retaining terms up to z_0^3 to get

$$(c_3 - (2/3)a_{01})z_0^3 = 0. \tag{9.39}$$

This determines the unknown coefficient c_3:

$$c_3 = \frac{2}{3}a_{01}. \tag{9.40}$$

Substitution of this in (9.38) yields the expression for z valid up to z_0^3:

$$z = z_0(1 + (2/3)a_{01}z_0^2). \tag{9.41}$$

Energy is obtained by using the expression for z above in (9.35) and retaining terms up to z_0^3:

$$U = \frac{3}{5}\bar{N}\epsilon_F \left(1 + (a_{11} - (5/3)a_{01})z_0^2 \right). \tag{9.42}$$

Invoking the expressions in (9.33) for a_{01}, a_{11} with $\zeta(2) = \pi^2/6$, we have

$$a_{01} = \frac{\pi^2}{8}, \qquad a_{11} = \frac{5\pi^2}{8} \tag{9.43}$$

so that

$$U = \frac{3\bar{N}\epsilon_F}{5}\left[1 + \frac{5\pi^2}{12}\left(\frac{k_BT}{\epsilon_F}\right)^2\right]. \tag{9.44}$$

Invoking (9.37), the chemical potential is given by

$$\mu = \epsilon_F\left[1 - \frac{\pi^2}{12}\left(\frac{k_BT}{\epsilon_F}\right)^2\right]. \tag{9.45}$$

We have thus at hand the expressions for energy and chemical potential to the lowest order in temperature relative to the Fermi temperature. We derive below other thermodynamic characteristics of the Fermi gas to the said order of approximation.

1. Pressure P may be obtained by using the relation (8.130) which determines it in terms of U and V. On substituting in that relation the expression (9.44) for U follows the equation of state

$$PV = \frac{2\bar{N}\epsilon_F}{5}\left[1 + \frac{5\pi^2}{12}\left(\frac{k_BT}{\epsilon_F}\right)^2\right]. \tag{9.46}$$

The first term on the right side above gives the equation of state (9.7) for the Fermi gas at zero temperature whereas the second term is the lowest order correction to the zero temperature term.
2. The specific heat at constant volume is given by

$$\frac{C_V}{\bar{N}k_B} = \frac{1}{k_B\bar{N}}\frac{\partial U}{\partial T} = \frac{\pi^2}{2}\frac{k_BT}{\epsilon_F}. \tag{9.47}$$

This shows that the specific heat goes to zero linearly as $T \to 0$.
3. Invoking Euler's relation, the entropy of the Fermi gas in low-temperature limit may be shown to be given by

$$\frac{S}{\bar{N}k_B} = \frac{\pi^2}{2}\frac{k_BT}{\epsilon_F}. \tag{9.48}$$

This equation rightly predicts $S \to 0$ as $T \to 0$.
4. Recalling (9.8) whereby $\epsilon_F \sim V^{-2/3}$, it follows that T and V in an adiabatic process ($S = $ constant) are related by

$$V^{2/3}T = \text{constant}. \tag{9.49}$$

The relation between P and V in an adiabatic process can be found by expressing PV in (9.46) in terms of S (see (9.68)). Since $\epsilon_F \sim V^{-2/3}$, the said equation for constant S yields

$$PV^{5/3} = \text{constant}. \tag{9.50}$$

On combining (9.49) and (9.50) we get $TP^{-2/5} = \text{constant}$. The adiabatic relations between the pairs of P, V, T derived above for the Fermi gas in low temperature limit are same as the corresponding ones for an ideal monoatomic gas. However, whereas the exponents in the said relations in the classical gas are related with C_P/C_V, the exponents in the present case, though their values are same as the corresponding ones in the classical gas, have no relation with C_P/C_V (see Ex. 9.8).

Second-Order Correction

The second-order correction is obtained by terminating the series in (9.36) at z_0^4:

$$z = z_0(1 + c_3 z_0^2 + c_5 z_0^4). \tag{9.51}$$

This determines z up to z_0^5. Substitute this in (9.34) retaining terms to order z_0^5 to get

$$(c_3 - (2/3)a_{01})z_0^2 + D_5 z_0^4 = 0, \tag{9.52}$$

where

$$D_5 = c_5 - \left(2c_3 a_{01} + \frac{2}{3}a_{02} - \frac{5}{9}a_{01}^2\right). \tag{9.53}$$

The coefficient of z_0^2 is zero due to (9.40). Equating to zero the coefficient of z_0^4 determines c_5. The value of a_{01} needed for evaluating c_5 is given in (9.43) and that of a_{02} obtained using (9.33), with $\zeta(4) = \pi^4/90$, is

$$a_{02} = \frac{7\pi^4}{640}. \tag{9.54}$$

The value of c_5 turns out to be given by

$$c_5 = \frac{7\pi^4}{360}. \tag{9.55}$$

Recalling (9.37), the expression for the chemical potential reads

$$\mu = \epsilon_F \left[1 - \frac{\pi^2}{12}\left(\frac{k_B T}{\epsilon_F}\right)^2 - \frac{\pi^4}{80}\left(\frac{k_B T}{\epsilon_F}\right)^4\right]. \tag{9.56}$$

The average energy is evaluated by substituting (9.51) in (9.35) to obtain

$$U = \frac{3\bar{N}\epsilon_F}{5}\left[1 + \frac{5\pi^2}{12}\left(\frac{k_BT}{\epsilon_F}\right)^2 + I_4 z_0^4\right], \tag{9.57}$$

where

$$I_4 = a_{12} - \frac{a_{11}c_3}{2} + \frac{35c_3^2}{8} - \frac{5c_5}{2}. \tag{9.58}$$

The a_{11} is as in (9.43). Using (9.33), the values of a_{12} may be shown to be given by

$$a_{11} = \frac{5\pi^2}{8}, \qquad a_{12} = -\frac{7\pi^4}{384}. \tag{9.59}$$

It then follows that

$$U = \frac{3\bar{N}\epsilon_F}{5}\left[1 + \frac{5\pi^2}{12}\left(\frac{k_BT}{\epsilon_F}\right)^2 - \frac{\pi^4}{16}\left(\frac{k_BT}{\epsilon_F}\right)^4\right]. \tag{9.60}$$

The equations above provide means of evaluating various thermodynamic quantities.

Exercises

Ex. 9. 7. Show that the fundamental energetic equation for the gas of Fermions, in the first order of low-temperature approximation is

$$U = \frac{3\bar{N}\epsilon_F}{5}\left[1 + \frac{5}{3\pi^2}\left(\frac{S}{\bar{N}k_B}\right)^2\right]. \tag{9.61}$$

Hint: Use (9.48) to express k_BT/ϵ_F in terms of S.

Ex. 9. 8. (a) Show that for the gas of Fermions in the first order of low temperature approximation,

$$P\left(\frac{\partial V}{\partial T}\right)_P = \frac{2}{5}\left[1 + \frac{\pi^2}{3}\left(\frac{k_BT}{\epsilon_F}\right)^2\right]C_V. \tag{9.62}$$

(b) Use the result above to show that

$$\frac{C_P}{C_V} = 1 + \frac{\pi^2}{3}\left(\frac{k_BT}{\epsilon_F}\right)^2. \tag{9.63}$$

Hint: (a) Differentiate (9.46) with respect to T at constant P using $(\partial \epsilon_F / \partial T)_P = (\partial V / \partial T)_P (d\epsilon_F / dV)$ and recall from (9.8)) that $\epsilon_F = CV^{-2/3}$ (C is a constant) so that

$$\frac{d\epsilon_F}{dV} = -\frac{2\epsilon_F}{3V}. \tag{9.64}$$

Recall also the expression (9.47) for C_V. (b) Use first law of thermodynamics and the relation (8.130) to show that

$$C_p = \frac{5P}{2} \left(\frac{\partial V}{\partial T} \right)_P. \tag{9.65}$$

Ex. 9. 9. Show that the isothermal compressibility of the gas of Fermions in the first order of low-temperature approximation is given by

$$\kappa_T = \frac{3V}{2\bar{N}\epsilon_F} \left[1 - \frac{\pi^2}{12} \left(\frac{k_B T}{\epsilon_F} \right)^2 \right], \tag{9.66}$$

where κ_T is defined in (2.110). Hint: Differentiate (9.46) with respect to V at constant T using (9.64).

Ex. 9. 10. Show that the isentropic compressibility of the gas of Fermions in the first order of low-temperature approximation is given by

$$\kappa_S = \frac{3V}{2\bar{N}\epsilon_F} \left[1 - \frac{5\pi^2}{12} \left(\frac{k_B T}{\epsilon_F} \right)^2 \right], \tag{9.67}$$

where κ_S is defined in (2.103). Hint: Use (9.48) to express $k_B T / \epsilon_F$ in the expression (9.46) for PV in terms of S:

$$PV = \frac{2\bar{N}\epsilon_F}{5} \left[1 + \frac{5}{3\pi^2} \left(\frac{S}{\bar{N}k_B} \right)^2 \right]. \tag{9.68}$$

Differentiate this with respect to V at constant S using (9.64).

Chapter 10
Ideal Bose Gas

In this chapter we investigate properties of the gas of free non-interacting Bosons. We will see that at low temperatures it exhibits the phenomenon of condensation of macroscopic number of Bosons to the zero-energy state, called the Bose–Einstein condensation. The relation of the phenomenon of the Bose–Einstein condensation with that of the phase transitions is explored.

10.1 Bose Gas

Since $z < 1$ for an ideal Bose gas, its average particle density is given by (8.141) with $\eta = 1$ therein:

$$\bar{n} \equiv \frac{\bar{N}}{V} = \lambda_T^{-3} g_{3/2}(z) + \frac{1}{V}\frac{z}{1-z} \equiv \bar{n}_e + \bar{n}_0, \qquad (10.1)$$

where $g_{3/2}^{(1)}(z)$ has been written as $g_{3/2}(z)$ for convenience and

$$\bar{n}_e = \lambda_T^{-3} g_{3/2}(z), \qquad \bar{n}_0 = \frac{1}{V}\frac{z}{1-z}. \qquad (10.2)$$

The $\bar{n}_0$ is the number density in the zero-energy ground state, and $\bar{n}_e$ stands for the number density in the excited states. We address below the question of inverting (10.1) to obtain z in terms of $\bar{n}$ and T.

If the second term in (10.1) is ignored then $\bar{n}_e = \bar{n}$ so that

$$\lambda_T^3 \bar{n} = g_{3/2}(z). \qquad (10.3)$$

© The Author(s), under exclusive license to Springer Nature Switzerland AG 2024
R. R. Puri, *Modern Thermodynamics and Statistical Mechanics*, Undergraduate Lecture
Notes in Physics, https://doi.org/10.1007/978-3-031-54310-4_10

Fig. 10.1 $g_{3/2}(z)$ and $g_{5/2}(z)$ as functions of z

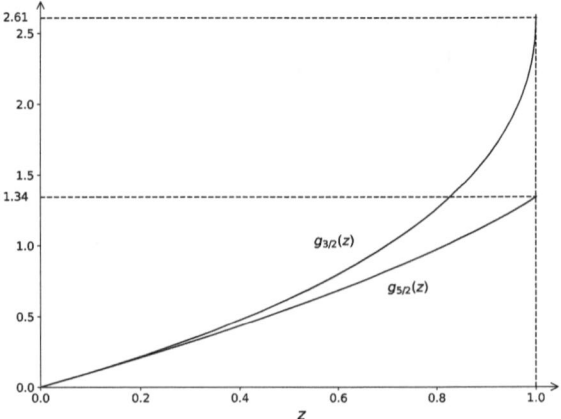

The function $g_{3/2}(z)$ is monotonically increasing and is bounded for $0 \le z \le 1$ (see Fig. 10.1), its maximum value in the said interval being

$$g_{3/2}(1) = 2.612\ldots \tag{10.4}$$

Equation (10.3) determines the fugacity z for specified values of $\bar{n}$ and T. The maximum value $z = 1$ will be reached for such values of $\bar{n}$ and λ_T^3 for which the product $\bar{n}\lambda_T^3$ attains the value $g_{3/2}(1)$, called the critical value $(\bar{n}\lambda_T^3)_c$:

$$(\bar{n}\lambda_T^3)_c = g_{3/2}(1). \tag{10.5}$$

Since $g_{3/2}(z)$ is monotonically increasing, (10.3) shows that if $\bar{n}\lambda_T^3$ is increased then z must increase. Since the right side of (10.3) attains maximum value for $z = 1$, it follows that there is no value of z for which (10.3) can be satisfied if the values of $\bar{n}$ and λ_T^3 are such that $\bar{n}\lambda_T^3 > g_{3/2}(1)$. This is clearly an unacceptable situation because, in principle, $\bar{n}\lambda_T^3$ can be made arbitrarily large. We will see that inclusion of the second term in (10.1), ignored in writing (10.3), provides correct description.

To that end, recall that (10.3) is obtained if the discrete sum in the expression (8.45) for the grand partition function is replaced by integration. Such a replacement is justified for high values of energy. Hence, if the population in lower energy levels is insignificant, the sum may be replaced by the integral. In order to incorporate the situations in which lower energy levels may be significantly populated, we evaluate that sum by splitting the energy levels into three groups: (i) zero-energy ground level, (ii) starting from the first excited state, the group consisting of a finite number P of levels and (iii) the rest of the levels to write the average particle number as

$$\bar{N} = \bar{N}_0 + \sum_{k=1}^{P} \bar{N}_k + \sum_{k>P} \bar{N}_k, \tag{10.6}$$

where $\bar{N}_0$ is average number of particles in the state of zero energy and $\bar{N}_k$ is that in the state of energy ϵ_k. The last sum can be replaced by an integral whose contribution will be same as in (10.3) because removal of a small number of points does not change the integral so that

$$\bar{n} = \frac{1}{\lambda_T^3} g_{3/2}(z) + \frac{\bar{N}_0}{V} + \frac{1}{V} \sum_{k=1}^{P} \bar{N}_k. \tag{10.7}$$

Now, from (8.38) we know that (with $\bar{n}_{kB} \rightarrow \bar{N}_k$)

$$\bar{N}_0 = \frac{z}{1-z}, \qquad \bar{N}_k = \frac{z \exp(-\beta \epsilon_k)}{1 - z \exp(-\beta \epsilon_k)}, \qquad \epsilon_k \neq 0. \tag{10.8}$$

The average occupation numbers $\{\bar{N}_k\}$ corresponding to $\epsilon_k \neq 0$ are clearly finite for any z. Hence the last term in (10.7) would approach zero as $V \rightarrow \infty$. However, the average occupation number $\bar{N}_0$ of the ground state increases indefinitely as $z \rightarrow 1$ and may become so large as to make the number density in the ground state,

$$\bar{n}_0 = \frac{\bar{N}_0}{V} = \frac{z}{V(1-z)}, \tag{10.9}$$

finite even for macroscopic volumes. It is therefore sufficient to retain only the $\bar{N}_0$ term from the discrete summation part reducing it to the anticipated form (10.1), rewritten as

$$\bar{n}\lambda_T^3 = g_{3/2}(z) + \lambda_T^3 \bar{n}_0, \qquad \bar{n}_0 = \frac{\bar{N}_0}{V} = \frac{z}{V(1-z)}. \tag{10.10}$$

This determines z for a specified value of $\bar{n}\lambda_T^3$. For the values of $\bar{n}\lambda_T^3 < g_{3/2}(1)$, we know that $z < 1$. The factor $z/(1-z)$ in that case is finite and hence $z/(V(1-z)) \rightarrow 0$ in the thermodynamic limit $V \rightarrow \infty$ so that $\bar{n}_0 = 0$, $\bar{n}_e = \bar{n}$. The gas is then said to be in the *normal phase*:

$$\begin{aligned} \bar{n}_0 = 0, \qquad & \bar{n}_e = \bar{n} \\ \bar{n}\lambda_T^3 = g_{3/2}(z), \qquad & \bar{n}\lambda_T^3 < g_{3/2}(1), \quad \text{normal phase.} \end{aligned} \tag{10.11}$$

The thermodynamic properties of the gas in the normal phase are determined by evaluating z by solving the second of the equations above for the specified values of $\bar{n}$ and T.

The factor $z/(V(1-z))$ would contribute significantly when

$$\frac{z}{1-z} \sim V, \quad \text{which implies} \quad z \sim 1 - \frac{1}{V}. \tag{10.12}$$

The values of $z \sim 1 - 1/V$ are said to be in the vicinity of $z = 1$. We know that when $\bar{n}\lambda_T^3 \to (\bar{n}\lambda_T^3)_c$, $z \to 1$ so that z lies in the vicinity of $z = 1$, making the ground state number density $\bar{n}_0$ non-zero. The z remains in the vicinity of 1 even as $\bar{n}\lambda_T^3$ increases beyond $(\bar{n}\lambda_T^3)_c$. Since $z \approx 1$ in the vicinity of $z = 1$, $g_{3/2}(z) \approx g_{3/2}(1)$. Hence, when $\bar{n}\lambda_T^3 > (\bar{n}\lambda_T^3)_c$ the occupation number of the ground state, determined by letting $g_{3/2}(z) = g_{3/2}(1)$ in (10.10), is

$$\frac{\bar{n}_0}{\bar{n}} = 1 - (\bar{n}\lambda_T^3)^{-1}g_{3/2}(1), \qquad \bar{n}\lambda_T^3 > g_{3/2}(1). \tag{10.13}$$

This shows that, out of total number density $\bar{n}$, the number density $\bar{n}_e$ in the excited state is

$$\frac{\bar{n}_e}{\bar{n}} = (\bar{n}\lambda_T^3)^{-1}g_{3/2}(1), \qquad \bar{n}\lambda_T^3 > g_{3/2}(1), \tag{10.14}$$

The last two equations give the fraction of molecules in the zero energy, and the excited states when $\bar{n}\lambda_T^3 > g_{3/2}(1)$.

We see that at the critical point, the ground state number density becomes finite and keeps increasing as the value of the parameter $\bar{n}\lambda_T^3$ increases beyond $g_{3/2}(1)$. The phenomenon of the Bose gas making transition from the normal state in which $\bar{n}_0 = 0$ to the one in which $\bar{n}_0$ acquires macroscopic value is called the *Bose–Einstein condensation* (BEC). The region defined by the condition $\bar{n}\lambda_T^3 > g_{3/2}(1)$ is referred to as the *condensation region*. As $\bar{n}\lambda_T^3 \to \infty$, $\bar{n}_e/\bar{n} \to 0$ and $\bar{n}_0/\bar{n} \to 1$, i.e. all the molecules condense to the zero-energy state. The gas is then said to be in the *condensed phase*. For finite values of $n\lambda_T^3 > g_{3/2}(1)$, it is a mixture of the condensed and the normal phase, called the *condensate*.

Summary

The value of the parameter $\bar{n}\lambda_T^3$ defines two phases of the Bose gas. When $\bar{n}\lambda_T^3 < g_{3/2}(1)$, the ground state of energy zero has no macroscopic number of molecules in it. This is called the normal phase. When $\bar{n}\lambda_T^3 = g_{3/2}(1)$, called the critical value of $\bar{n}\lambda_T^3$ denoted by $(\bar{n}\lambda_T^3)_c$, the ground state starts occupying macroscopic population which continues to increase as the value of $\bar{n}\lambda_T^3$ increases beyond $g_{3/2}(1)$. The phenomenon of the Bose gas making transition from the normal phase in which the ground state population density $\bar{n}_0 = 0$ to the one in which $\bar{n}_0$ acquires macroscopic value is called the Bose–Einstein condensation. The entire number density resides in the ground state as $\bar{n}\lambda_T^3 \to \infty$. This is called the condensed phase. The phase of the gas corresponding to finite values of $\bar{n}\lambda_T^3 > g_{3/2}(1)$ is called the condensate. The equations determining the number of molecules in various phases are:

Normal Phase $\bar{n}\lambda_T^3 < g_{3/2}(1)$

$$\bar{n}_0 = 0, \qquad \bar{n}_e = \bar{n}, \qquad \bar{n}\lambda_T^3 = g_{3/2}(z). \tag{10.15}$$

Condensate $\bar{n}\lambda_T^3 > g_{3/2}(1)$

$$\frac{\bar{n}_0}{\bar{n}} = 1 - (\bar{n}\lambda_T^3)^{-1}g_{3/2}(1), \qquad \frac{\bar{n}_e}{\bar{n}} = (\bar{n}\lambda_T^3)^{-1}g_{3/2}(1). \qquad (10.16)$$

Condensed Phase $\bar{n}\lambda_T^3 \to \infty$

$$\frac{\bar{n}_0}{\bar{n}} = 1, \qquad \frac{\bar{n}_e}{\bar{n}} = 0. \qquad (10.17)$$

Let us determine how $\bar{n}_0$ varies when the gas is compressed at constant temperature and how it varies when it is cooled keeping density fixed.

1. If gas in the normal phase is compressed keeping temperature fixed at T then $\bar{n}$ would increase till its critical value $\bar{n}_c$ at which $(\bar{n}\lambda_T^3)_c = \bar{n}_c\lambda_T^3 \equiv g_{3/2}(1)$ so that

$$\bar{n}_c = \lambda_T^{-3}g_{3/2}(1). \qquad (10.18)$$

The $\bar{n}_c$ is called the *critical density* corresponding to a specified T. The gas is in the normal phase for $\bar{n} < \bar{n}_c$ so that the value of $\bar{n}_0$ in the normal phase given by the first equation in (10.15) holds for $\bar{n} < \bar{n}_c$:

$$\bar{n}_0 = 0, \qquad \bar{n} < \bar{n}_c, \qquad \text{normal phase}. \qquad (10.19)$$

For $\lambda_T^3 > \bar{n}_c$, the ground state occupancy is determined by (10.16) which, on using (10.18), reads

$$\frac{\bar{n}_0}{\bar{n}} = 1 - \frac{\bar{n}_c}{\bar{n}}, \qquad \bar{n} > \bar{n}_c, \qquad \text{condensate}. \qquad (10.20)$$

In terms of the number $\bar{N}_0$ of molecules in the ground state, and their total number $\bar{N}$, the equation above assumes the form

$$\frac{\bar{N}_0}{\bar{N}} = 1 - \frac{\bar{n}_c}{\bar{n}}, \qquad \bar{n} > \bar{n}_c, \qquad \text{condensate}. \qquad (10.21)$$

Equations (10.19) and (10.20) determine the ground state population as a function of density at a fixed temperature. We see that the ground state population remains zero for densities less than the critical density but starts acquiring finite values at the critical value and keeps increasing as the density is increased further.

2. If the gas in the normal phase is cooled keeping the density fixed at $\bar{n}$ it will remain in normal phase till temperature attains the value T_c, called the *critical temperature* for the specified value of $\bar{n}$, at which $(\bar{n}\lambda_T^3)_c = \bar{n}\lambda_{T_c}^3 \equiv g_{3/2}(1)$, i.e. when

$$\lambda_{T_c}^3 = \bar{n}^{-1}g_{3/2}(1). \qquad (10.22)$$

Hence, due to (10.15),

$$\bar{n}_0 = 0, \quad T > T_c, \qquad\qquad \text{normal phase.} \qquad\qquad (10.23)$$

For $\lambda_T^3 > \lambda_{T_c}^3$, the ground state occupancy is determined by (10.16) which, on using (10.22) reads

$$\frac{\bar{n}_0}{\bar{n}} = 1 - \left(\frac{T}{T_c}\right)^{3/2}, \quad T < T_c, \qquad \text{condensate.} \qquad\qquad (10.24)$$

In terms of the number $\bar{N}_0$ of particles in the ground state, and their total number N, the equation above reads

$$\frac{\bar{N}_0}{\bar{N}} = 1 - \left(\frac{T}{T_c}\right)^{3/2}, \quad T < T_c, \qquad \text{condensate.} \qquad\qquad (10.25)$$

Equations (10.23) and (10.25) determine the ground state population as a function of temperature above and below the critical temperature at a fixed density. We see that the ground state population remains zero for temperatures higher than the critical temperature but starts acquiring finite values at the critical value and keeps increasing as temperature is lowered further.

In Sect. 10.1.3 we will see that the BEC may be viewed as a phenomenon of phase transition.

The BEC is the phenomenon of condensation of the molecules in the momentum space. However, falling under gravity, the condensed phase would separate from the normal phase in real space. See [1] and references therein for the theory of Bose gas under gravity.

Though predicted in 1924, the experimental realization of BEC had to wait for technological advancements to create low density gases and low temperatures. First BEC was observed in 1995 in the gas of Rb-87 atoms at the critical temperature 170nK at the number density of about 2.5×10^{12}/c.c. [2]. It was soon followed by its observation in Na atoms at the critical temperature 2μK at the number density of about 10^{14}/c.c. [3].

10.1.1   Conditions for BEC

The occurrence of BEC is the result of non-solvability of the (10.3) for $g_{3/2}(z)$ in terms of the parameter $\bar{n}\lambda_T^3$ under the restriction $0 \leq z \leq 1$ on the values of z. The reason for it is that $g_{3/2}(z)$ is monotonically increasing and bounded at $z = 1$. As a result, $g_{3/2}(z)$ has maximum value $g_{3/2}(1)$ in the desired range $0 \leq z \leq 1$ whereas there is no bound on the values of $\bar{n}\lambda_T^3$. No such problem would arise if in place of $g_{3/2}(z)$

there were another function unbounded at $z = 1$. To look for situations in which said possibility may be realized, recall that the relation (10.3) is for three-dimensional non-relativistic Bose gas. Let us examine the equations for $\bar{n}\lambda_T^3$ in different dimensional spaces obeying different energy-momentum relations.

To that end, recall that the average number of particles in d-dimensional Bose gas in which energy is related with momentum by the relation $\epsilon = C_n p^n$ is given by (8.150):

$$\bar{N} = \frac{A_{nd}}{\beta^{d/n}} \Gamma(d/n) g_{d/n}(z),$$

where A_{nd} is as in (8.90). The BEC will occur if $g_{d/n}(1)$ is bounded. If $g_{d/n}(1)$ is not bounded, $\bar{N}$ in the equation above can take values without restriction so that BEC will not occur if $g_{d/n}(1)$ is unbounded. We know that $g_{d/n}(1)$ is bounded when $d/n > 1$ else it is not. We thus see that the condition for BEC to occur is $d/n > 1$. It is, of course, satisfied for 3-dimensional non-relativistic Bose gas. However, for two-dimensional non-relativistic Bose gas $d = 2, n = 2$ so that $d/n = 1$. Hence BEC will not occur in a two-dimensional non-relativistic Bose gas. On the other hand, for two-dimensional ultra-relativistic Bose gas $d = 2, n = 1$ so that $d/n > 1$ and hence such a gas will exhibit BEC.

For Bose gas of photons in a blackbody cavity $d = 3, n = 1$ so that the condition $d/n > 1$ for BEC to occur is satisfied. It, however, can not exhibit BEC for different reasons. For, recall that the fugacity is $z = \exp(\alpha)$ where α is the Lagrange multiplier for imposing the restriction on the average number of particles while maximizing entropy. In case the average number of particles cannot be specified, the corresponding Lagrange multiplier α is zero so that $z = 1$. For such gases therefore there is no question of determining z as a function of temperature and density and hence those gases will not exhibit BEC. That is the case with the photon gas in a blackbody cavity because the photons in the cavity are created and absorbed continuously by the oscillators in the walls of the cavity due to which it is not possible to specify the photon number. Hence photon gas in a blackbody cannot exhibit BEC.

However, though the restriction on the average photon number does not apply to photons in a blackbody, it is possible to create conditions in which that restriction applies making z vary. Such conditions have been conceptualized and BEC of gas of photons realized experimentally. See for example [4, 5].

10.1.2 Thermodynamic Properties

We showed that the state of the Bose gas makes transition from normal to the condensate phase when its temperature and density are such that the parameter $\bar{n}\lambda_T^3$ exceeds the value $g_{3/2}(1)$. We now study its thermodynamic properties in those two phases. To that end, since $z < 1$ for the Bose gas, we use the expressions for various thermodynamic quantities derived in Sect. 8.7 for $z < 1$ corresponding to $\eta = 1$ with $g_p^{(1)}(z) \to g_p(z)$.

1. Consider the expression (8.139) for pressure:

$$P = \frac{1}{\beta \lambda_T^3} g_{5/2}(z) - \frac{1}{\beta V} \ln(1 - z). \qquad (10.26)$$

When the gas is in the normal phase, z is not in the vicinity of 1 so that the second term in (10.26) goes to zero in the thermodynamic limit. On the other hand, when z is in the vicinity of 1, $(1/V)\ln(1 - z) \to 0$ as $V \to \infty$ whereby the second term in (10.26) vanishes in the condensate as well. Recalling that $z = 1$ in the condensate, (10.26) reduces to

$$P = \frac{1}{\beta \lambda_T^3} g_{5/2}(z), \qquad \lambda_T^3 \bar{n} < g_{3/2}(1), \quad \text{normal phase,}$$

$$P = \frac{1}{\beta \lambda_T^3} g_{5/2}(1), \qquad \lambda_T^3 \bar{n} > g_{3/2}(1), \quad \text{condensate,} \qquad (10.27)$$

where from the tabulated values

$$g_{5/2}(1) = 1.341 \ldots \qquad (10.28)$$

Equation (10.27) shows that the pressure of the condensate is independent of $\bar{n}$. Consequently, its isothermal compressibility κ_T (defined in (2.110)) is infinity in the condensate including at the critical point when approached from the condensate side. It is shown in Ex. 10.4 that $\kappa_T \to \infty$ even when the critical point is approached from the normal phase side.

In the normal phase, dependence of pressure on $\bar{n}$ comes from the dependence of z on $\bar{n}$ (see Ex. 10.3).

On using the expression (7.6) for λ_T, the second equation in (10.27) may be rewritten as

$$P = (k_B T)^{5/2} \left(\frac{2\pi m}{h^2} \right)^{3/2} g_{5/2}(1), \qquad \text{condensate.} \qquad (10.29)$$

This is the equation of state for the BEC.

2. On combining (10.15) and (10.27) we obtain

$$P = \frac{N k_B T}{V} \frac{g_{5/2}(z)}{g_{3/2}(z)}, \qquad \text{normal phase.} \qquad (10.30)$$

Since $g_{5/2}(z) \leq g_{3/2}(z)$ (see Fig. 10.1), we see that the pressure exerted by the Bose gas in the normal phase is always less than that exerted by the classical gas at same density and temperature. Recall that we arrived at the same conclusion by expressing P as a power series in $\bar{n}\lambda_T^3$ in (8.163). In particular, at the critical temperature T_c,

$$P = \frac{Nk_BT_c}{V}\frac{g_{5/2}(1)}{g_{3/2}(1)} \approx 0.513\frac{Nk_BT_c}{V}. \tag{10.31}$$

This shows that the pressure of the Bose gas at the critical temperature is about half of that of the classical gas at the same temperature and density.

3. Recall the expression (8.140) for the internal energy U:

$$U = \frac{3V}{2\beta\lambda_T^3}g_{5/2}(z). \tag{10.32}$$

By the same arguments as were used to arrive at (10.27) we get

$$U = \frac{3V}{2\beta\lambda_T^3}g_{5/2}(z), \qquad \lambda_T^3\bar{n} < g_{3/2}(1), \quad \text{normal phase,}$$

$$U = \frac{3V}{2\beta\lambda_T^3}g_{5/2}(1), \qquad \lambda_T^3\bar{n} > g_{3/2}(1), \quad \text{condensate.} \tag{10.33}$$

4. Consider the Bose gas at fixed number density $\bar{n}$ and vary its temperature. The transition between normal phase and the condensate then occurs at the critical temperature T_c. Using (10.33), the heat capacity at constant volume and number in the normal phase turns out to be

$$C_V = \frac{3k_BV}{2\lambda_T^3}\left(\frac{5}{2}g_{5/2}(z) + \frac{g_{3/2}(z)}{k_B\beta}\frac{1}{z}\left(\frac{\partial z}{\partial T}\right)_V\right), \quad T > T_c, \tag{10.34}$$

where we have used the identity (8.143). Use (10.15), and (10.54) to reduce the equation above to the form

$$\frac{C_V}{k_BN} = \frac{3}{4}\left(\frac{5g_{5/2}(z)}{g_{3/2}(z)} - \frac{3g_{3/2}(z)}{g_{1/2}(z)}\right), \quad T > T_c. \tag{10.35}$$

The value of C_V at the critical point is obtained by taking the limit $z \to 1$ of the expression above. Since $g_p(1)$ is finite for $p > 1$ and $g_{1/2}(1) \to \infty$ as $z \to 1$, the expression above yields

$$\left.\frac{C_V}{k_BN}\right|_{T_c+0} = \frac{15}{4}\frac{g_{5/2}(1)}{g_{3/2}(1)} \approx 1.925. \tag{10.36}$$

This is the value of the specific heat at the critical temperature when approached from above.

Next, the derivative of C_V with respect to T may be seen to be given by

$$\frac{1}{k_B N}\frac{\partial C_V}{\partial T} = \left[\frac{15}{4}\left(1 - \frac{g_{5/2}(z)g_{1/2}(z)}{g_{3/2}^2(z)}\right)\right.$$

$$\left. -\frac{9}{4}\left(1 - \frac{g_{3/2}(z)g_{-1/2}(z)}{g_{1/2}^2(z)}\right)\right]\frac{1}{z}\left(\frac{\partial z}{\partial T}\right)_V. \qquad (10.37)$$

Using (10.54) and the fact that $g_p(1)$ is finite if $p > 1$, $g_{1/2}(1) \to \infty$, the first term in (10.37) in the limit $z \to 1$ reduces to

$$\frac{15}{4}\left(1 - \frac{g_{5/2}(z)g_{1/2}(z)}{g_{3/2}^2(z)}\right)\frac{1}{z}\left(\frac{\partial z}{\partial T}\right)_V \to \frac{45}{8T_c}\frac{g_{5/2}(1)}{g_{3/2}(1)} \approx \frac{2.888}{T_c}. \qquad (10.38)$$

In the limit $z \to 1$, the second term in (10.37) reads

$$-\frac{9}{4}\left(1 - \frac{g_{3/2}(z)g_{-1/2}(z)}{g_{1/2}^2(z)}\right)\frac{1}{z}\left(\frac{\partial z}{\partial T}\right)_V$$

$$\to -\frac{27T_c}{8}Lt_{z\to 1}\left(\frac{g_{3/2}^2(z)g_{-1/2}(z)}{g_{1/2}^3(z)}\right). \qquad (10.39)$$

The functions $g_{\pm 1/2}(1)$ are not finite. We can, however, evaluate the expression above by using the asymptotic forms of those functions:

$$g_{1/2}(\exp(\alpha)) = \sqrt{\pi}(-\alpha)^{-1/2},$$

$$g_{-1/2}(\exp(\alpha)) = \frac{\sqrt{\pi}}{2}(-\alpha)^{-3/2}, \quad \alpha \to 0. \qquad (10.40)$$

It is readily seen that the expression (10.39) then reduces to

$$-\frac{27T_c}{8}Lt_{z\to 1}\left(\frac{g_{3/2}^2(z)g_{-1/2}(z)}{g_{1/2}^3(z)}\right) = -\frac{27}{16\pi}g_{3/2}^2(1) \approx -3.665. \qquad (10.41)$$

On substituting (10.38) and (10.41) in (10.37) we obtain

$$\frac{1}{k_B N}\frac{\partial C_V}{\partial T}\bigg|_{T_c+0} \approx -\frac{0.778}{T_c}. \qquad (10.42)$$

This is the slope of the C_V curve as a function of temperature at $T = T_c$ on the normal side of the gas.

5. We now evaluate C_V when $T < T_c$. Invoking (10.33) for $\lambda_T^3 \bar{n} < g_{3/2}(1)$ it is straightforward to see that

$$\frac{C_V}{k_B N} = \frac{15}{4\bar{n}\lambda_T^3}g_{5/2}(1). \quad T \le T_c. \qquad (10.43)$$

Since $\lambda_T \sim T^{-1/2}$, this shows that $C_V \sim T^{3/2}$ as $T \to 0$. Recall from (9.47) that in a Fermi gas $C_V \sim T$ as $T \to 0$.

At the critical point, $(\bar{n}\lambda_T)_c = g_{3/2}(1)$. Hence, the value of C_V at T_c on the condensate side is given by

$$\left.\frac{C_V}{k_B N}\right|_{T_c-0} = \frac{15}{4}\frac{g_{5/2}(1)}{g_{3/2}(1)} \approx 1.925. \tag{10.44}$$

This is the value of specific heat at the critical temperature when approached from below. It is same as that in (10.36) for C_V at $T_c + 0$. Thus the specific heat is continuous at the critical point.

On differentiating (10.43) with respect to T we obtain

$$\frac{1}{k_B N}\frac{\partial C_V}{\partial T} = \frac{45}{8T\bar{n}\lambda_T^3}g_{5/2}(1), \qquad T \le T_c. \tag{10.45}$$

Since $(\bar{n}\lambda_T)_c = g_{3/2}(1)$, at the transition point,

$$\left.\frac{1}{k_B N}\frac{\partial C_V}{\partial T}\right|_{T_c-0} = \frac{45}{8T_c}\frac{g_{5/2}(1)}{g_{3/2}(1)} \approx \frac{2.89}{T_c}. \tag{10.46}$$

On comparing this with (10.42) we see that

$$\left.\frac{\partial C_V}{\partial T}\right|_{T_c-0} \neq \left.\frac{\partial C_V}{\partial T}\right|_{T_c+0}. \tag{10.47}$$

This shows that the slope of the specific heat curve as a function of T is not continuous at the transition point as is exhibited also by the plot of C_V/Nk_B as a function of T/T_c in Fig. 10.2.

6. Invoking the identity (2.118) for the ratio of the specific heat at constant pressure to that at constant volume and recalling the expressions (10.57), (10.58) for isothermal and isentropic compressibilities we obtain

Fig. 10.2 Specific heat of Bose gas as a function of temperature

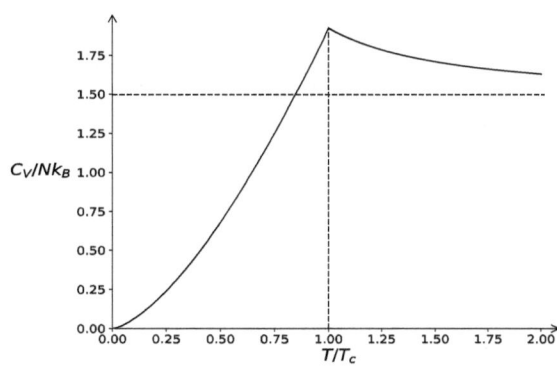

$$\frac{C_P}{C_V} = \frac{\kappa_T}{\kappa_S} = \frac{5}{3} \frac{g_{5/2}(z)g_{1/2}(z)}{g_{3/2}^2(z)}. \tag{10.48}$$

In the classical limit characterized by $z \ll 1$ it is straightforward to see by keeping the lowest order term in the summation defining $g_p(z)$ that $\Gamma = 5/3$ which is the known classical value.

7. The relations derived in Ex. 10.7 between pairs of (P, V, T) for the Bose gas undergoing adiabatic transformation are same as those for the classical gas. However, the exponents therein are not related with C_P/C_V, derived in (10.48) for the Bose gas, as they are in the classical gas (see also the discussion circa (8.151)).

8. Recalling (2.76) and (8.130), the grand potential for the Bose gas reads

$$\Omega(T, V, \mu) = -\frac{2}{3}U. \tag{10.49}$$

Invoking the expression (10.33) for U, we obtain

$$\Omega = -\frac{V}{\beta\lambda_T^3}g_{5/2}(z), \quad \lambda_T^3\bar{n} < g_{3/2}(1), \quad \text{normal phase,}$$

$$\Omega = -\frac{V}{\beta\lambda_T^3}g_{5/2}(1), \quad \lambda_T^3\bar{n} > g_{3/2}(1), \quad \text{condensate.} \tag{10.50}$$

We use these results to derive expression for the entropy of the Bose gas.

9. Recall from (2.78) the relation,

$$S = -\left(\frac{\partial\Omega}{\partial T}\right)_{V,\mu}, \tag{10.51}$$

to show using the expression (10.50) for $\Omega(T, V, \mu)$ that

$$\frac{S}{k_B N} = \frac{1}{\bar{n}\lambda_T^3}\left(\frac{5}{2}g_{5/2}(z) - \mu\beta g_{3/2}(z)\right) = \frac{5}{2}\frac{g_{5/2}(z)}{g_{3/2}(z)} - \mu\beta,$$

$$= \frac{5}{2}\frac{g_{5/2}(z)}{g_{3/2}(z)} - \ln(z) \quad \text{normal phase,}$$

$$\frac{S}{k_B N} = \frac{5}{2\bar{n}\lambda_T^3}g_{5/2}(1) \quad \text{condensate.} \tag{10.52}$$

We leave the derivation of the expressions above as an exercise. It is readily seen that S in (10.52) in the condensate can be written as

$$ST = \frac{5}{2}PV = \frac{5}{3}U, \quad \text{condensate.} \tag{10.53}$$

Since $\mu = 0$ in the condensate, the relation above is consistent with Euler's equation.

Exercises

Ex. 10.1. Show that

$$\frac{1}{z}\left(\frac{\partial z}{\partial T}\right)_{V,N} = -\frac{3g_{3/2}(z)}{2T g_{1/2}(z)}. \tag{10.54}$$

Hint: Differentiate second equation in (10.15) with respect to T.

Ex. 10.2. Show that

$$\frac{1}{z}\left(\frac{\partial z}{\partial T}\right)_{P,N} = -\frac{5}{2T}\frac{g_{5/2}(z)}{g_{3/2}(z)}. \tag{10.55}$$

Hint: Use (10.27).

Ex. 10.3. Show that

$$\frac{1}{z}\left(\frac{\partial z}{\partial V}\right)_{T,N} = -\frac{g_{3/2}(z)}{V g_{1/2}(z)}. \tag{10.56}$$

Hint: Use (10.15).

Ex. 10.4. Show that the isothermal compressibility κ_T (defined in (2.110)) for a Bose gas in normal phase is given by

$$\kappa_T = \frac{\beta}{\bar{n}}\frac{g_{1/2}(z)}{g_{3/2}(z)}. \tag{10.57}$$

Since $g_{1/2}(z) \to \infty$ as $z \to 1$, this shows that $\kappa_T \to \infty$ as the transition point is approached from the normal phase side. It has been shown circa (10.28) that $\kappa_T \to \infty$ as the transition point is approached from the condensate side as well. Hint: Differentiate (10.27) with respect to V at constant N, T and use (10.56).

Ex. 10.5. Show that the isentropic compressibility κ_S (defined in (2.103)) of Bose gas in normal phase is given by

$$\kappa_S = \frac{3\beta}{5\bar{n}}\frac{g_{3/2}(z)}{g_{5/2}(z)}. \tag{10.58}$$

Hint: Since S is constant, (10.52) shows that, at constant N, $z = $ constant for the gas in the normal phase. Hence partial derivatives at constant S and N are equivalent with those at constant z. Eliminate β between (10.15) and (10.27) to rewrite the expression for P in terms of V and z:

$$P = \left(\frac{h^2}{2\pi m}\right)^{2/3} \left(\frac{N}{V g_{3/2}(z)}\right)^{5/3} g_{5/2}(z). \qquad (10.59)$$

Use also (10.56).

Ex. 10.6. Derive (10.52) using (10.51). Hint: Note that the derivative in (10.51) is to be carried keeping μ fixed. Since $z = \exp(\mu\beta)$,

$$\left(\frac{\partial g_p(z)}{\partial T}\right)_{V,\mu} = -k_B\beta^2 \left(\frac{\partial g_p(z)}{\partial z}\right)_{V,\mu} \left(\frac{\partial z}{\partial \beta}\right)_{V,\mu} = -k_B\beta^2 \mu g_{p-1}.$$
$$(10.60)$$

Ex. 10.7. Show that with S given by (10.52), $T(\partial S/\partial T)_{N,V}$ leads to the same expression for U as in (10.33).

Ex. 10.8. Show that the equations of state for an adiabatic transformation of the Bose gas are

$$TP^{-2/5} = \text{constant}, \qquad TV^{2/3} = \text{constant}. \qquad (10.61)$$

Hint: Since S and N are constants in an adiabatic process, (10.52) shows that $z = $ constant for the gas in normal state. It then follows from (10.27) that $TP^{-2/5} = $ constant in both phases. Recall (8.151) $(PV^{5/3} = $ constant) to derive the other result above.

Ex. 10.9. Consider a two-dimensional ultra-relativistic Bose gas having average number $\bar{n}$ of particles per unit area. Show that, unlike the case of two-dimensional non-relativistic gas, the two-dimensional ultra-relativistic gas exhibits BEC and find the critical temperature T_c.

Hint: The number density per unit area in two-dimensional ultra-relativistic Bose gas, given by (8.150) corresponding to $d = 2, n = 1$ is

$$\bar{n} = \frac{2\pi}{h^2 c^2 \beta^2} g_2(z) \qquad (10.62)$$

The maximum possible value 1 of z shall be reached when

$$\frac{\bar{n} h^2 c^2 \beta^2}{2\pi} = g_2(1) = \frac{\pi^2}{6}. \qquad (10.63)$$

This will therefore have no solution when $\bar{n} h^2 c^2 \beta^2 / 2\pi > \pi^2/6$.

10.1.3   BEC as a Phenomenon of Phase Transition

Note: Reader may skip this subsection in first reading and return to it after going through the chapter on phase transitions.

The BEC exhibits the features of second as well as first-order phase transitions. For, the behavior of the population in the ground state in a BE gas remaining finite below a critical temperature and vanishing suddenly thereat to remain zero for higher temperatures is reminiscent of the behavior of the order parameter characteristic of a second-order phase transition. The compressibility of the BE gas going to infinity at the transition temperature (See Ex. 10.4) is also a characteristic of the second-order phase transition.

However, as we will show below, the BE transition involves absorption or release of the latent heat, a characteristic of the first-order phase transition. To that end we investigate the characteristics of $P - v$ ($v = 1/\bar{n}$) isotherms and the $P - T$ coexistence curve.

Consider the behavior of P as a function of the specific volume $v \equiv 1/\bar{n}$ at some fixed temperatures T. We know that the condensation region is characterized by such values of v, T for which $\lambda_T^3/v \geq g_{3/2}(1)$. Hence, for a given T, the specific volume v in the condensation region is such that $v \leq g_{3/2}^{-1}(1)\lambda_T^3$ whereby the maximum value $v_{\max}$ turns out to be

$$v_{\max} = g_{3/2}^{-1}(1)\lambda_T^3. \tag{10.64}$$

Now, as the expression (10.27) shows, pressure in the condensate for a given T does not depend on v. Hence an isotherm on the $P - v$ plot in the condensation region will be a line parallel to the v-axis extending from $v = 0$ to $v_{\max}$. We can express $v_{\max}$ in terms of P by eliminating β between (10.64) and the second equation for P in (10.27), valid in the condensation region, leading to

$$P v_{\max}^{5/3} = \frac{h^2}{2\pi m} \frac{g_{5/2}(1)}{g_{3/2}^{5/3}(1)}. \tag{10.65}$$

Thus, at a given temperature, the $P - v$ plot in the condensation region at any pressure P extends from $v = 0$ to $v_{\max}$ where $v_{\max}$ is determined by (10.65). This is depicted in Fig. 10.3 where the dashed curve is the plot of $v_{\max}$ given by (10.65). It is called the coexistence curve because it is the locus of the meeting points of the isotherms in the normal and the condensate phases. Referring to the $T = T_1$ isotherm in the figure, the system is in condensed phase at A where $v = 0$ and in the normal phase at B where $v = v_{\max}$. It is a mixture of the two phases, which we have been calling condensate, for any other v between A and B. The part of the isotherm corresponding to $v > v_{\max}$ is determined by the first equation in (10.27) in which z is determined implicitly by the (10.15). The part of an isotherm in the condensate region, such as AB in Figure 10.3, resembles the plot of liquid–gas phase transition (see Fig. 12.4) wherein the part of the isotherms in the region where liquid and gas phases coexist is a straight line parallel to the v-axis.

We can determine the fraction of the condensed and the normal phases for any v in the condensate region by the lever rule (11.7). According to that rule, if v_1 is the specific volume of one phase, and v_2 that of the other at a given temperature then the

Fig. 10.3 $P - v$ diagram of
Bose gas. The dashed curve
is the plot of v_{max} given by
(10.65)

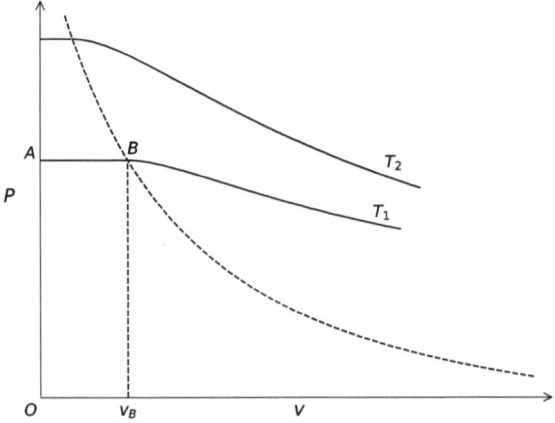

fractions of the two phases having combined specific volume v is given by

$$f_1 = \frac{v_2 - v}{v_2 - v_1}, \qquad f_2 = \frac{v - v_1}{v_2 - v_1}.$$

To apply this rule in the present case, consider the $P - v$ isotherm at temperature
T_1 in Figure 10.3. Let the index 1 in the expression above stand for the condensed
phase and 2 for the normal phase so that

$$f_{cond} = \frac{v_B - v}{v_B - v_A}, \quad f_{normal} = \frac{v - v_A}{v_B - v_A}, \quad v_B = v_{max} = \frac{\lambda_T^3}{g_{3/2}(1)}, \quad (10.66)$$

and $v_A = 0$. It then follows that

$$f_{cond} = \frac{v_B - v}{v_B}, \qquad f_{normal} = \frac{v}{v_B}. \tag{10.67}$$

We can evaluate also the difference in entropies at the points A and B noting that
entropy S_A in the condensed phase at A is zero, and entropy S_B in the normal phase
at B, evaluated using (10.52), is given by

$$\frac{S_B}{k_B N} = \frac{5 v_{max}}{2 \lambda_T^3} g_{5/2}(1) = \frac{5}{2} \frac{g_{5/2}(1)}{g_{3/2}(1)}. \tag{10.68}$$

The second equation above is due to (10.64). Hence difference between the normal
and the condensed state entropies per molecule is

$$\Delta s = (S_B - S_A)/N = \frac{5}{2} \frac{g_{5/2}(1)}{g_{3/2}(1)} k_B. \tag{10.69}$$

Since the transition occurs at a fixed temperature, the change in entropy is accompanied by the heat exchange called the latent heat L given by

$$L = T \Delta s = \frac{5}{2} \frac{g_{5/2}(1)}{g_{3/2}(1)} k_B T. \tag{10.70}$$

The presence of latent heat is a signature of first-order phase transition. To establish the consistency, we determine the latent heat also using the Clapeyron–Clausius equation (11.26) which relates latent heat with the slope of the $P - T$ coexistence curve.

The functional relation between P and T at the transition point defines the $P - T$ coexistence curve. The Clapeyron–Clausius equation (11.26) relating latent heat to the slope of the $P - T$ coexistence curve in the preset case reads

$$L = T (v_B - v_A) \left(\frac{dP}{dT} \right)_{\text{coex}}. \tag{10.71}$$

The slope of the $P - T$ coexistence curve is determined by (10.27) relating P and T in the condensate:

$$\frac{dP}{dT} = \frac{5k_B}{2\lambda_T^3} g_{5/2}(1). \tag{10.72}$$

Combine this with (10.71) with $v_A = 0$, $v_B = v_{\text{max}}$ to get (10.70). This establishes the consistency of the interpretation of the $P - v$ isotherms and the $P - T$ coexistence curve in BEC as the first-order phase transition.

10.2 Gas of Photons

By the gas of photons we mean quanta of electromagnetic (e.m.) radiation inside a blackbody at some temperature T. A photon is the quantum of a mode of the em field wave characterized by the wave vector $\mathbf{k}$ and one of the two directions of polarization $\mathbf{e}_{1\mathbf{k}}, \mathbf{e}_{2\mathbf{k}}$ for each value of $\mathbf{k}$. The state of a photon is therefore characterized by the parameter $k \equiv (\mathbf{k}, \mathbf{e}_{i\mathbf{k}})$ ($i = 1, 2$). We denote by ϵ_k the energy of the photon in the state characterized by k. The equilibrium state of the gas of photons inside a blackbody is characterized by average energy without any restriction on the number of photons. For, photons in the cavity are continuously being absorbed and emitted by the atoms in the walls of the cavity due to which their number cannot be specified, neither as exact nor as an average. The spin quantum number of a photon is 1. Hence photons are Bosons. The partition function Z_P of the gas of photons is therefore Bosonic given by (8.45) with $\alpha = 0$:

$$\ln(Z_P) = -\sum_k \ln\{1 - \exp(-\beta\epsilon_k)\}. \qquad (10.73)$$

The parameter α is put equal to zero because it is the Lagrange multiplier corresponding to the constraint on the average number of particles which is absent in the present case. The probability of having n_k photons of energy ϵ_k is given by (8.46):

$$p(n_k) = (1 - \exp(-\beta\epsilon_k)) \exp(-\beta\epsilon_k n_k). \qquad (10.74)$$

Photons are massless obeying the energy-momentum relation

$$\epsilon = pc. \qquad (10.75)$$

We therefore convert the sum to an integral by using the expression (8.77) for the density of states for a massless particle by the correspondence

$$\sum_k f(\epsilon(k)) \rightarrow \int_0^\infty f(\epsilon)\, D(\epsilon)\, d\epsilon, \qquad D(\epsilon) = 2\frac{4\pi V \epsilon^2}{h^3 c^3}. \qquad (10.76)$$

The expression for the density of states $D(\epsilon)$ above differs from the one given in (8.77) by multiplication by the factor of two to account for two directions of the field polarization.

In the following we use the correspondence above to investigate the thermodynamic properties of the gas of photons. See also [6].

10.2.1   Thermodynamic Properties

In Appendix F we have derived thermodynamic properties of the blackbody radiation by starting from the phenomenological Stefan–Boltzmann law. In the following we derive those properties by using the formalism of statistical mechanics and find that, it not only reproduces the results derived using the said law, but also yields expression for the phenomenological Stefan–Boltzmann constant in terms of the fundamental constants.

1. Following the procedure outlined in Sect. 8.7, we convert the sum in (10.73) defining $\ln(Z_P)$ to the integral with $D(\epsilon)$ given by (10.76), perform the integration by parts and transform the variable of integration ϵ to $x = \beta\epsilon$ to obtain

$$\ln(Z_P) = \frac{8\pi V}{3h^3 c^3 \beta^3} \int_0^\infty \frac{x^3\, dx}{\exp(x) - 1}. \qquad (10.77)$$

On using (H.12) to carry the integral above we get

$$\ln(Z_P) = \frac{8\pi^5 V}{45 h^3 c^3 \beta^3}. \tag{10.78}$$

2. The pressure of the photon gas is given by

$$P = \frac{1}{\beta V}\ln(Z_P) = \frac{8\pi^5 k_B^4}{45 h^3 c^3}T^4. \tag{10.79}$$

This shows that, since $(\partial P/\partial V)_T = 0$, the isothermal compressibility κ_T, defined in (2.110), is infinitely large for photon gas, or as is argued in [6], one must say it does not exist because the identity $(\partial V/\partial P)_T = 1/(\partial P/\partial V)_T$ does not hold when its left side is zero.
3. The internal energy of the photon gas is

$$U = -\frac{\partial \ln(Z_P)}{\partial \beta} = \frac{8\pi^5 k_B^4 V}{15 h^3 c^3}T^4. \tag{10.80}$$

The intensity I of radiation coming out from a small opening from the blackbody cavity is $I = cU/4V$ so that, with U given by (10.80), we have

$$I = \sigma(k_B T)^4, \qquad \sigma = \frac{2\pi^5}{15 h^3 c^2}. \tag{10.81}$$

This is the well-known Stefan–Boltzmann law where σ is Stefan–Boltzmann constant. That constant is included in thermodynamics as an unknown to be determined experimentally. In the statistical mechanical formalism we see that it emerges naturally and is a function only of the fundamental constants c and h.
4. On comparing (10.79) and (10.80) it is straightforward to see that

$$PV = \frac{1}{3}U. \tag{10.82}$$

It is same as the relation arrived in (F.5) by following the Stefan–Boltzmann law. It is also consistent with the general formula (8.147) as in the present case the system is three-dimensional and due to the energy-momentum relation (10.75) for photons, $n = 1$. Compare (10.82) with the corresponding relation (8.130) for massive quantum particles to note the difference of the factor of 2. On substituting in (10.82) the expression (10.80) for U, we recover the expression (10.79) for P.
5. The adiabatic equation of state for the photon gas is given by (8.151) with $d = 3$ and due to (10.75), $n = 1$ therein:

$$PV^{4/3} = \text{constant.} \tag{10.83}$$

6. Invoking (10.80), the specific heat at constant volume turns out to be

$$\frac{C_V}{k_B V} = \frac{1}{k_B V}\frac{\partial U}{\partial T} = \frac{32\pi^5 k_B^3}{15 h^3 c^3} T^3.$$ (10.84)

Note that in the photon gas, which is the gas of massless bosons, the specific heat $C_V \sim T^3$ as $T \to 0$ whereas $C_V \sim T^{3/2}$ for a massive Bosonic gas (see (10.43)).

7. Using the identity (H.1), the average number of photons in the level of energy ϵ_k may be shown to be given by

$$\bar{N}_k = \sum_{n_k=1}^{\infty} n_k p(n_k) = \frac{1}{\exp(\beta \varepsilon_k) - 1}.$$ (10.85)

Average number of photons per unit volume in all the states is

$$\frac{\bar{N}}{V} \equiv \frac{1}{V}\sum_k \frac{\exp(-\beta \epsilon_k)}{1 - \exp(-\beta \varepsilon_k)} = \frac{8\pi k_B^3 T^3}{h^3 c^3} \int_0^{\infty} \frac{x^2 dx}{\exp(x) - 1}.$$ (10.86)

Invoking the identity (H.12) we get

$$\frac{\bar{N}}{V} = \frac{8\pi k_B^3 T^3}{h^3 c^3}\Gamma(3)\zeta(3) = 1.202\frac{16\pi k_B^3}{h^3 c^3} T^3.$$ (10.87)

8. The expression for entropy, evaluated using Euler's equation with $\mu = 0$ therein for the gas of photons, reads

$$\frac{S}{k_B V} = \frac{32\pi^5 k_B^3}{45 h^3 c^3} T^3.$$ (10.88)

The T^3 dependence of S/V in the equation above is consistent with that arrived in (F.4) by Stefan–Boltzmann law.

9. To determine fluctuations in the number of photons in the mode k we evaluate $\langle N_k^2 \rangle$ using the identity (H.1) to get

$$\langle N_k^2 \rangle = \sum_{n_k=0}^{\infty} n_k^2 p(n_k) = \frac{\exp(-\beta \epsilon_k)(1 + \exp(-\beta \epsilon_k))}{(1 - \exp(-\beta \epsilon_k))^2}.$$ (10.89)

Derivation of the equation above is left as an exercise. It then follows that the variance in the number of photons in the state k is

$$\langle N_k^2 \rangle - \langle N_k \rangle^2 = \frac{\exp(-\beta \epsilon_k)}{(1 - \exp(-\beta \epsilon_k))^2}.$$ (10.90)

The variance in total number of photons is

$$\langle N^2 \rangle - \langle N \rangle^2 = \frac{8\pi k_B^3 T^3 V}{h^3 c^3} \int_0^\infty \frac{x^2 \exp(-x) \mathrm{d}x}{(1 - \exp(-x))^2}. \tag{10.91}$$

Invoking the identity (H.13), we obtain

$$\langle N^2 \rangle - \langle N \rangle^2 = \frac{8\pi^3 k_B^3 V}{3 h^3 c^3} T^3. \tag{10.92}$$

This shows that the fluctuations in the number of photons is finite. Now recall the result obtained following (10.79) namely κ_T is infinitely large (or does not exist) for the photon gas using which we see that the equation (6.77) relating number fluctuations with isothermal compressibility predicts that number fluctuations must be infinitely large (or do not exit). The question of the origin of incompatibility of the said results has been traced in [6] to the inapplicability of the derivation of (6.77) when applied to the photon gas. For, note that the derivation of (6.77) involves differentiating average number $\bar{N}$ of particles with respect to α which, being zero for photon gas, is put equal to zero while evaluating $\bar{N}$. Consequently $\bar{N}$ has no α dependence left in it. Alternatively, note that we get the factor $1/(\partial P/\partial \alpha)_{V,T}$ in (6.76) on way to the derivation of (6.77). Since $(\partial P/\partial \alpha)_{V,T} = 0$ for photon gas, the derivation becomes questionable for photon gas. For further discussion see [6].

References

1. R.K. Bhaduri, W. van Dijk, Phys. Lett. A **380**, 2480 (2016)
2. M.H. Anderson, J.R. Ensher, M.R. Mathews, C.E. Wieman, E.A. Cornell, Science **269**, 198 (1995)
3. K.B. Davis, M.-O. Mewes, M.R. Andrews, N.J. van Druten, D.S. Durfee, D.M. Kurn, W. Ketterle, Phys. Rev. Lett. **75**, 3969 (1995)
4. J. Klaers, J. Schmitt, F. Vewinger, M. Weitz, Nature **468**, 545 (2010)
5. J. Klaers, M. Weitz, (2012) arXiv:1210.7707v1 [cond-mat.quant-gas]
6. H.S. Leff, Am. J. Phys. **83**, 362 (2015)

Chapter 11
Phase Transitions and Critical Phenomena

Transformation of a substance between its solid, liquid, and gaseous forms is a common experience. In this chapter, we study the thermodynamic description of the said transformations called the phenomenon of phase transitions. Introduced also is the concept of the critical phenomenon.

11.1 Phase Equilibrium

It is observed that at certain values of pressure and temperature a homogeneous substance may separate into two homogeneous parts of different densities in contact with each other. The coexisting parts of a substance in contact with each other having different densities are called its *phases*. The gaseous, liquid, and solid forms of matter in that sense are its different phases. The phenomenon of separation of a homogeneous substance into its different phases is called the phenomenon of *phase transition*. The conditions under which the phase transition may occur can be determined as follows.

We label the two coexisting phases as phases 1 and 2. Since they are in equilibrium, their pressure, and temperature, denoted by (P_0, T_0), must be same. In addition, since the molecules are exchanged between two phases, the condition of equilibrium requires their chemical potentials also to equalize leading to the equation

$$\mu_1(P_0, T_0) = \mu_2(P_0, T_0), \tag{11.1}$$

where μ_i is the chemical potential of the ith phase ($i = 1, 2$). Comparing this with (2.66), we see that the relation above implies

$$g_1(P_0, T_0) = g_2(P_0, T_0), \tag{11.2}$$

© The Author(s), under exclusive license to Springer Nature Switzerland AG 2024
R. R. Puri, *Modern Thermodynamics and Statistical Mechanics*, Undergraduate Lecture
Notes in Physics, https://doi.org/10.1007/978-3-031-54310-4_11

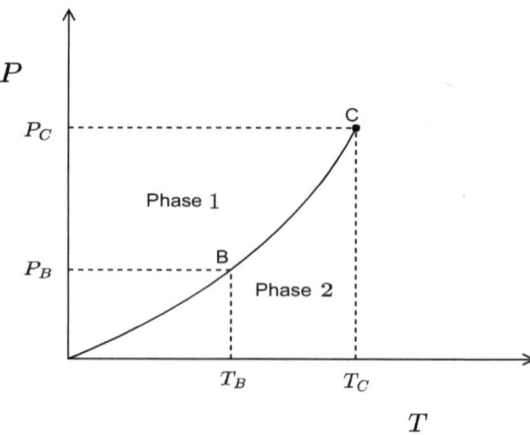

Fig. 11.1 P-T coexistence curve

where

$$g_i(P, T) = \frac{G_i(P, T)}{N_i}, \qquad (11.3)$$

is Gibbs potential per molecule of the ith phase. The two phases coexist only at the values of pressure and temperature related by (11.2). It describes a curve in the $P - T$ plane along which the two phases coexist. It is called the $P - T$ coexistence curve, depicted in Fig. 11.1. Though the $P - T$ coexistence curve is shown as terminating in the figure, it may not. When it does, the point of its termination is called the *critical point*. The temperature T_c at that point is called the critical temperature, and the corresponding pressure P_c the critical pressure. To see when the $P - T$ coexistence curve can terminate, consider the coexistence curve in Fig. 11.1 terminating at C. The substance is homogeneous at pressures and temperatures greater than P_c and T_c. This can happen if the two phases have same symmetry in their microscopic structure. On the other hand, if the two phases are of different symmetry, they can not constitute a homogeneous system thereby ruling out terminating $P - T$ coexistence curve. An example of non-terminating $P - T$ curve is that of the solid-liquid phase transition because molecules in a solid are arranged periodically but not in the liquid. The liquid-gas phase transition, on the other hand, is described by a terminating $P - T$ coexistence curve because both have non-periodic arrangement of molecules.

The $P - T$ coexistence curve therefore either does not terminate, or it terminates at the point of intersection of coexistence curves of other phases.

11.1.1   Triple Point

We have obtained the condition under which two phases coexist. We can similarly derive the conditions required for the existence of three phases in equilibrium. Those

Fig. 11.2 P-T coexistence
curves for three phases

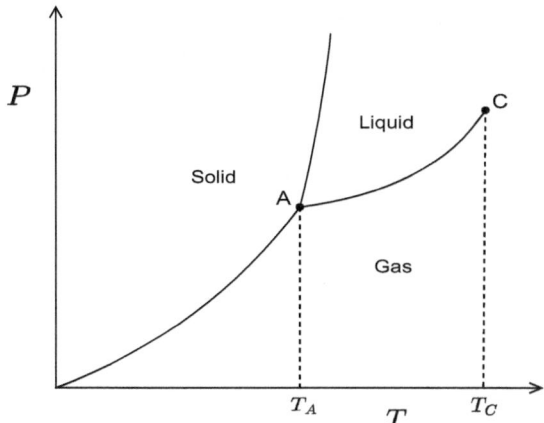

conditions evidently are same pressure P_0 and temperature T_0 in the three phases
and the equality of their chemical potentials:

$$\mu_1(P_0, T_0) = \mu_2(P_0, T_0) = \mu_3(P_0, T_0). \tag{11.4}$$

These are two simultaneous equations in two unknowns and hence define a point in the
$P - T$ plane. At the said point, called the *triple point*, three phases coexist. Clearly,
there can not be more than three coexisting phases of a substance as that would require
simultaneously satisfying three equations in two unknowns, the problem which has
no solution. Fig. 11.2 depicts coexistence curves for solid–liquid–gas transition, each
for equilibrium between two of the three phases. The gas–liquid coexistence curve
terminates at the critical point C but the liquid–solid curve does not except at the
triple point A.

The critical point of water occurs approximately at 373.9 °C, and 217.7 atm. Its
triple point occurs approximately at 0.01 °C, and 6.03×10^{-3} atm.

We construct next the $P - v$ coexistence curve and determine the fraction of each
phase in the coexistence region.

11.1.2 P − v Isotherms in Coexistence Region

We restrict our attention to the case when the $P - T$ coexistence curve terminates.
Consider the point B on the $P - T$ coexistence curve in Fig. 11.1 having coordinates
(T_B, P_B). We examine the nature of the $P - v$ ($v = V/N$ is the specific volume)
isotherm at temperature T_B. To that end, consider the point A_1 on the said isotherm
on the $P - v$ diagram in Fig. 11.3. As pressure is decreased keeping temperature
fixed at T_B, the $P - v$ values would trace the dashed curve on which there is one-to-
one relation between P and v and therefore the system is homogeneous till pressure

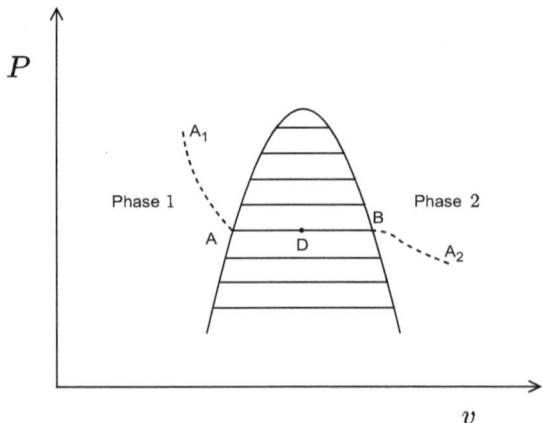

Fig. 11.3 Solid curve in the plot is the $P - v$ coexistence curve

reaches the value P_B (point A on the dashed isotherm) which is the value of the pressure corresponding to the temperature T_B on the $P - T$ coexistence curve. We know that the equation of state corresponding to (T_B, P_B) is solved by two values of v. Hence, at A the system separates into two parts of different specific volumes v_1, and v_2 where v_1 is the specific volume at the point A and v_2 that at the point B at the same pressure P_B. For $v > v_2$, the system follows the dashed curve on which there is one-to-one relation between P and v. Hence, for each temperature on the $P - T$ coexistence curve, there is a pair of points, one in the state of phase 1 and the other in phase 2. Join the points in phase 1 on different isotherms, and also those in phase 2. At the critical point on the $P - T$ coexistence curve, (point C in Fig. 11.1), the two solutions merge. Hence, the said two curves in the $P - v$ plot merge at the temperature T_C corresponding to the critical point C. The result is the solid curve depicted in Fig. 11.3, called the $P - v$ coexistence curve. Similar curve is obtained for the $T - v$ plot called the $T - v$ coexistence curve. The region bound by the coexistence curve is called the region of coexistence of the two phases.

As described above, the curve $A_1 A B A_2$ in Fig. 11.3 depicts the part of the isotherm corresponding to some temperature on the $P - T$ isotherm. The parts $A_1 A$, and $B A_2$ are determined by the given equation of state and describe homogeneous system. The straight line part $A B$ is the part of the isotherm in the coexistence region. The system at A is in phase 1 having specific volume v_1 and is in phase 2 at B having specific volume v_2. It is in mixed state for the values of v lying between v_1, and v_2. We will address the question of finding the equation of state in the coexistence region in Sect. 11.2. For now, we determine the fraction of the two phases when the specific volume of the system is in the coexistence region.

11.1.3 Lever Rule

In order to find the proportion of the amount of the two phases in the mixed state, let v be the specific volume of the system, represented by the point D on the line AB in the coexistence region. Let V be the volume of the system, and N the number of particles in it. Let V_i and N_i be the volume and the number of molecules in the ith phase at D so that $v_i = V_i/N_i$ is the specific volume of the ith phase with $V = V_1 + V_2$, $N = N_1 + N_2$. We then have

$$v \equiv \frac{V}{N} = \frac{V_1 + V_2}{N} = f_1 v_1 + f_2 v_2, \tag{11.5}$$

where $f_i = N_i/N$ is the fraction of the substance in the ith phase with

$$f_1 + f_2 = 1, \quad f_i = \frac{N_i}{N}. \tag{11.6}$$

Equations (11.5) and (11.6) lead to the following expression for f_i for the state of specific volume v:

$$f_1 = \frac{v_2 - v}{v_2 - v_1}, \quad f_2 = \frac{v - v_1}{v_2 - v_1}. \tag{11.7}$$

This is called the *lever rule*. It determines the fraction of the number of molecules in the phases 1, and 2 when their combined specific volume is v.

Using the lever rule, we can evaluate energy and entropy in the region of coexistence as follows.

Referring to Fig. 11.3, we evaluate energy of the system in the state described by the point D. Let U be the energy of the system so that its specific energy is $u = U/N$. Let U_i be the energy of the phase i component at the point D so that the specific energy of the ith phase component in the mixture of the phases at D is $u_i = U_i/N_i$. Due to additivity of energy, it is straightforward to show that

$$u = f_1 u_1 + f_2 u_2, \tag{11.8}$$

where f_i is given by the lever rule (11.7).

Like energy, due to additivity of entropy, the specific entropy of the system is given by

$$s = f_1 s_1 + f_2 s_2, \tag{11.9}$$

where s_i is specific entropy of the ith phase component.

11.2   Equation of State in Coexistence Region

An equation of state is the relation between P, v, T of a single phase system. The system in the coexistence region, on the other hand, is a mixture of two phases. In this section, we outline the procedure for obtaining the equation of state, energy, and entropy for the system in the coexistence region.

To determine the equation of state in the region of coexistence, referring to Fig. 11.3, let A_1ABA_2 therein be the $P - v$ isotherm corresponding to temperature T meeting the coexistence curve at the points A and B. Since pressure is same at all points on AB, the equation of state in the region of coexistence corresponding to temperature T would be

$$P(v, T) = \pi(T), \tag{11.10}$$

where

$$\pi(T) = P(v_1(T), T) = P(v_2(T), T). \tag{11.11}$$

Determination of the equation of state therefore requires knowledge of the specific volumes $v_1(T), v_2(T)$ corresponding to the points A and B at which the isotherm in question meets the coexistence curve. Those are determined by two equations relating $v_1(T)$, and $v_2(T)$. One is their defining relation: the equality of pressure at A and B,

$$P(v_1, T) = P(v_2, T), \tag{11.12}$$

and the other is the equality of the chemical potentials at A and B: $\mu_A = \mu_B$. Note that, since the system is in single phase at A, and at B where its specific volumes are v_1, v_2, the $P(v_i, T)$ in (11.12) is determined by the single phase equation of state.

The second relation between $v_1(T)$, and $v_2(T)$ emerging from the equality of μ_A and μ_B is obtained by integrating the Gibbs–Duhem relation (2.49) along the isotherm joining A and B on which it reduces to $d\mu = vdP$ so that

$$\mu_B - \mu_A = \int_A^B vdP. \tag{11.13}$$

Since $\mu_A = \mu_B$, the equation above yields

$$\int_A^B vdP \equiv \int_{v_1}^{v_2} v\frac{dP}{dv}dv = 0. \tag{11.14}$$

The expressions (11.12) and (11.14) are evaluated using single phase given equation of state. They determine two unknowns, v_1 and v_2, as the function of T and hence, due to (11.11), $\pi(T)$ which on substitution in (11.10) gives the equation of state in the region of coexistence.

The evaluation of v_1, v_2 using (11.14) thus requires knowledge of single phase equation of state. In Chap. 12, we will outline method of solving (11.14) when the equation of state is the van der Waals equation. We will introduce the concept of critical point in Sect. 11.4 and find the equation of state in the region of coexistence in its vicinity.

11.3 First-Order Phase Transition

We have seen that the intensive variables, P, T, μ, are same for the two coexisting phases but their specific volumes are different. Since, due to (2.66), $\mu = g$ where g is specific Gibbs potential, on replacing μ by g in the Gibbs–Duhem relation (2.49) or by invoking (2.68), we have

$$v = \left(\frac{\partial g}{\partial P}\right)_T, \qquad s = -\left(\frac{\partial g}{\partial T}\right)_P. \tag{11.15}$$

Since the value of v, and as we will see, also of s, for the two coexisting phases are different, it follows that, though $g(T, P)$ is continuous at the transition points on the coexistence curve, its first derivatives are not. The phase transition, characterized by continuity of the specific Gibbs potential but discontinuity of its first derivatives, is called *first-order phase transition*. Discontinuity in entropy is manifested, as we will see in the Sect. 11.3.3, in absorption or release of heat, called the latent heat, when the substance transforms from one phase to the other.

In Sect. 11.4, we will come across transitions in which specific Gibbs potential as well as its first derivatives are continuous but some of its second derivatives exhibit singularity. Such a transition is called *second-order phase transition*. Note that, since the first derivatives of $g(T, P)$ are continuous, the entropy of the coexisting phases is same and hence there is no transfer of latent heat. In view of this, the phase transitions which involve transfer of latent heat are called first order, and all the others *continuous-phase transitions*.

11.3.1 Entropy Discontinuity

We determine the difference in entropy of the two coexisting phases. To that end, consider the system at fixed pressure P_0 on its $P - T$ coexistence curve. Let $\mu_1(T)$ be the chemical potential of phase 1, and $\mu_2(T)$ that of phase 2 at temperature T. Let T_0 be the temperature corresponding to P_0 on the $P - T$ coexistence curve. The condition of coexistence implies $\mu_1(T_0) = \mu_2(T_0)$. We know that the equilibrium state for given P, T corresponds to the minimum of the Gibbs potential (see Sect. 2.6.2).

Hence, due to equality of specific Gibbs potential and chemical potential, it follows that the equilibrium state is one in which μ is minimum. Since, at the given pressure P_0, the system is in phase 1 below T_0, and in phase 2 above it, and the equilibrium state is one of lower chemical potential, we have

$$
\begin{aligned}
\mu_1(T_0 - \delta T) - \mu_2(T_0 - \delta T) &< 0, \quad \text{in phase 1,} \\
\mu_2(T_0 + \delta T) - \mu_1(T_0 + \delta T) &< 0, \quad \text{in phase 2.}
\end{aligned}
\tag{11.16}
$$

Since pressure is assumed to be constant, these equations imply

$$
-\left(\frac{\partial \mu_1}{\partial T}\right)_{P_0} < -\left(\frac{\partial \mu_2}{\partial T}\right)_{P_0}.
\tag{11.17}
$$

Due to the second equation in (11.15), the inequality above is same as

$$
s_1 < s_2,
\tag{11.18}
$$

where s_i is the entropy per molecule in the ith phase. This shows that the specific entropy in the phase at lower temperature is less than that at higher temperature close to the coexistence curve. This is what is expected intuitively too.

11.3.2 Energy Discontinuity

The internal energy also undergoes jump at phase transition. It can be evaluated by invoking the entropic Gibbs–Duhem equation (2.50) for $d(\mu/T)$ and noting that equality of μ on two sides of a point on the $P - T$ coexistence curve also implies equality of μ/T so that

$$
u_1 d(1/T) + v_1 d(P/T) = u_2 d(1/T) + v_2 d(P/T).
\tag{11.19}
$$

This leads to the equation

$$
u_2 - u_1 = (v_2 - v_1)\left[T\left(\frac{dP}{dT}\right)_{\text{coex}} - P\right],
\tag{11.20}
$$

which expresses jump in energy at the phase transition point in terms of the slope of the $P - T$ coexistence curve.

11.3.3 Latent Heat

We will show that the system absorbs or releases heat during first-order phase transition. To that end, note that since phase transition takes place at a constant temperature, the amount of heat L absorbed per molecule by the system while transforming from phase 1 to phase 2, obtained by integrating the equation $dq = T ds$ is given by

$$L = T(s_2 - s_1), \qquad L \equiv q_2 - q_1. \qquad (11.21)$$

Since, due to (11.18), $s_1 < s_2$, we see that $L > 0$, i.e. heat is absorbed while transforming from phase 1 to phase 2 when phase 2 is on the higher side of temperature of the $P - T$ coexistence curve. Since temperature of the system does not change while it absorbs the said heat, it is called the *latent heat*. Conversely, while transforming from phase 2 to phase 1, $L < 0$, i.e. latent heat is released. Hence, for example, heat is absorbed when liquid changes to gas, and is released in the reverse process.

11.3.4 Clapeyron–Clausius Equation

The Clapeyron–Clausius equation expresses slope of the $P - T$ coexistence curve in terms of the latent heat, and the difference in the specific volumes of the substance in two phases. It enables the determination of change in transition temperature due to the change in pressure in terms of the said measurable quantities. Referring to Fig. 11.4, let A_1 and A_2 be the points at (T, P) on the $P - T$ coexistence curve, one on its phase 1 side, and the other on the side of the phase 2. Similarly, let B_1 and B_2 be the points at $(T + dT, P + dP)$ on its two sides. Due to (11.1), we have

$$\mu_{A_1}(P, T) = \mu_{A_2}(P, T),$$
$$\mu_{B_1}(P + dP, T + dT) = \mu_{B_2}(P + dP, T + dT). \qquad (11.22)$$

Fig. 11.4 Plot for deriving Clapeyron–Clausius equation

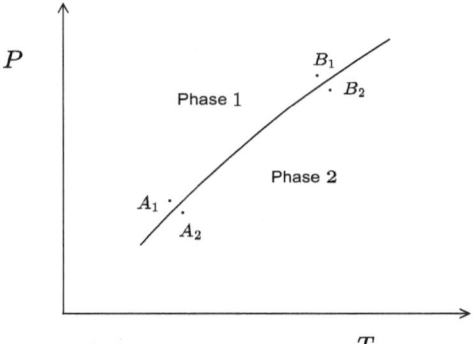

On subtracting the two equations above, we obtain

$$d\mu_{A_1}(P, T) = d\mu_{A_2}(P, T). \tag{11.23}$$

Invoking Gibbs–Duhem relation this leads to

$$-s_1 dT + v_1 dP = -s_2 dT + v_2 dP. \tag{11.24}$$

This relates the slope of the $P - T$ curve with the change in entropy of the two phases:

$$s_2 - s_1 = (v_2 - v_1) \left(\frac{dP}{dT}\right)_{coex}. \tag{11.25}$$

On substituting in this the expression (11.21) relating change in entropy with the latent heat follows the *Clapeyron–Clausius equation*

$$\left(\frac{dP}{dT}\right)_{coex} = \frac{L}{T(v_2 - v_1)}. \tag{11.26}$$

This relates latent heat to the change in pressure and transition temperature of the two coexisting phases. In particular, if phase 2 is gas, and phase 1 is liquid then, as argued circa (11.21), latent heat is positive, and so is $v_2 - v_1$ because the specific volume of gas is always greater than that of the liquid. Hence, $dP/dT > 0$. This means the transition temperature from liquid to gas will increase as pressure increases.

11.4   Critical Phenomenon

Fig. 11.5 depicts some $P - v$ isotherms. An isotherm for $T > T_c$ describes single homogeneous state. It is therefore a continuous curve each point of which is a point of stable equilibrium conforming to the stability requirement $(\partial P/\partial v)_T < 0$. On the other hand, a part of an isotherm for $T < T_c$ lies in one phase and the other in the other phase. As an example, consider the isotherm in the Fig. 11.5 for temperature $T_1 < T_c$. Its part in phase 1 meets the coexistence curve at the point A, and that in phase 2 meets the coexistence curve at B. The said points are at same pressure and temperature but have different specific volumes, v_1 and and v_2, respectively. Since the points A and B are on the stable parts of isotherms describing one homogeneous system or the other, $(\partial P/\partial v)_T < 0$ at A as well as at B. The part of the isotherm at A should therefore continue to fall for $v \geq v_A$, and that at B should continue to rise for $v \leq v_B$. Since the points A and B to be joined by the isotherm have same value of P, the falling part of the isotherm inside the region of coexistence should reach a minimum, and the rising one a maximum to enable them to join continuously. Let A_1 and B_1 in the figure be the points where $(\partial P/\partial v)_T = 0$. The isotherm has minimum

Fig. 11.5 P-v isotherms,
coexistence curve (solid),
and spinodal curve (dashed)

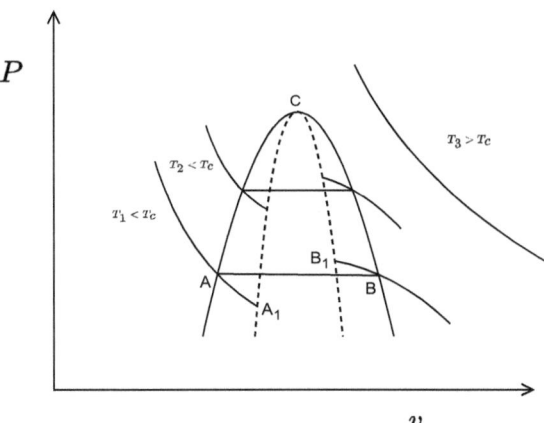

at A_1 and the maximum at B_1. A part of the dashed curve in Fig. 11.5 joins the points of minima, and the other joins the points of maxima of different isotherms. The two meet at the critical point, marked C in the figure, which is at the maximum of the coexistence curve. The dashed curve is called the *spinodal curve*. We examine the stability of the system on the spinodal curve on which, by construction,

$$\left(\frac{\partial P}{\partial v}\right)_T = 0, \quad \text{on spinodal curve.} \tag{11.27}$$

Note that, since C is the point of maximum of the spinodal curve,

$$\left(\frac{\partial^2 P}{\partial v^2}\right)_T = 0, \quad \text{at } C. \tag{11.28}$$

To examine stability, we first choose an appropriate thermodynamic potential to describe the system. Since the control parameters presently are temperature and pressure, we look upon the system as interacting with the heat reservoir at temperature T_0, and the pressure reservoir at pressure P_0. As shown in Sect. 2.13.2, the state of equilibrium of the system in this case is the one at which G_0 attains minimum value. It has been shown therein that, at equilibrium, the temperature and pressure of the system attain the same values as those of the respective reservoirs, and that the equilibrium state is stable if $\delta^2 G_0 > 0$ where $\delta^2 G_0$ is second-order variation in G_0 given, to second order in the variation in V, T by (2.217). Since we are interested in the condition of stability as we move along an isotherm the stability condition (2.217) when $\delta T = 0$ reads

$$-\left(\frac{\partial P}{\partial V}\right)_T (\delta V)^2 > 0. \tag{11.29}$$

Since our interest is in the stability at the points where $\partial P/\partial V = 0$, the condition above is not satisfied at the said points. We therefore need to examine the higher order terms. The next higher terms, namely the terms of order $(\delta V)^3$, and $(\delta V)^4$ have been evaluated in (2.219) under the conditions $\partial P/\partial V = 0$, and $\delta T = 0$ leading to the following stability criterion:

$$\left[\frac{1}{3!} \left(\frac{\partial^2 P}{\partial V^2} \right)_T + \frac{1}{4!} \left(\frac{\partial^3 P}{\partial V^3} \right)_T (\delta V) \right] (\delta V)^3 < 0. \tag{11.30}$$

The sign of the first term on the left side in the inequality above depends on the sign of δV and hence it can not be satisfied except when the condition

$$\left(\frac{\partial^2 P}{\partial V^2} \right)_T = 0, \tag{11.31}$$

also holds in which case (11.30) reduces to

$$\left(\frac{\partial^3 P}{\partial V^3} \right)_T (\delta V)^4 < 0, \tag{11.32}$$

which is independent of the sign of δV. The point where, along with $\partial P/\partial V = 0$, (11.31) holds is the critical point, which is point C in Fig. 11.5. It is the point of inflexion of the isotherm corresponding to the critical temperature T_c. It follows from (11.32) that the critical point will be stable if

$$\left(\frac{\partial^3 P}{\partial V^3} \right)_T < 0. \tag{11.33}$$

To summarize: the critical point is characterized by following conditions:

$$\left(\frac{\partial P}{\partial V} \right)_T = 0, \quad \left(\frac{\partial^2 P}{\partial V^2} \right)_T = 0, \quad \left(\frac{\partial^3 P}{\partial V^3} \right)_T < 0, \quad \text{at } T = T_c, \ V = V_c. \tag{11.34}$$

Using the conditions above we examine the behavior of the thermodynamic quantities in the vicinity of the critical point.

11.4.1 Critical Exponents

To examine the behavior of the system in the vicinity of the critical point, we define the scaled variables τ, η, p corresponding to T, v, P by the relations

$$\tau = \frac{T - T_c}{T_c}, \quad \eta = \frac{v - v_c}{v_c}, \quad p = \frac{P - P_c}{P_c}, \tag{11.35}$$

where x_c denotes the value of x at the critical point. In terms of the scaled variables, the equations in (11.34) assume the form:

$$\left(\frac{\partial p}{\partial \eta}\right)_\tau = 0, \quad \left(\frac{\partial^2 p}{\partial \eta^2}\right)_\tau = 0, \quad \left(\frac{\partial^3 p}{\partial \eta^3}\right)_\tau < 0, \quad \text{at } \tau = 0, \eta = 0. \tag{11.36}$$

The critical exponents determine how the value of a thermodynamic quantity varies in the neighborhood of the critical point. For example, let $f(\tau)$ be the thermodynamic quantity expressed as a function of deviation τ of temperature from its critical value. Then λ defined by (see [Stanley])

$$\lambda = \mathrm{Lt}_{\tau \to 0} \frac{\ln(f(\tau))}{\ln(\tau)}, \tag{11.37}$$

is called the *critical point exponent* associated with $f(\tau)$. The relation (11.37) is often written as

$$f(\tau) \sim \tau^\lambda. \tag{11.38}$$

The critical point exponents showing dependence of a thermodynamic quantity on the deviation of a specific volume from that at the critical point is similarly defined as

$$f(\eta) \sim \eta^\mu. \tag{11.39}$$

The interest in critical point exponents is due to the fact that they possess certain universal properties independent of the detailed nature of interaction or the equation of state.

Some critical exponents of interest in fluids are defined as follows (in the following $p = 0$ except in the last equation):

$$
\begin{aligned}
c_v &\sim \tau^{-\alpha}, & \tau &> 0, \quad \eta = 0, \\
c_v &\sim (-\tau)^{-\alpha'} & \tau &< 0, \quad \eta = 0 \\
\eta_g - \eta_l &\sim (-\tau)^\beta, & \tau &< 0, \\
\kappa_T &\sim (\tau)^{-\gamma}, & \tau &> 0, \quad \eta = 0, \\
\kappa_T &\sim (-\tau)^{-\gamma'}, & \tau &< 0, \quad \eta = \eta_g \text{ or } \eta = \eta_l, \\
p &\sim \eta^\delta, & \tau &= 0.
\end{aligned}
\tag{11.40}
$$

In the following, we discuss a simple approach to the theory of critical exponents (see [Landau and Lifshitz] and [Stanley] for further details).

To evaluate the exponents defined above, consider pressure a function of v, T and expand it in Taylor series at the critical point v_c, T_c. Retaining the lowest order contribution, we have

$$p(\eta, \tau) = A\tau - 2B\tau\eta - (4/3)C\eta^3, \tag{11.41}$$

based on the following rationale: Due to the first two conditions in (11.36), there is no term in η and η^2. Retained in (11.41) is the lowest power η^3 of η. The other third-order terms, namely $\eta\tau^2$ and $\tau\eta^2$, not retained in writing (11.41) is due to the assumption that they are smaller than the $\tau\eta$ term in (11.41), respectively, by the smallness factors τ and η (see [Landau and Lifshitz]). However, in the study of the equation of state of the van der Waals gas we will see that the term $\eta^2\tau$ ignored within the same order of approximation, as also the higher order term $\eta^3\tau$, contribute to the observables to the same order in powers of η as does (11.41).

The chosen form of the constants is for later convenience. Next, we determine the signs of the constants in (11.41). Due to the last relation in (11.36), we must have

$$C > 0. \tag{11.42}$$

Also, the system is homogeneous when $\tau > 0$; hence, we must have $\partial p/\partial \eta < 0$ for all $\tau > 0$ which implies

$$B > 0. \tag{11.43}$$

We have shown circa (11.58) that

$$A > 0. \tag{11.44}$$

We derive next the expressions for specific energy, and entropy corresponding to $p(\eta, \tau)$ in (11.41).

To evaluate energy, rewrite (2.232) expressing energy in terms of pressure and temperature in the present notation:

$$u(\eta, \tau) = P_c v_c \int \left\{ (1 + \tau) \left(\frac{\partial p}{\partial \tau} \right)_\eta - (p + 1) \right\} d\eta + \phi(\tau). \tag{11.45}$$

It then follows that

$$u(\eta, \tau) = P_c v_c \left\{ (A - 1)\eta - B\eta^2 + \frac{C\eta^4}{3} \right\} + \phi(\tau). \tag{11.46}$$

To evaluate entropy, recall (2.235) and rewrite it in the present notation:

$$s(\eta, \tau) = \frac{P_c v_c}{T_c} \int \left(\frac{\partial p}{\partial \tau}\right)_\eta d\eta + \psi(\tau), \tag{11.47}$$

where, due to (2.236),

$$\frac{d\phi(\tau)}{d\tau} = T_c(1 + \tau)\frac{d\psi(\tau)}{d\tau}. \tag{11.48}$$

Evaluation of (11.47) yields

$$s(\eta, \tau) = \frac{P_c v_c}{T_c} \left(A\eta - B\eta^2\right) + \psi(\tau). \tag{11.49}$$

The expressions above are for single phase region. The equation of state and other thermodynamic quantities in the two-phase region are derived next.

1. The equation of state in the region of coexistence corresponding to that in the single phase region is given by (11.10). In terms of the scaled variables, it reads

$$p(\eta, \tau) = p(\eta_g, \tau) \equiv p(\eta_l, \tau), \tag{11.50}$$

where η_l, η_g are the scaled specific volumes of the two phases in equilibrium: we have changed the nomenclature of the two phases, labeling the lower density state as g in place of 2, and the higher density state l in place of 1, with l indicating the liquid phase and g the gas phase. Equation (11.50) requires determination of η_g, η_l using the conditions (11.12) and (11.14). The condition (11.14) in the present case reads

$$\int_A^B v \, dP = v_c P_c \int_A^B (1 + \eta) dp = v_c P_c \int_A^B \eta \, dp = 0, \tag{11.51}$$

where the second equation is due to the fact that the value of P at the limits of integration are same. Now, with $p(\eta, \tau)$ given by (11.41),

$$\int_A^B \eta \, dp = \int_{\eta_l}^{\eta_g} \eta \frac{dp}{d\eta} d\eta = -\int_{\eta_l}^{\eta_g} \eta \left(2B\tau + 4C\eta^2\right) d\eta = 0. \tag{11.52}$$

This yields

$$(\eta_g^2 - \eta_l^2) \left(B\tau + C(\eta_g^2 + \eta_l^2)\right) = 0. \tag{11.53}$$

The second equation resulting from (11.50) is

$$(\eta_g - \eta_l) \left[B\tau + \frac{2C}{3} \left(\eta_g^2 + \eta_l^2 + \eta_g \eta_l\right)\right] = 0. \tag{11.54}$$

The $\eta_g \neq \eta_l$ solution of (11.53) is

$$\eta_l = -\eta_g. \tag{11.55}$$

This shows that the specific volumes at which phase transition occurs are placed symmetrically about the critical specific volume v_c. On substituting (11.55) in (11.54), we obtain

$$B\tau + \frac{2C}{3}\eta_g^2 = 0. \tag{11.56}$$

This implies

$$\eta_g^2 = -\frac{3B}{2C}\tau, \qquad \tau < 0, \tag{11.57}$$

and

$$p(\eta_g, \tau) = p(\eta_l, \tau) = A\tau. \tag{11.58}$$

Note that $\tau < 0$, $p < 0$ in the coexistence region. Hence, for the equation above to hold, we must have $A > 0$. On combining the results above with (11.50) follows the equation of state in the region of coexistence:

$$p(\eta, \tau) = A\tau. \tag{11.59}$$

2. If η is the specific volume in the coexistence region, then the fractions f_g, f_l of the substance in gas and liquid states are given by the lever rule (11.7) which in terms of the scaled variables reads

$$f_l = \frac{\eta_g - \eta}{\eta_g - \eta_l}, \qquad f_g = \frac{\eta - \eta_l}{\eta_g - \eta_l}. \tag{11.60}$$

3. Invoking (11.8), energy in the coexistence region for system of total specific volume v is given by

$$u = f_l u_l + f_g u_g, \tag{11.61}$$

where the fraction f_i of the substance in the ith phase is given by (11.60), u_l, u_g are given by (11.46), with $\eta \to \eta_l$, and $\eta \to \eta_g$, respectively.

4. Invoking (11.9), entropy in the coexistence region for system of total specific volume v is given by

$$s = f_l s_l + f_g s_g, \tag{11.62}$$

where the fraction f_i of the substance in the ith phase is given by (11.60), s_l, s_g are given by (11.49) with $\eta \to \eta_l$, $\eta \to \eta_g$, respectively.

The critical exponents may now be evaluated as follows.

1. Let the critical point be approached along the isochore (the path of constant v) $\eta = 0$ from above. The internal energy in that case is given by (11.46), and the specific heat at constant volume c_v at $\eta = 0$ turns out to be given by

$$c_v \equiv \left(\frac{\partial u}{\partial T} \right)_{v=v_c} = \frac{1}{T_c} \left(\frac{\partial u}{\partial \tau} \right)_{\eta=0} = \frac{1}{T_c} \frac{d\phi(\tau)}{d\tau}, \qquad T > T_c. \qquad (11.63)$$

The behavior of c_v in the vicinity of the critical point is thus determined by the functional form of $\phi(T)$. For several gases $\phi(T) \sim T$ in which case c_v is a constant and hence $\alpha = 0$.

2. To evaluate α' we compute c_v using the expression (11.61) for energy in the coexistence region by taking $\eta = 0$ therein. With $\eta_l = -\eta_g$ (see (11.55)), the expressions in (11.60) for the fractions in the two phases in equilibrium at $\tau = 0$ reduce to

$$f_l = f_g = \frac{1}{2}. \qquad (11.64)$$

Equation (11.61) for energy then reads

$$u = (u_l + u_g)/2. \qquad (11.65)$$

Evaluation of u_l, u_g using (11.46) with $\eta \to \eta_l$, $\eta \to \eta_g$, respectively, and also the relations (11.55), and (11.57) yields

$$u = \frac{3B^2 P_c v_c}{2C} \left(\tau + \frac{1}{2}\tau^2 \right) + \phi(\tau). \qquad (11.66)$$

Hence

$$c_v = \frac{1}{T_c} \left\{ \frac{3B^2 P_c v_c}{2C} (1+\tau) + \frac{d\phi(\tau)}{d\tau} \right\}, \qquad T < T_c. \qquad (11.67)$$

On comparing this with the expression (11.63) for c_v for $\tau > 0$, we see that c_v exhibits finite jump at $\tau = 0$. If $\phi(T) \sim T$, then the expression above shows that $\alpha' = 0$.

We will see that the equation of state of van der Waals gas can be approximated by the form in (11.41) in which A, B, C are given by (12.69). We will also investigate its critical properties in higher order approximation. The c_v in the coexistence region, obtained in higher order approximation, will be shown to agree with the expression (11.67) of c_v only up to the terms of order τ^0. This suggests that contribution to the linear terms in τ in c_v arises also from some higher order terms ignored in approximating $p(\eta, \tau)$ by (11.41).

3. Evaluation of β, defined in (11.40), requires determination of $\eta_g - \eta_l$. In the present case, recalling (11.55) and (11.57),

$$\eta_g - \eta_l = 2\eta_g = 2\sqrt{\frac{3B}{2C}}\sqrt{-\tau}. \tag{11.68}$$

This shows that $\beta = 1/2$.

4. To determine γ, we have

$$\kappa_T = -\frac{1}{v}\left(\frac{\partial v}{\partial P}\right)_T = -\frac{1}{P_c(1+\eta)}\left(\frac{\partial \eta}{\partial p}\right)_\tau. \tag{11.69}$$

Using (11.41), we get

$$\left(\frac{\partial \eta}{\partial p}\right)_\tau = -\frac{1}{2B\tau + 4C\eta^2}. \tag{11.70}$$

On substituting this in (11.69), it follows that, at the critical point $\eta = 0$,

$$\kappa_T = \frac{1}{2BP_c}(\tau)^{-1}, \qquad T > T_c. \tag{11.71}$$

Hence $\gamma = 1$. This shows that κ_T diverges at the critical point when it is approached from its higher temperature side.

5. To find γ', we need to evaluate (11.69) at $\eta = \eta_g$. With $(\partial \eta/\partial p)_\tau$ given by (11.70) we have at $\eta = \eta_g$,

$$\left(\frac{\partial \eta}{\partial p}\right)_\tau = \frac{1}{4B\tau}, \tag{11.72}$$

where (11.57) has been used. Substitution of this in (11.69) yields

$$\kappa_T = \frac{1}{4BP_c(1+\eta_g)}(-\tau)^{-1}, \qquad \tau < 0. \tag{11.73}$$

This shows that $\gamma' = 1$ i.e. κ_T diverges at the critical point also when it is approached from its lower temperature side.

Furthermore, since $\eta_g \to 0$ as $\tau \to 0$, we see that

$$\frac{\kappa_T(T > T_c)}{\kappa_T(T < T_c)} = 2. \tag{11.74}$$

6. To evaluate δ, note from (11.41) that, at the critical point $\tau = 0$,

$$p = -\frac{4C}{3}\eta^3. \qquad (11.75)$$

This shows that $\delta = 3$.

The theory of critical exponents discussed above is independent of any particular equation of state. In Chap. 12, we will evaluate those exponents for the system described by van der Waals equation and show that they are same as the ones obtained above.

The values of the critical exponents derived above are not in agreement with their observed values. The reason behind that failure can be understood best in the framework of statistical mechanics. To that end, recall (6.77) which relates fluctuations in the number of particles, $\Delta N/N$, with the isothermal compressibility κ_T. As shown in equations (11.71) and (11.73), κ_T diverges at the critical point which means there are infinitely large fluctuations in the particle number. We know that the thermodynamic quantities are averages over statistical mechanical probability distributions. The thermodynamic relations, including the equations of state are the relations between the averages. They hold as long as the number fluctuations are negligibly small. Since the number fluctuations are infinitely large at the critical point, the applicability of the equations of state, including that of (11.41) on which the theory of critical phenomena outlined above is based, is questionable. The failure of the theory above of critical indices is therefore not surprising. The theory which describes the critical phenomena accurately is the renormalization group theory due to K. G. Wilson for which he was awarded the Nobel Prize in 1982. Its discussion is beyond the scope of this book.

Chapter 12
Interacting Classical Gas

The assumption that the molecules of a gas do not interact though models a number of systems, there are situations of practical interest which the assumption in question leaves out. For example, the phenomenon of phase transition, discussed in Chap. 11, can not be explained by the equation of state of a non-interacting gas. In this chapter, we introduce a systematic way of treating interaction between molecules of a classical gas by expressing its equation of state as an expansion in terms of its number density. We restrict our attention to terms up to second order in the number density. The first-order terms describe the familiar ideal gas. The equation of state obtained by keeping terms up to second order in the number density turns out to be the van der Waals equation. We will see how that equation captures the essence of the gas–liquid phase transition.

12.1 Virial Expansion

Consider a gas of N atoms contained in the volume V at temperature T. The expression of pressure P as an expansion in powers of N/V in the form

$$P = \frac{Nk_{\mathrm{B}}T}{V}\left\{1 + \frac{N}{V}B_2(T) + \left(\frac{N}{V}\right)^2 B_3(T) + \cdots\right\}, \qquad (12.1)$$

is known as the *virial expansion*. The temperature-dependent coefficient $B_i(T)$ is known as the ith *virial coefficients*. The first term in the expansion gives the equation of state of an ideal gas. The virial expansion is the expansion in powers of the number density. For small densities, the lowest order correction to the ideal gas equation is obtained by ignoring the $B_3(T)$ and higher coefficient terms so that

$$P = \frac{Nk_{\mathrm{B}}T}{V}\left(1 + \frac{N}{V}B_2(T)\right). \qquad (12.2)$$

© The Author(s), under exclusive license to Springer Nature Switzerland AG 2024
R. R. Puri, *Modern Thermodynamics and Statistical Mechanics*, Undergraduate Lecture Notes in Physics, https://doi.org/10.1007/978-3-031-54310-4_12

At the temperature T_B at which $B_2(T) = 0$ the equation above reduces to the ideal gas equation. It is called the *Boyle temperature*:

$$B_2(T_B) = 0, \qquad T_B: \text{Boyle temperature.} \tag{12.3}$$

Statistical mechanics determines the virial coefficients in terms of the microscopic interaction between the atoms. In the following, we outline a formal approach to seek that end.

To obtain the virial expansion, recall that the partition function of a classical gas of N particles is given by (7.5) in which $\mathcal{V}(\mathbf{r})$ is the potential energy of the particles. We assume that the inter-particle potential $\mathcal{V}(\{\mathbf{r}_i\}_N)$ is a function only of the distances between the atoms expressible in the form

$$\mathcal{V}(\{\mathbf{r}_i\}_N) = \sum_{i<j=1}^{N} \mathcal{V}(|\mathbf{r}_i - \mathbf{r}_j|) \equiv \sum_{i<j=1}^{N} \mathcal{V}_{ij}, \qquad \mathcal{V}_{ij} \equiv \mathcal{V}(|\mathbf{r}_i - \mathbf{r}_j|). \tag{12.4}$$

Substitution of the equation above in (7.5) yields

$$\ln(Z_N) - \ln(Z_{NF}) = \ln\left(\frac{1}{V^N} \int \prod_{i<j=1}^{N} \exp\left(-\beta \mathcal{V}_{ij}\right) d^{3N}\mathbf{r} \right), \tag{12.5}$$

where Z_{NF} stands for the free particle partition function given in (7.18). Introduce the so-called *Meyer function*

$$f_{ij} = \exp\left(-\beta \mathcal{V}_{ij}\right) - 1, \tag{12.6}$$

to rewrite (12.5) as

$$\ln(Z_N) - \ln(Z_{NF}) = \ln\left\langle \prod_{i<j=1}^{N} (1 + f_{ij}) \right\rangle, \tag{12.7}$$

where $\langle A(\mathbf{r}) \rangle$ denotes the average of $A(\mathbf{r})$ defined by

$$\langle A(\mathbf{r}) \rangle = \frac{1}{V^N} \int A(\mathbf{r}) \, d^{3N}\mathbf{r}. \tag{12.8}$$

Clearly, if c is a constant then $\langle c \rangle = c$. Write (12.7) in the form

$$\left\langle \prod_{i<j=1}^{N} (1 + f_{ij}) \right\rangle = 1 + \left\langle \sum_{i<j=1}^{N} f_{ij} \right\rangle + x, \tag{12.9}$$

where x represents the average over products of the $f'_{ij}s$:

$$x = \left\langle \sum_{i<j} \sum_{k<l} f_{ij} f_{kl} + \sum_{i<j} \sum_{k<l} \sum_{m<n} f_{ij} f_{kl} f_{mn} + \cdots \right\rangle. \tag{12.10}$$

Using (12.9) rewrite (12.7) as

$$\ln(Z_N) - \ln(Z_{NF}) = \ln\left(1 + \left\langle \sum_{i<j=1}^{N} f_{ij} \right\rangle + x \right). \tag{12.11}$$

We assume that the interaction between atoms is very weak and is effective only when two atoms come very close to each other and that $V_{ij} \to 0$ as $|\mathbf{r}_i - \mathbf{r}_j| \to \infty$. As a consequence, we see that $f_{ij} \to 0$ as $|\mathbf{r}_i - \mathbf{r}_j| \to \infty$. Furthermore, the gas is assumed to be of so low density that not more than two atoms at a time are close enough for effective interaction. The contribution of the products of f_{ij} terms is consequently small compared with the single f_{ij} term in (12.11). We therefore use the expansion $\ln(1 + y) = y - y^2/2 + \cdots$ and ignore the terms involving products of the $f'_{ij}s$ to arrive at the following approximate form of (12.11):

$$\ln(Z_N) - \ln(Z_{NF}) = \left\langle \sum_{i<j=1}^{N} f_{ij} \right\rangle$$

$$= \frac{1}{V^N} \sum_{i<j=1}^{N} \int \{\exp(-\beta V_{ij}) - 1\} \prod_i d^3\mathbf{r}_i. \tag{12.12}$$

Since V_{ij} depends only on the positions $\mathbf{r}_i$ and $\mathbf{r}_j$, we can separate the integration over the positions $\mathbf{r}_i$ and $\mathbf{r}_j$ of the pair of atoms at $\mathbf{r}_i$ and $\mathbf{r}_j$ from the rest to write

$$\prod_k d^3 r_k = d^3 r_i d^3 r_j \prod_{k \neq i,j} d^3 \mathbf{r}_k. \tag{12.13}$$

Furthermore, since V_{ij} depends only on $|\mathbf{r}_i - \mathbf{r}_j|$ and the positions are variables of integration, the integral (12.12) is same for all pairs of atoms. The summation over i, j in (12.12) is therefore over all possible pairs. There being N atoms, the number of independent pairs that can be formed, assuming $N \gg 1$, are $N(N-1)/2 \approx N^2/2$. Hence (12.12) assumes the form

$$\ln(Z_N) - \ln(Z_{NF})$$

$$= \frac{N^2}{2V^N} \int \{\exp(-\beta V(|\mathbf{r}_i - \mathbf{r}_j|)) - 1\} d^3 r_i d^3 r_j \prod_{k \neq i,j} d^3 r_k$$

$$= \frac{N^2}{2V^2} \int \{\exp(-\beta V(|\mathbf{r}_i - \mathbf{r}_j|)) - 1\} d^3 r_i d^3 r_j, \tag{12.14}$$

where in writing the last line we have used the fact that the product over $k \neq i, j$ excludes two positions $\mathbf{r}_i$ and $\mathbf{r}_j$ leaving integration over remaining $N - 2$ positions as a result of which

$$\int \prod_{k \neq i, j} d^3\mathbf{r}_k = V^{N-2}. \tag{12.15}$$

To evaluate the remaining integral in (12.14), transform $(\mathbf{r}_i, \mathbf{r}_j)$ to the relative and center of mass coordinates $(\mathbf{r}, \mathbf{R})$ with $\mathbf{r} = \mathbf{r}_i - \mathbf{r}_j$, $\mathbf{R} = (\mathbf{r}_i + \mathbf{r}_j)/2$ to rewrite (12.14) as

$$\ln(Z_N) - \ln(Z_{NF}) = \frac{N^2}{2V^2} \int \left\{ \exp(-\beta \mathcal{V}(r)) - 1 \right\} d^3\mathbf{r} d^3\mathbf{R}. \tag{12.16}$$

The integration over R gives V so that

$$\ln(Z_N) - \ln(Z_{NF}) = -\frac{N^2}{V} B_2(T), \tag{12.17}$$

where $B_2(T)$ is the second virial coefficient:

$$B_2(T) = 2\pi \int_0^\infty \left\{ 1 - \exp(-\beta \mathcal{V}(r)) \right\} r^2 dr. \tag{12.18}$$

That $B_2(T)$ defined above is the second virial coefficient may be ascertained by differentiating (12.17) with respect to V keeping T constant and recalling the expression (6.39) for pressure and (7.18) for Z_{NF} to get (12.2) confirming thereby that $B_2(T)$ in (12.18) is indeed the second virial coefficient. We have thus obtained the expression for the second virial coefficient in terms of the interaction potential between the atoms when the gas is sufficiently rare. The higher order contributions are obtained by the method of so-called cluster expansion discussion of which is beyond the scope of this book (see [Landau and Lifshitz], [Huang], [Pathria]).

The detailed functional form of $B_2(T)$ depends, of course, on the functional form of $\mathcal{V}(r)$. However, certain characteristic features of $\mathcal{V}(r)$ are: to prevent collapse of atoms in to each other, the interaction should be repulsive at short separation between atoms, going to infinity as $r \to 0$. The interaction is attractive after certain distance between atoms, going to zero as their separation increases and has a minimum at the stable equilibrium separation of the atoms. A typical form of $\mathcal{V}(r)$ is shown in the Fig. 12.1.

For $\mathcal{V}(r)$ depicted in Fig. 12.1, we rewrite $B_2(T)$ in (12.18) as

$$B_2(T) = 2\pi \left(\int_0^d + \int_d^\infty \right) \left\{ 1 - \exp(-\beta \mathcal{V}(r)) \right\} r^2 dr. \tag{12.19}$$

Fig. 12.1 Typical
inter-atomic potential

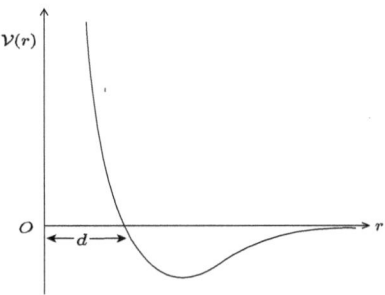

If temperature is high such that $\beta|V(r)| \ll 1$ for $r > d$, the integrand in the second integral above is close to zero. The dominant contribution to $B_2(T)$ then comes from the first integral which is positive due to the fact that $V(r) > 0$ for $r < d$. Hence, $B_2(T) > 0$ at high temperatures. On the other hand, if T is small then $\beta V(r)$ is large and positive for $r < d$ but large and negative for $r > d$. The dominant contribution to $B_2(T)$ in this case therefore comes from the second integral in (12.19) which is negative. Hence $B_2(T) < 0$ at low temperatures. Since $B_2(T)$ changes sign as T varies from low to high values, $B_2(T_B) = 0$ for some temperature T_B. This shows that the gas has a Boyle temperature as defined in (12.3).

As regards explicit functional forms of potentials, a widely used inter-atomic potential is the so-called *Lennard–Jones potential*:

$$V(r) = 4V_0 \left[\left(\frac{d}{r}\right)^{12} - \left(\frac{d}{r}\right)^{6} \right]. \tag{12.20}$$

The properties of $B_2(T)$ as a function of T for different choices of the parameters d, V_0 have been studied in the literature.

Widely used also is the simpler form of $V(r)$, the *hardcore potential* depicted in Fig. 12.2:

$$V(r) = \infty, \quad r \le d, \quad V(r) = -\mathcal{U}(r), \quad r \ge d, \tag{12.21}$$

where $\mathcal{U}(r) \ge 0$ and approaches zero as $r \to \infty$. The d is the distance of the closest approach of atoms.

On using this in (12.19), we get

$$B_2(T) = b - 2\pi \int_d^\infty \left\{ \exp(\beta \mathcal{U}(r)) - 1 \right\} r^2 dr, \tag{12.22}$$

$$b = \frac{2\pi d^3}{3}. \tag{12.23}$$

Fig. 12.2 Hardcore
potential (12.21)

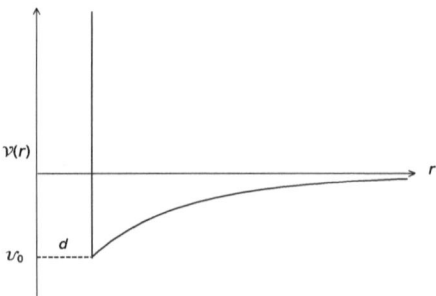

We evaluate $B_2(T)$ making further assumptions about $\mathcal{U}(r)$ leading to the van der Waals equation of state.

12.2   Van der Waals Equation of State

Assuming $\mathcal{U}(r)\beta \ll 1$, expand the exponential in (12.22) to first order in $\mathcal{U}(r)$ to obtain

$$B(T) = b - 2\pi\beta \int_d^\infty \mathcal{U}(r)r^2 dr. \qquad (12.24)$$

Assuming the integral in the equation above is finite:

$$\int_d^\infty \mathcal{U}(r)r^2 dr = \frac{a}{2\pi}, \qquad (12.25)$$

we get

$$B_2(T) = b - \beta a. \qquad (12.26)$$

Equation (12.17) then reads

$$\ln(Z_N) - \ln(Z_{NF}) = \frac{N^2}{V}(-b + a\beta). \qquad (12.27)$$

The parameter b may be related with the volume occupied by a molecule. For, if we think of the molecules as hard spheres of radii r_0, then the distance of the closest approach between two molecules, denoted in the definition (12.21) of the intermolecular potential by d, would be $d = 2r_0$. In terms of r_0, b in (12.23) is $b = 16\pi r_0^3/3$ which is four times the volume occupied by a molecule which is known to be the average excluded volume per molecule in the gas of a large number of molecules.

We could ignore the contribution from the products of the f_{ij} under the assumption that the gas is sufficiently rare. It means the volume per atom V/N is much larger than the molecular volume $\sim b$, i.e. $Nb/V \ll 1$ so that the approximation $-Nb/V \approx \ln(1 - Nb/V)$ holds enabling us to rewrite (12.27) as

$$\ln(Z_N) - \ln(Z_{NF}) = N\ln(1 - Nb/V) + \frac{N^2}{V}\beta a. \qquad (12.28)$$

In terms of the specific volume $v = V/N$, this reads

$$\ln(Z_N) - \ln(Z_{NF}) = N\left[\ln(v - b) - \ln(v) + \frac{a\beta}{v}\right]. \qquad (12.29)$$

This is the canonical partition function Z_N for an interacting gas under the approximations detailed while arriving at it. Since, as we will show, this leads to the van der Waals equation of state, we call it the partition function for the van der Waals gas. Using Z_{NF} the expression (7.18) for the N-particle partition function for a non-interacting gas, along with the use of Stirling's approximation, (12.29) reads

$$\ln(Z_N) = N\left[\ln(v - b) - \frac{3}{2}\ln(\beta) + \frac{a\beta}{v} - \frac{3}{2}\ln(w) + 1\right], \qquad (12.30)$$

where

$$w = \frac{h^2}{2\pi m}. \qquad (12.31)$$

The parameter a, defined by (12.25), depends on the specific form of $\mathcal{U}(r)$. A form of $\mathcal{U}(r)$ often used is the van der Waals potential:

$$\mathcal{U}(r) = \mathcal{U}_0\left(\frac{r_0}{r}\right)^6. \qquad (12.32)$$

The $-\mathcal{U}(r)$ is the attractive dipole interaction potential between neutral atoms. The hardcore potential (12.21) with $\mathcal{U}(r)$ given by (12.32) is also called the *Sutherland potential*.

We study next the thermodynamics of the system using the expression (12.30) for Z_N.

1. Using (6.39) with Z_N given by (12.29), the equation of state turns out to be given by

$$P = \frac{k_B T}{v - b} - \frac{a}{v^2}, \qquad v \ge b. \qquad (12.33)$$

This is the van der Waals equation for a non-ideal gas derived here from first principles. A phenomenological derivation of this equation has been outlined in Sect. 1.6.

2. The expression for specific energy reads

$$u = \frac{3}{2}k_B T - \frac{a}{v}. \tag{12.34}$$

This reduces to the ideal gas equation, as it must, when $a = 0$, i.e. when there is no interaction.

Recall that, starting from van der Waals equation of state, we derived in (2.258) the expression for energy using thermodynamic relations. That equation has in it the unknown function $\Phi(T)$ which can not be determined by thermodynamics. We, however, could deduce it by demanding that the limit in which van der Waals gas reduces to the ideal gas, so should its specific heat. We see that the result so arrived at in (2.259) is in agreement with (12.34) which is an outcome of the microscopic theory.

3. The expression for specific entropy is

$$s = \frac{k_B}{N} \left(\beta U + \ln(Z_N) \right)$$
$$= k_B \left[\ln(v - b) - \frac{3}{2}\ln(\beta) - \frac{3}{2}\ln(w) + \frac{5}{2} \right]. \tag{12.35}$$

4. The specific Gibbs potential, which is same as the chemical potential, is

$$\mu = G(P, T)/N = -\frac{k_B T}{N}\ln(Z_N) + Pv$$
$$= k_B T \left[\frac{v}{v - b} - \ln(v - b) + \frac{3}{2}\ln(\beta) + \frac{3}{2}\ln(w) - 1 \right] - \frac{2a}{v}, \tag{12.36}$$

where (6.114) has been recalled to express $G(P, T)$ in terms of Z_N.

Next, we study consequences of the equation of state (12.33). See also [1].

12.3 Critical Point

Recall from Sect. 11.4 that the point v_c, P_c on the isotherm corresponding to temperature T_c which is such that

$$(\partial P/\partial v)_{T_c} = (\partial^2 P/\partial v^2)_{T_c} = 0, \tag{12.37}$$

is called the critical point. The temperature for which the equations above hold is called the critical temperature T_c. We know that there is no such point for an ideal gas. We show that the critical point exists for the van der Waals equation of state.

To that end, use the equation of state (12.33) to evaluate the derivatives in the two equations in (12.37) to get

$$\frac{\partial P}{\partial v} = -\frac{k_B T}{(v-b)^2} + \frac{2a}{v^3} = 0, \qquad \frac{\partial^2 P}{\partial v^2} = \frac{2k_B T}{(v-b)^3} - \frac{6a}{v^4} = 0. \quad (12.38)$$

It is straightforward to solve the equations above. Let the solution be denoted by (v_c, T_c), which on substitution in (12.33) determines P_c resulting in the expressions:

$$T_c = \frac{8}{27}\frac{a}{k_B b}, \quad v_c = 3b, \quad P_c = \frac{a}{27b^2}. \quad (12.39)$$

It may be verified that

$$Z \equiv \frac{P_c v_c}{k_B T_c} = \frac{3}{8}. \quad (12.40)$$

The Z is known as the *compression parameter*. We see that, according to van der Waals model, Z has same value for all gases. However, the experimental values of Z are somewhat lower than the predicted value $3/8$ and not same for all gases (see [Stanley] for tabulated values of Z for various gases).

The critical values depend on the parameters (a, b) which are different for different systems. We show below that the equation of state assumes the form which is independent of system-specific parameters (a, b) if written in terms of the quantities scaled by respective critical values.

12.3.1 Law of Corresponding States

Define $(\widetilde{P}, \widetilde{v}, \widetilde{T})$ by the relations

$$\widetilde{P} = \frac{P}{P_c}, \quad \widetilde{v} = \frac{v}{v_c}, \quad \widetilde{T} = \frac{T}{T_c}, \quad (12.41)$$

where (T_c, v_c, P_c) are as in (12.39). On using (12.41) and (12.39) the van der Waals equation (12.33) assumes the form

$$\widetilde{P} = \frac{8}{3}\frac{\widetilde{T}}{\widetilde{v} - 1/3} - \frac{3}{\widetilde{v}^2}, \quad \widetilde{v} \geq 1/3. \quad (12.42)$$

The equation of state above, written in terms of the scaled variables, has no material-specific parameter. It is known as the *law of corresponding states*.

Exercises

Ex. 12.1. In terms of the reduced variables defined in (12.41) show that the partition function Z_N and energy assume the forms:

$$\ln(Z_N) = N\left[\ln(3\tilde{v} - 1) + \frac{3}{2}\ln(\tilde{T}) + \frac{9}{8}\frac{1}{\tilde{v}\tilde{T}} + C\right], \qquad (12.43)$$

where

$$C = \ln(v_c/3) + \frac{3}{2}\ln(k_B T_c) - \frac{3}{2}\ln(w) + 1, \qquad (12.44)$$

$$u = \frac{3k_B T_c}{2}\left(-\frac{3}{4\tilde{v}} + \tilde{T}\right). \qquad (12.45)$$

12.4   *P − v* Isotherms

Some isotherms in the $P - v$ plane obeying the van der Waals equation (12.33) are displayed in Fig. 12.3. Note that the isotherms for temperatures $T > T_c$ are monotonically decreasing functions of v, whereas each isotherm for $T < T_c$ exhibits a minimum and a maximum. The isotherm corresponding to $T = T_c$ has a point of inflexion at (v_c, P_c). In the following, we extract the said features of the isotherms from the equation of state.

To that end, it is convenient to work with the scaled form (12.42) of the equation of state, rewritten in terms of the variable x related with $\tilde{v}$ by

$$x = \tilde{v} - 1/3, \qquad x \geq 0, \qquad (12.46)$$

where the condition on x is due to that on $\tilde{v}$ (see (12.42)). The equation of state (12.42) then reads

$$\tilde{P} = \frac{8}{3}\frac{\tilde{T}}{x} - \frac{3}{(x + 1/3)^2}. \qquad (12.47)$$

Fig. 12.3 $P - v$ isotherms
obeying van der Waals
equation

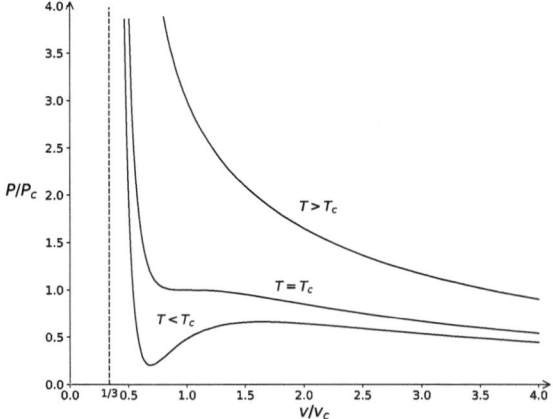

For a fixed $\widetilde{T}$, position of the extrema of a $P - v$ curve is determined by the roots of $d\widetilde{P}/dx = 0$ which is the cubic equation

$$x^3 + (1 - \alpha)x^2 + \frac{x}{3} + \frac{1}{27} = 0, \qquad (12.48)$$

where

$$\alpha = \frac{9}{4\widetilde{T}}. \qquad (12.49)$$

Let the roots of (12.48) be x_1, x_2, x_3. The roots have the following properties:

1. We know that, if z is a root of a real polynomial then so is z^*. It implies that, since (12.48) is real, either x_1, x_2, x_3 are all real or one of them is real and the other two complex conjugate of each other.
2. The roots obey the following relations,

$$x_1 + x_2 + x_3 = -\text{coefficient of } x^2 = \alpha - 1,$$
$$x_1(x_2 + x_3) + x_2 x_3 = \text{coefficient of } x = \frac{1}{3}$$
$$x_1 x_2 x_3 = -\text{coefficient of } x^0 = -\frac{1}{27}. \qquad (12.50)$$

The formulas for finding the roots of a cubic are summarized in the Appendix E. We use the results presented therein to find the roots of (12.48).

1. Comparing (12.48) with the standard form (E.1) of the cubic we have

$$a_2 = 1 - \alpha \equiv 1 - \frac{9}{4\widetilde{T}}, \quad a_1 = \frac{1}{3}, \quad a_0 = \frac{1}{27}. \qquad (12.51)$$

Hence the parameters p and q, defined in (E.2), are

$$p = a_1 - \frac{1}{3}a_2^2 = \frac{\alpha}{3}(2 - \alpha), \tag{12.52}$$

and

$$q = a_0 - \frac{1}{3}a_1 a_2 + \frac{2}{27}a_2^3$$
$$= -\frac{\alpha}{27}(2\alpha^2 - 6\alpha + 3). \tag{12.53}$$

2. On substituting the expressions (12.52), (12.53) for p, q in the definition (E.4) of the discriminant Δ, which determines the nature of the roots,

$$\Delta = \frac{1}{4}q^2 + \frac{1}{27}p^3, \tag{12.54}$$

we get

$$\Delta = \frac{\alpha^2}{12}\left(1 - \frac{1}{\tilde{T}}\right) = \frac{27}{64\tilde{T}^2}\left(1 - \frac{1}{\tilde{T}}\right). \tag{12.55}$$

Recall from the Appendix E that

$\Delta < 0 \implies$ roots are real

$\Delta = 0 \implies$ roots are real, two roots being equal

$\Delta > 0 \implies$ one root is real and other two are complex conjugate pair.

We discuss the question of locating roots for each of the cases above separately.

$\Delta = 0$: **Critical Temperature Isotherm**

With Δ given by (12.55), we see that

$$\Delta = 0 \text{ when } \tilde{T} = 1 \implies T = T_c. \tag{12.56}$$

Hence $\alpha \equiv 9/(4\tilde{T}) = 9/4$, which on substitution in (12.51), (12.52) and (12.53) gives

$$a_2 - \frac{5}{4}, \quad p = -\frac{3}{16}, \quad q = \frac{1}{32}. \tag{12.57}$$

To evaluate the roots, we compute first the quantities A, B defined in (E.3):

$$A = B \equiv -\frac{q}{2} = -\frac{1}{64}. \tag{12.58}$$

The roots, found using (E.5) are given by

$$x_1 = 2A^{1/3} - \frac{1}{3}a_2 = -\frac{1}{12}$$
$$x_2 = x_3 = -A^{1/3} - \frac{1}{3}a_2 = \frac{2}{3} \qquad (12.59)$$

The root x_1, being negative, falls outside the acceptable values of x which must be positive. Since $x = \tilde{v} - 1/3$, it follows that the roots x_2, x_3 correspond to $\tilde{v} = 1$, i.e. $v = v_c$.

We see that one of the roots is negative and the other two are positive and equal to each other, located at $v = v_c$. This implies, since the admissible values of x are positive, the $P - v$ isotherm corresponding to temperature $T = T_c$ has $\partial P / \partial v = 0$ only at $v = v_c$. The said root being a repeated root, implies $\partial^2 P / \partial v^2 = 0$ at $v = v_c$. Thus, the $P - v$ isotherm corresponding to temperature $T = T_c$ must have point of inflexion at $v = v_c$. This explains the behavior of the isotherm at $T = T_c$ in Fig. 12.3.

$\Delta > 0 : T > T_c$

The expression (12.55) shows that $\Delta > 0$ if $\tilde{T} > 1$, i.e. if $T > T_c$. As discussed in the Appendix E, for $\Delta > 0$, one of the roots, say x_1, is real and the others are complex conjugate of each other. The expression in (12.50) for the product of the roots shows that $x_1 x_2 x_3 = -1/27 < 0$ which, due to $x_3 = x_2^*$, means $x_1 |x_2|^2 = -1/27 < 0$, i.e. the real root is negative which is unphysical. Thus none of the three roots is physically acceptable. Consequently the $P - v$ plot is expected to be monotonic. This is consistent with the $P - v$ isotherms for temperatures $T > T_c$ depicted in Fig. 12.3.

$\Delta < 0 : T < T_c$

Equation (12.55) shows that $\Delta < 0$ if $\tilde{T} < 1$, i.e. if $T < T_c$. Since $\Delta < 0$, the roots are all real. Due to (12.50), $x_1 x_2 x_3 = -1/27 < 0$. Hence at least one of the roots, say x_1, is negative and the roots x_2, x_3 can either be both positive or both negative. Being negative, x_1 is unphysical. Furthermore, since $\tilde{T} < 1$, $\alpha - 1 = 9/4\tilde{T} - 1 > 0$. It therefore follows from (12.50) that the sum of the roots is positive. Hence all the roots can not be negative implying thereby that, since $x_1 < 0$, we must have $x_2, x_3 > 0$.

We thus see that the $P - v$ isotherms corresponding to $T < T_c$ have extrema at physically acceptable positions x_2, x_3. Now, from the expression (12.47) for $\tilde{P}$ as a function of x we see that (i) $\tilde{P}$ is a continuous function of x, (ii) $\tilde{P} \to \infty$ as $x \to 0+$, (iii) $\tilde{P} \to 0$ as $x \to \infty$. Hence, one of the extrema should be a minimum and the other a maximum. Consequently the $P - v$ isotherms for $T < T_c$ should be as shown in the Fig. 12.3. In Sect. 12.5 we discuss how these isotherms predict gas–liquid transition.

12.5 Gas–Liquid Transition

We know that the ideal gas equation does not predict the familiar phenomenon of phase transition. We will show that it is predicted by the van der Waals equation even though it is derived assuming smallness of the molecular density, a condition not satisfied by the liquid state.

The expectation that the van der Waals equation can predict phase transition is supported by the fact that the $P - v$ isotherms of van der Waals gas in Fig. 12.3 are qualitatively the same as those in Fig. 11.5. The van der Waals gas serves as a microscopic model for the thermodynamics of phase transition. Since the van der Waals model is not universal, the quantitative results derived from it too are limited in their scope.

To understand the phenomenon of phase transition predicted by the van der Waals model, refer to Fig. 12.3. We see that $dP/dv < 0$ on the isotherms for $T > T_c$. This behavior is analogous to that exhibited by an ideal gas and is consistent with the principles of thermodynamics.

On the other hand, the isotherms for $T < T_c$ exhibit a minimum and a maximum. One such isotherm is drawn in Fig. 12.4. The parts AC and EG in it exhibit physically acceptable variation of pressure with volume, namely, $dP/dv < 0$. However, $dp/dv > 0$ on the segment CE which is an unacceptable behavior and hence system can not be stable in the states lying on the said segment. The segment CE is therefore said to be an unstable branch of the $P - v$ isotherm.

We see that, for P such that $P_C < P < P_A$ (P_X denotes pressure at the point X on the $P - v$ diagram), there are three values of v on the same $T < T_c$ isotherm, one on the unstable branch CE and the other two on the stable branches AC and EG. This is illustrated in Fig. 12.4 in which the points B, D, F on the same isotherm

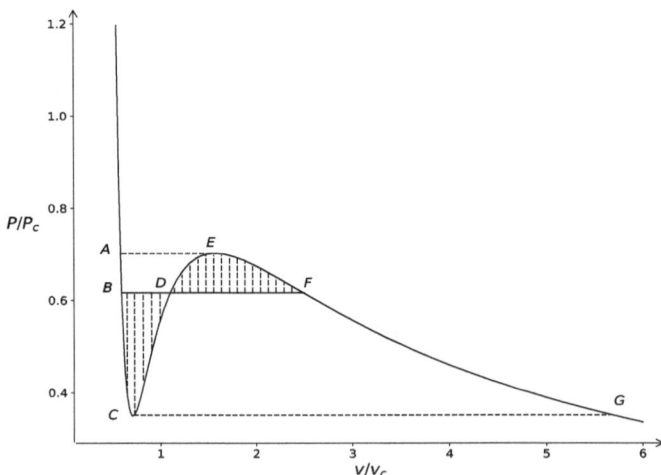

Fig. 12.4 Maxwell construction

are at the same pressure P_B. The three values of the specific volume solving van der Waals equation for the pressure P_B are: v_D corresponding to the point D on the unstable branch CE; v_B and v_F corresponding to the points B and F on the two stable branches AC and EG. Since the point D is on the unstable branch, the stable equilibrium states of the system at pressure P_B correspond to the points B and F. We can therefore say that the system separates in to two homogeneous components or phases, one having specific volume v_B and the other v_F (v_X denotes specific volume at the point X on the $P - v$ diagram). However, the two phases can not be at equilibrium with each other at all such pairs of points as it would mean that the two phases can coexist at any P_B lying in (P_C, P_A) for a given T which is absurd. There can be only one value of pressure for a given T at which the two phases can coexist. To find the said pressure, we recall that the coexistence of the phases requires not only equality of pressure and temperature, but that of their chemical potentials too. Hence only such pairs of points B, F can be in equilibrium at which the chemical potential is same.

To locate the pair of points at which the chemical potential is same, referring to Fig. 12.4, assume that the pair of points (B, F) on it is the pair in thermodynamic equilibrium. The equality of chemical potentials at such two points has been shown to lead to the equation (11.14) determining the specific volumes in equilibrium. We outline a graphical method of solving that equation.

The condition (11.14) of thermodynamic equilibrium in the present case between the coexisting phases at B and F reads

$$\int_{v_B}^{v_F} v \frac{dP}{dv} dv = 0. \tag{12.60}$$

The integration is to be carried along the isotherm on which the points lie, including its unstable part. In the present case it is along the curve $BCDEF$. The equation above can be solved graphically by noting that it implies equality of areas of the two shaded regions in the plot. To see that, carry the integration in (12.60) by parts to get $(P_F = P_B)$

$$\int_{v_B}^{v_F} v \frac{dP}{dv} dv = vP \Big|_B^F - \int_{v_B}^{v_F} P dv = P_B(v_F - v_B) - \int_{v_B}^{v_F} P dv$$

$$= \left[P_B(v_D - v_B) - \int_{v_B}^{v_D} P dv \right] + \left[P_B(v_F - v_D) - \int_{v_D}^{v_F} P dv \right] = 0. \tag{12.61}$$

Hence

$$P_B(v_D - v_B) - \int_{v_B}^{v_D} P dv = \int_{v_D}^{v_F} P dv - P_B(v_F - v_D). \tag{12.62}$$

Clearly, the two sides in the equation above are the two shaded areas in Fig. 12.4.

Fig. 12.5 The solid curve is
the coexistence curve and the
dashed curve is the spinodal
curve. The region bound by
the solid curve is the
coexistence region

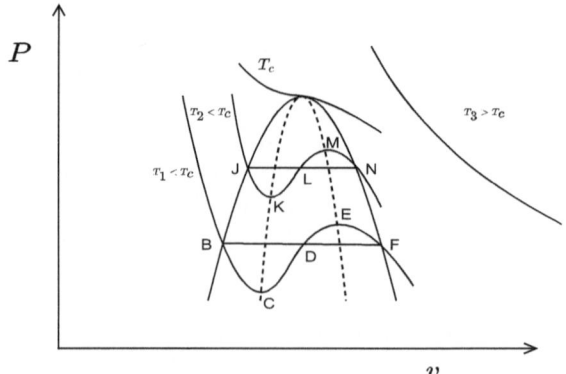

The procedure outlined above to find the pressure at which two phases coexist is
called *Maxwell construction*. The pair of points so identified are the ones at which
the two phases coexist (points B and F in Fig. 12.4). They are called the *transition
points*. Note that (i) the specific volume v_B at B on the stable part AC is closer to the
distance of closest approach b between the atoms than the specific volume v_F at the
point F at the same pressure. Hence the state of the system represented by the points
on the segment AC is relatively dense, (ii) the $\partial P/\partial v$ is very steep on the segment
AC suggesting that points on the said segment represent the state of the system which
is not easily compressible. These are the characteristics of a liquid. Hence the points
on the segment AC are said to represent the liquid phase. On the other hand, (i) the
specific volume v_F of the system at the point F on the segment EG, at which the
pressure is same as that at B, is much higher than v_B, (ii) the $\partial P/\partial v$ is relatively
small suggesting that the system in the said state can be compressed relatively easily.
These are the characteristics of a gas. The state of the system on the points on the
segment EG is therefore said to be gaseous phase. The van der Waals equation thus
predicts the liquid phase even though it is derived assuming the system to be of low
density.

At F the system is completely in the gaseous state, whereas at B it is completely in
the liquid state. We denote by (v_l, v_g) the specific volumes, respectively, in the liquid
and the gas phases in equilibrium with each other. At any other specific volume v
between the two, the system is a mixture of the liquid and the gas phases. The fractions
f_l, f_g of the phases, respectively, in the liquid and the gas state is given by the lever
rule:

$$f_l = \frac{v_g - v}{v_g - v_l}, \qquad f_g = \frac{v - v_l}{v_g - v_l}. \qquad (12.63)$$

We can use the Maxwell construction to identify for each isotherm the pair of
points at which the fluid can coexist in liquid and gas phases. In Fig. 12.5 we have
drawn a couple of isotherms and joined the pairs of points on each of them corresponding to the coexisting phases by straight lines. The two phases merge in to one

at the critical temperature T_c. We also join the transition points in the liquid phase and those in the gas phase by solid curves which meet at the critical point. It is the coexistence curve predicted in Sect. 11.1.2 based on thermodynamic considerations. The region bound by the curve is the region of coexistence in which the gas and the liquid phases coexist.

Going back to Fig. 12.4 in which points B and F constitute the pair in equilibrium, note that there is part BC of the isotherm below the point B on the liquid side and the part EF above the point F on the gaseous side on which, in accordance with thermodynamics, P is a monotonically decreasing function of v. These parts are, however, metastable. For, recall that the point B and F, respectively, on the parts AC and EG on the isotherm are the only points of stable equilibrium on the said segments. If the liquid is expanded sufficiently slowly near B, one can go past it. The liquid is then said to be *superheated*. Similarly, one can go past F if gas is compressed sufficiently slowly near it. The gas is then said to be *supercooled*. As argued above, being metastable, the superheated and supercooled states are short lived.

Join the points of minima of different isotherms (for example points B and J corresponding to T_1 and T_2 in Fig. 12.5), and the points of maxima (for example points E and M corresponding to T_1 and T_2 in the figure). The curve joining those points is shown by the dotted line. It is the spinodal curve predicted based on general thermodynamic considerations as well (see Sect. 11.4).

We thus see that the qualitative features of phase transition of the van der Waals gas agree with those predicted based on general thermodynamic considerations. We derive next the critical exponents based on the van der Waals equation of state.

Exercises

Ex. 12.2. Show that the pressure given by the van der Waals equation of state (12.47) is positive for all T and v if

$$T > \frac{27}{32} T_c. \tag{12.64}$$

The pressure can become negative for certain values of T, v for T not obeying (12.64). Note that negative pressure is predicted for the values of T lower than T_c. Hence the arguments leading to the identification of the parts of the $P - v$ plot for $T < T_c$ as metastable hold also for showing that the negative pressure part is metastable. Hint: Rewrite (12.47) as

$$\tilde{P} = \frac{8\tilde{T}}{3x(x+1/3)^2} \left(x^2 + ((2/3) - (9/8\tilde{T}))x + 1/9 \right). \tag{12.65}$$

$\tilde{P}$ will be positive if the quadratic in the equation above is positive which will be the case if the roots of the quadratic are complex, i.e. if its discriminant is negative.

Ex. 12.3. Show that the integral in (12.25) will be finite if $\mathcal{U}(r)$ decreases faster
than $1/r^3$ for large r. Hint: Let $\mathcal{U}(r) \sim 1/r^n$ for large r and carry the
integral to get the desired result.

12.6   Critical Exponents for van der Waals Fluid

Whereas the theory of phase transition in Chap. 11 does not assume any particular
form of the equation of state, the theory of critical exponents developed in (11.4) is
based on the approximate form (11.41) of the equation of state in the vicinity of the
critical point. We show below that the van der Waals equation can be reduced to the
said form. As a result the critical exponents computed in 11.4 hold for the van der
Walls fluid. We confirm those results without reducing the van der Waals equation
of state to the form (11.41).

In terms of the variables defined in (11.35), the equation of state (12.42) assumes
the form

$$p + 1 = \frac{4(1 + \tau)}{1 + 3\eta/2} - \frac{3}{(1 + \eta)^2}. \tag{12.66}$$

On expanding the right hand side of (12.66) about $\eta = 0$, $\tau = 0$ and ignoring the $\tau \eta^2$
term for the reasons mentioned circa (11.41), the expression for p up to η^3 reduces
to

$$p(\eta, \tau) = 4\tau \left(1 - \frac{3}{2}\eta\right) - \frac{3}{2}\eta^3. \tag{12.67}$$

Similarly, in terms of the variables defined in (11.35) and invoking also (12.39), u in
(12.34) up to η^4 reads

$$
\begin{aligned}
u &= -\frac{a}{v} + \frac{3k_B T}{2} = -\frac{9}{8} k_B T_c (1 + \eta)^{-1} + \frac{3k_B T}{2} \\
&= \frac{9}{8} k_B T_c (\eta - \eta^2 + \eta^3 - \eta^4) + \frac{3}{2} k_B T_c \tau + \frac{3}{8} k_B T_c. \tag{12.68}
\end{aligned}
$$

Equation (12.66) for $p(\tau, \eta)$ is of the form (11.41) with the parameters A, B, C
therein given by

$$A = 4, \quad B = 3, \quad C = 9/8. \tag{12.69}$$

The results derived in Sect. 11.4.1 are therefore applicable to the present system With
A, B, C as in (12.69), the said results yield:

1. Equation (11.46) for u in single phase region in the present case yields

$$u(\eta, \tau) = \frac{9k_B T_c}{8} \left\{ \eta - \eta^2 + \frac{\eta^4}{8} \right\} + \phi(\tau), \qquad (12.70)$$

where we take

$$\phi(\tau) = \frac{3}{2} k_B T_c \tau + \frac{3}{8} k_B T_c. \qquad (12.71)$$

However, comparison of (12.70) with (12.68) shows that they agree only up to η^2. The reason for said discrepancy lies in ignoring $\tau \eta^k$ ($k = 2, 3$) terms, linear in τ, in the expansion (12.67). As can be seen from (11.45), they contribute η^k ($k = 3, 4$) terms to energy. When added to already present η^4 term in (12.70) and the addition of the new η^3 term to it gives results in agreement with (12.68). For consistency we must therefore retain $\tau \eta^2$, $\tau \eta^3$ terms in the expansion (12.67).

We proceed nevertheless by assuming (12.67) to hold. Hence the results derived below are reliable only up to η^2. We will outline method of obtaining results reliable up to η^4 in the sequel. We will see that the inclusion of additional terms though changes numerical value of the heat capacity, it does not alter the critical exponents.

2. Equations (11.63) for c_v when $T > T_c$, (11.67) for c_v when $T < T_c$, (11.68) for $\eta_g - \eta_l$, (11.71) for κ_T when $T > T_c$, (11.73) for κ_T when $T < T_c$ and (11.75) for variation of p as a function of η on the critical isotherm in the present case read

$$c_v = \frac{3k_B}{2}, \qquad T > T_c,$$

$$c_v = 6k_B - \frac{9k_B}{2} |\tau|, \qquad T < T_c,$$

$$\eta_g - \eta_l = 4\sqrt{-\tau}, \qquad T < T_c$$

$$\kappa_T = \frac{1}{6P_c} (\tau)^{-1}, \qquad T > T_c,$$

$$\kappa_T = \frac{1}{12P_c} (-\tau)^{-1}, \qquad T < T_c,$$

$$p = -\frac{3}{2} \eta^3, \qquad \tau = 0, \qquad (12.72)$$

so that

$$\alpha = \alpha' = 0, \quad \beta = \frac{1}{2}, \quad \gamma = \gamma' = 1, \quad \delta = 3. \qquad (12.73)$$

Due to the reasons mentioned following (12.70), the coefficient of τ in the expression of c_v when $T < T_c$ is not reliable. Higher order approximation predicts c_v as in (12.91) which has same constant term as c_v for $T < T_c$ in (12.72) but different coefficient for τ.

3. Equations (11.55) and (11.57) for η_g, η_l in the present case read

$$\eta_g^2 = -4\tau, \qquad \eta_l = -\eta_g. \tag{12.74}$$

4. Using (11.59), the equation of state in the coexistence region is

$$p(\eta, \tau) = 4\tau. \tag{12.75}$$

The results above are based on the expansion (12.67). Its fallacies have been pointed out in the discussion circa (12.70). To make amends, we address afresh the question of determining η_g, η_l. A method of evaluating $\eta_g - \eta_g$ in terms of difference in entropy per molecule in the coexisting liquid and gas phases is described in [2]. We however follow the method of [3] to evaluate thermodynamic observables in the coexistence region.

In [3], the quantities $\tilde{\rho}_g$ and $\tilde{\rho}_l$ defined below have been evaluated as a series in $\sqrt{-\tau}$ where

$$\tilde{\rho} = \frac{1}{\tilde{v}} \equiv \frac{1}{1+\eta}, \tag{12.76}$$

$$\tilde{\rho}_g = \frac{1}{\tilde{v}_g} \equiv \frac{1}{1+\eta_g}, \qquad \tilde{\rho}_l = \frac{1}{\tilde{v}_l} \equiv \frac{1}{1+\eta_l}. \tag{12.77}$$

It has been shown that, with

$$x = \sqrt{-\tau}, \qquad \tau \le 0, \tag{12.78}$$

to order x^4,

$$\begin{aligned} \tilde{\rho}_g &= 1 + \alpha_1 x + \alpha_2 x^2 + \alpha_3 x^3 + \alpha_4 x^4, \\ \tilde{\rho}_l &= 1 - \alpha_1 x + \alpha_2 x^2 - \alpha_3 x^3 + (\alpha_4 - \beta)x^4, \end{aligned} \tag{12.79}$$

where

$$\alpha_1 = -2, \quad \alpha_2 = \frac{2}{5}, \quad \alpha_3 = \frac{13}{25}, \quad \alpha_4 = .207, \quad \beta = .092. \tag{12.80}$$

We use the expressions above to derive various thermodynamic quantities in the coexistence region.

1. We compute first the internal energy. In terms of the scaled density $\tilde{\rho}$ defined in (12.76), the expression (12.45) for internal energy reads

$$u(T, \rho) = -\frac{9k_B T_c}{8}\tilde{\rho} + \frac{3k_B T}{2}. \tag{12.81}$$

As a consequence, the internal energy at gas–liquid equilibrium state is

$$u(T, \rho) = f_g u_g + f_l u_l = -\frac{9k_B T_c}{8}\left(f_g \tilde{\rho}_g + f_l \tilde{\rho}_l\right) + \frac{3k_B T}{2}, \tag{12.82}$$

where f_g, f_l are determined by the lever rule (zzlivr4). We assume that the total specific volume v is equal to the critical volume v_c, i.e. $\rho = 1/v_c \equiv \rho_c$. The f_g, f_l in (11.7) in that case, rewritten in terms of the density, are

$$f_g = \frac{v_c - v_l}{v_g - v_l} = \frac{1 - \tilde{v}_l}{\tilde{v}_g - \tilde{v}_l} = \rho_g \frac{\tilde{\rho}_l - 1}{\tilde{\rho}_l - \tilde{\rho}_g},$$

$$f_l = \frac{v_g - vc}{v_g - v_l} = \frac{\tilde{v}_g - 1}{\tilde{v}_g - \tilde{v}_l} = \rho_l \frac{1 - \tilde{\rho}_g}{\tilde{\rho}_l - \tilde{\rho}_g}. \tag{12.83}$$

On using the identities above it is straightforward to see that

$$f_g \tilde{\rho}_g + f_l \tilde{\rho}_l = \tilde{\rho}_g + \tilde{\rho}_l - \tilde{\rho}_g \tilde{\rho}_l. \tag{12.84}$$

On using this, the expression (12.82) for internal energy for $\rho = \rho_c$ reads

$$u(\rho_c, T) = -\frac{9k_B T_c}{8}\left(\tilde{\rho}_g + \tilde{\rho}_l - \tilde{\rho}_g \tilde{\rho}_l\right) + \frac{3k_B T}{2}. \tag{12.85}$$

Recalling (12.79) we obtain

$$\begin{aligned}
\tilde{\rho}_g \tilde{\rho}_l &= \left\{\left(1 + \alpha_2 x^2 + \alpha_4 x^4\right) + \left(\alpha_1 x + \alpha_3 x^3\right)\right\} \\
&\quad \times \left\{\left(1 + \alpha_2 x^2 + \alpha_4 x^4\right) - \left(\alpha_1 x + \alpha_3 x^3\right) - \beta x^4\right\} \\
&= \left(1 + \alpha_2 x^2 + \alpha_4 x^4\right)^2 - \left(\alpha_1 x + \alpha_3 x^3\right)^2 \\
&\quad - \beta x^4 \left\{\left(1 + \alpha_2 x^2 + \alpha_4 x^4\right) + \left(\alpha_1 x + \alpha_3 x^3\right)\right\}.
\end{aligned} \tag{12.86}$$

On retaining terms up to x^4, the equation above reduces to

$$\tilde{\rho}_g \tilde{\rho}_l = 1 + (2\alpha_2 - \alpha_1^2)x^2 + (2\alpha_4 + \alpha_2^2 - 2\alpha_1\alpha_3 - \beta)x^4. \tag{12.87}$$

Also

$$\tilde{\rho}_g + \tilde{\rho}_l = 2(1 + \alpha_2 x^2 + \alpha_4 x^4) - \beta x^4. \tag{12.88}$$

Substitution of (12.87) and (12.88) in (12.85) gives

$$u(T, \rho_c) = -\frac{9k_B T_c}{8}\left(1 + \alpha_1^2 x^2 - (\alpha_2^2 - 2\alpha_1\alpha_3)x^4\right) + \frac{3k_B T}{2}. \tag{12.89}$$

On using (12.80) the equation above reads

$$u(T, \rho_c) = \frac{9k_B T_c}{2}\left(\tau + \frac{14}{25}\tau^2\right) + \frac{3k_B T_c}{2}\tau + \frac{3k_B T_c}{8} \tag{12.90}$$

This is internal energy in the coexistence region when total volume of the fluid is equal to the critical volume.

2. The specific heat per molecule for $v = v_c$ in the coexistence region is given by

$$c_v \equiv \frac{1}{T_c}\frac{du(T, \rho_c)}{d\tau} = k_B\left(6 + \frac{126}{25}\tau\right)$$

$$= k_B\left(6 - \frac{126}{25}|\tau|\right), \quad \tau \le 0. \tag{12.91}$$

The constant term in the equation above is same as that in the expression for c_v for $T < T_c$ in (12.72) but the τ-dependent terms in the two equations are different.

3. To evaluate compressibility κ_T for $T < T_c$, differentiate (12.42) and rewrite it in terms of $\tilde{\rho}$ to obtain

$$\left(\frac{\partial \tilde{P}}{\partial \tilde{v}}\right)_T = -\tilde{\rho}^2\left(\frac{8\tilde{T}}{3}(1 - 3\tilde{\rho})^{-2} - 6\tilde{\rho}\right). \tag{12.92}$$

We need to evaluate (12.92) for $\rho = \rho_g$. To that end, invoke (12.79) to write ρ_g to order x^2 as

$$\rho_g = 1 + f(x), \quad f(x) = -2x + \frac{2}{5}x^2, \tag{12.93}$$

so that, for $\tilde{\rho} = \tilde{\rho}_g$, (12.92) may be rewritten as

$$\left(\frac{\partial \tilde{P}}{\partial \tilde{v}}\right)_T = -6\rho_g^2\left(\frac{8\tilde{T}}{3}(1 - f(x)/2)^{-2} - (1 + f(x))\right)$$

$$= -12x^2. \tag{12.94}$$

It then follows that

$$\kappa_T \equiv -\frac{1}{P_c \tilde{v}} \left(\frac{\partial \tilde{v}}{\partial \tilde{P}} \right)_T = \frac{1}{12 P_c} (-\tau)^{-1}, \qquad T < T_c. \tag{12.95}$$

This is same as the expression for κ_T in (12.72) for $T < T_c$.

4. Invoking (12.79) it is straightforward to see that, to the leading order,

$$\eta_g - \eta_l = 4\sqrt{-\tau}, \qquad T < T_c. \tag{12.96}$$

This is same as the expression for $\eta_g - \eta_l$ in (12.72) for $T < T_c$.

Exercises

Ex. 12.4. Show that the equation of state of the van der Waals fluid in the gas–liquid coexistence region is given by

$$\tilde{P} = \tilde{\rho}_g \tilde{\rho}_l \left(3 - (\tilde{\rho}_g + \tilde{\rho}_l) \right). \tag{12.97}$$

Hint: Write the van der Waals equation of state in terms of the density:

$$\tilde{P}(\tilde{\rho}, \tilde{T}) = \frac{8 \tilde{T} \tilde{\rho}}{3 - \tilde{\rho}} - 3 \tilde{\rho}^2. \tag{12.98}$$

Use the equation $\tilde{P}(\tilde{\rho}_g, \tilde{T}) = \tilde{P}(\tilde{\rho}_l, \tilde{T})$ characterizing the gas–liquid coexistence region to express $\tilde{T}$ in terms of $\tilde{\rho}_l, \tilde{\rho}_g$:

$$\tilde{T} = (\tilde{\rho}_l + \tilde{\rho}_g)(3 - \tilde{\rho}_l)(3 - \tilde{\rho}_g). \tag{12.99}$$

Substitute this in (12.98) with $\tilde{\rho} = \tilde{\rho}_l$ or $\tilde{\rho} = \tilde{\rho}_g$ to get the desired result.

References

1. D.C. Johnson, 5 Feb. 2014. arXiv:1402.1205v1 [cond-mat.soft]
2. J. Lekner, Am. J. Phys. **50**, 161 (1982)
3. M.N. Berberan-Santos, E.N. Bodunov, L. Pogliani, J. Math. Chem. **43**, 1437 (2008)

Chapter 13
Density Operator Formalism

The occupation probabilities of energy levels of a macroscopic system in thermodynamic equilibrium are independent of time. The formalism of equilibrium statistical mechanics we presented could therefore be based on the occupation probabilities of energy levels. Those probabilities, however, change in time as the system evolves starting from an arbitrary initial state. That change takes place due to transitions between the energy levels. Hence time evolution of a system cannot be described only in terms of the probabilities of occupation of energy levels. In other words the probabilities of occupation of energy levels do not characterize the state of the system completely. A complete description of the state must include transition probabilities as well. We know that the state of an isolated system is characterized completely by a vector in the Hilbert space. The systems of interest in statistical mechanics are, however, not isolated; they interact with reservoirs. In this chapter we develop the formalism to characterize the state of an interacting system, called the density matrix formalism. We link the mechanical and the thermodynamic descriptions by quantum entropy. We show that the state of thermodynamic equilibrium so predicted is same as the one arrived at by working solely with energy levels occupation probabilities. The time-dependent aspects are discussed in Chap. 14.

The Sects. 13.1 and 13.2 essentially summarize the results the details of which may be found in [1].

13.1 Density Matrix

An isolated system in quantum mechanics is described by a vector in the Hilbert space whereas the state of a system interacting with other systems is described by the density matrix defined as follows (see [1] for the origin of the definition).

© The Author(s), under exclusive license to Springer Nature Switzerland AG 2024   341
R. R. Puri, *Modern Thermodynamics and Statistical Mechanics*, Undergraduate Lecture
Notes in Physics, https://doi.org/10.1007/978-3-031-54310-4_13

Consider an isolated system whose states are described by vectors in an n-dimensional Hilbert space, spanned by orthonormal basis vectors $|a_1\rangle, |a_2\rangle, \ldots,$ $|a_n\rangle$ so that any state vector $|\psi\rangle$ can be represented as a linear superposition,

$$|\psi\rangle = \sum_{i=1}^{n} c_i |a_i\rangle, \quad \langle a_i | a_j \rangle = \delta_{ij}, \quad c_i = \langle a_i | \psi \rangle. \qquad (13.1)$$

However, if the system interacts with other systems then its state is represented by $n \times n$ density matrix, also called density operator. Denoted by $\hat{\rho}$, it is a linear combination of the operators $|a_i\rangle\langle a_j|$ $(i, j = 1, 2, \ldots n)$:

$$\hat{\rho} = \sum_{i,j} \rho_{ij} |a_i\rangle\langle a_j|, \quad \rho_{ij} = \langle a_i | \hat{\rho} | a_j \rangle. \qquad (13.2)$$

Some properties of $\hat{\rho}$ are:

1. $\hat{\rho}$ is hermitian:

$$\hat{\rho} = \hat{\rho}^\dagger \implies \rho_{ij} = \rho_{ji}^*. \qquad (13.3)$$

2. As a consequence of its hermiticity, the eigenvalues λ_i, $(i = 1, 2, \ldots, n)$ of $\hat{\rho}$ are real and the corresponding eigenvectors orthonormal:

$$\hat{\rho}|\lambda_i\rangle = \lambda_i|\lambda_i\rangle, \quad \lambda_i = \lambda_i^*, \quad \langle \lambda_i | \lambda_j \rangle = \delta_{ij}. \qquad (13.4)$$

3. The trace of $\hat{\rho}$ is unity:

$$\mathrm{Tr}(\hat{\rho}) = 1, \qquad (13.5)$$

where $\mathrm{Tr}(\hat{A})$ stands for the trace of $\hat{A}$. Since the sum of the eigenvalues of a matrix equals its trace, we have

$$\sum_{i=1}^{n} \lambda_i = 1. \qquad (13.6)$$

4. The density matrix is positive:

$$\hat{\rho} \geq 0, \qquad (13.7)$$

in the sense that $\langle \psi | \hat{\rho} | \psi \rangle \geq 0$ for all $|\psi\rangle$. Since positivity of a matrix implies positivity of its eigenvalues, we have

$$\lambda_i \geq 0. \qquad (13.8)$$

Due to this and the relation (13.6), it follows that

$$0 \leq \lambda_i \leq 1. \qquad (13.9)$$

5. Since it is hermitian, $\hat{\rho}$ admits the spectral decomposition:

$$\hat{\rho} = \sum_{k=1}^{n} \lambda_k |\lambda_k\rangle\langle\lambda_k|. \tag{13.10}$$

Consequently, if $f(x)$ is a function such that $f(\lambda_k)$ is finite, then

$$f(\hat{\rho}) = \sum_{k=1}^{n} f(\lambda_k) |\lambda_k\rangle\langle\lambda_k|. \tag{13.11}$$

6. Due to (13.11) we have

$$\hat{\rho}^2 = \sum_{k=1}^{n} \lambda_k^2 |\lambda_k\rangle\langle\lambda_k|. \tag{13.12}$$

By virtue of (13.9), $\lambda_k^2 \leq \lambda_k$. Hence

$$\hat{\rho}^2 \leq \sum_{k=1}^{n} \lambda_k |\lambda_k\rangle\langle\lambda_k| = \hat{\rho}, \qquad \text{i.e.} \quad \hat{\rho}^2 \leq \hat{\rho}. \tag{13.13}$$

In particular this implies

$$\text{Tr}(\hat{\rho}^2) \leq 1. \tag{13.14}$$

The equality will hold if one of the eigenvalues of $\hat{\rho}$ is unity and all the others are zero. The $\hat{\rho}$ in that case is expressible in the form

$$\hat{\rho} = |\Psi\rangle\langle\Psi|, \tag{13.15}$$

where $|\Psi\rangle$ is normalized to unity. The system is then said to be in a *pure state*. If (13.13) does not hold with equality, the state of the system is said to be a *mixed state*.

7. Let the state of the system be described by the density matrix $\hat{\rho}$. The average value of a system operator $\hat{A}$, denoted by $\langle\hat{A}\rangle$, is given by

$$\langle\hat{A}\rangle = \text{Tr}(\hat{A}\hat{\rho}). \tag{13.16}$$

8. If the system is in the state described by the density matrix $\hat{\rho}$ then $\langle\psi|\hat{\rho}|\psi\rangle$ is the probability for it to be in the state $|\psi\rangle$.

9. Time evolution of $\hat{\rho}$ under the action of the Hamiltonian $\hat{H}$ is governed by the equation

$$i\hbar \frac{d\hat{\rho}(t)}{dt} = \left[\hat{H}, \hat{\rho}(t)\right], \tag{13.17}$$

known as the *Liouville–von Neumann equation*. Its formal solution reads

$$\hat{\rho}(t) = \hat{U}(t)\hat{\rho}(0)\hat{U}^{\dagger}(t), \quad \hat{U}(t) = \exp(-i\hat{H}t/\hbar). \tag{13.18}$$

The operator $\hat{U}(t)$ is the unitary time-evolution operator.

10. Consider a system composed of two subsystems A and B in which A lies in an n-dimensional Hilbert space and B in an m-dimensional one. Let $|a_1\rangle, |a_2\rangle, \ldots, |a_n\rangle$ be the orthonormal basis vectors of A and $|b_1\rangle, |b_2\rangle, \ldots, |b_m\rangle$ those of B. The density matrix of the combined system, denoted by $\hat{\rho}^{(A+B)}$, is given by

$$\hat{\rho}^{(A+B)} = \sum_{i,j=1}^{n} \sum_{k,l=1}^{m} c_{ik,jl}|a_i, b_k\rangle\langle a_j, b_l|, \tag{13.19}$$

where $|a, b\rangle$ stands for the tensor product of the state $|a\rangle$ of A and the state $|b\rangle$ of B.

11. If a system composed of two subsystems A and B is described by the density matrix $\hat{\rho}^{(A+B)}$ then the density matrix $\hat{\rho}^{(A)}$ of A alone and $\hat{\rho}^{(B)}$ of B alone are given by

$$\hat{\rho}^{(A)} = \mathrm{Tr}_B(\hat{\rho}^{(A+B)}), \quad \hat{\rho}^{(B)} = \mathrm{Tr}_A(\hat{\rho}^{(A+B)}), \tag{13.20}$$

where Tr_B is trace only over the states of the subsystem B and Tr_A is that only over the states of the subsystem A. They are called partial traces.

Next we introduce the notion of quantum entropy.

13.2   Quantum Entropy

Consider a system described by $n \times n$ density matrix $\hat{\rho}$. From the properties of $\hat{\rho}$ listed above, we know that the probability of finding the system in the state $|\lambda_k\rangle$ is $\langle\lambda_k|\hat{\rho}|\lambda_k\rangle$. We may consider the eigenvalues λ_k of $\hat{\rho}$ as random outcomes with probability $\langle\lambda_k|\hat{\rho}|\lambda_k\rangle$ of some hypothetical measurement process. The Shannon entropy of the distribution $\{\langle\lambda_k|\hat{\rho}|\lambda_k\rangle\}$ is evidently

$$S(\hat{\rho}) = -k_{\mathrm{B}} \sum_{k=1}^{n} \langle\lambda_k|\hat{\rho}|\lambda_k\rangle \ln(\langle\lambda_k|\hat{\rho}|\lambda_k\rangle) = -k_{\mathrm{B}} \sum_{k=1}^{n} \lambda_k \ln(\lambda_k), \tag{13.21}$$

where, due to (13.4), $\langle\lambda_k|\hat{\rho}|\lambda_k\rangle = \lambda_k$. Invoking (13.11), it is straightforward to see that the expression above for entropy may be written as

$$S(\hat{\rho}) = -k_{\mathrm{B}} \mathrm{Tr}\{\hat{\rho} \ln(\hat{\rho})\} \tag{13.22}$$

called *von Neumann entropy*.

Some properties of Von Neumann entropy are

1. Let $\hat{\tilde{\rho}}$ be obtained from $\hat{\rho}$ by a unitary transformation $\hat{U}$:

$$\hat{\tilde{\rho}} = \hat{U}^\dagger \hat{\rho} \hat{U}, \qquad \hat{U}\hat{U}^\dagger = \hat{U}^\dagger \hat{U} = I. \tag{13.23}$$

The entropy of the system in terms of $\hat{\tilde{\rho}}$ is

$$\tilde{S}(\hat{\tilde{\rho}}) = -k_B \text{Tr}(\hat{\tilde{\rho}} \ln(\hat{\tilde{\rho}})) = -k_B \text{Tr}\left\{ \hat{U}^\dagger \hat{\rho} \hat{U} \ln(\hat{U}^\dagger \hat{\rho} \hat{U}) \right\}. \tag{13.24}$$

Using the relation
$$F(\hat{U}^\dagger \hat{A} \hat{U}) = \hat{U}^\dagger F(\hat{A}) \hat{U}, \tag{13.25}$$

and the cyclic property of trace $(\text{Tr}(\hat{A}\hat{B}\hat{C}) = \text{Tr}(\hat{C}\hat{A}\hat{B}) = \text{Tr}(\hat{B}\hat{C}\hat{A}))$, (13.24) reduces to

$$\tilde{S}(\hat{\tilde{\rho}}) = -k_B \text{Tr}\left\{ \hat{U}^\dagger \hat{\rho} \ln(\hat{\rho}) \hat{U} \right\} = -k_B \text{Tr}\left\{ \hat{\rho} \ln(\hat{\rho}) \right\} = S(\hat{\rho}). \tag{13.26}$$

This shows that the entropy remains unchanged under unitary transformation of the density matrix.

2. If $\hat{\rho}$ represents a pure state $|\Psi\rangle$ then $\hat{\rho} = |\Psi\rangle\langle\Psi|$. In this case $\lambda_1 = 1$, $\lambda_k = 0$, $k \neq 1$ so that, on invoking (13.21),

$$S(\hat{\rho}) = 0. \tag{13.27}$$

Thus the entropy of a pure state is zero.

3. If $\hat{\rho} = I/n$, i.e. if all the states have equal probability of occupation without any correlation among them then (13.22) yields

$$S(\hat{\rho}) = k_B \ln(n). \tag{13.28}$$

It can be shown that $\hat{\rho} = I/n$ is the state of maximum entropy leading thereby to the inequality

$$0 \leq S(\hat{\rho}) \leq k_B \ln(n). \tag{13.29}$$

This is analogous to the property of Shannon entropy.

4. The *quantum joint entropy* of a system consisting of subsystems A and B is given by (13.22) with $\hat{\rho}$ identified as the density matrix $\hat{\rho}^{(A+B)}$ of the combined system of A and B.

5. If $\hat{\rho}^{(A)}$ is the density matrix of system A and $\hat{\rho}^{(B)}$ that of system B then

$$S(\hat{\rho}^{(A)} \otimes \hat{\rho}^{(B)}) = S(\hat{\rho}^{(A)}) + S(\hat{\rho}^{(B)}). \tag{13.30}$$

To prove this, assume that the system A is n-dimensional and that the eigenvalues of its density matrix $\hat{\rho}^{(A)}$ are $\lambda_1, \lambda_2, \ldots, \lambda_n$. Similarly, assume that the system B is m-dimensional and that the eigenvalues of its density matrix $\hat{\rho}^{(B)}$ are $\mu_1, \mu_2, \ldots, \mu_m$. The eigenvalues of the tensor product matrix $\hat{\rho}^{(A)} \otimes \hat{\rho}^{(B)}$ would be $\lambda_i \mu_j$, $(i = 1, 2, \ldots n; \ j = 1, 2, \ldots, m)$. The entropy of the combined system may then be written as

$$S\left(\hat{\rho}^{(A)} \otimes \hat{\rho}^{(B)}\right) = \sum_{j=1}^{n}\sum_{k=1}^{m} \lambda_j \mu_k \ln(\lambda_j \mu_k)$$

$$= \sum_{j=1}^{n} \lambda_j \ln(\lambda_j) \sum_{k=1}^{m}\mu_k + \sum_{k=1}^{m}\mu_k \ln(\mu_k) \sum_{j=1}^{n}\lambda_j$$

$$= \sum_{j=1}^{n} \lambda_j \ln(\lambda_j) + \sum_{k=1}^{m}\mu_k \ln(\mu_k), \tag{13.31}$$

where we have used the fact that the sum of the eigenvalues of a density matrix is unity. The last equation above is the sum of the entropies of the system A and B proving thereby (13.30).

6. It can be shown that $S(\hat{\rho})$ is concave in the sense that

$$S(f\hat{\rho}_1 + (1 - f)\hat{\rho}_2) \geq f S(\hat{\rho}_1) + (1 - f)S(\hat{\rho}_2). \tag{13.32}$$

In general

$$S\left(\sum_{i=1}^{k} f_i \hat{\rho}_i\right) \geq \sum_{i=1}^{k} f_i S\left(\hat{\rho}_i\right), \qquad \sum_{i=1}^{k} f_i = 1. \tag{13.33}$$

The quantum statistical mechanics is built on identifying von Neumann entropy as the statistical mechanical entropy.

13.3 Equilibrium Density Matrix

The properties of a system in thermodynamic equilibrium are described by the density matrix constructed according to the following postulate:

The density matrix of a system in thermodynamic equilibrium is one that maximizes the von Neumann entropy subject to the constraints on it.

We construct the equilibrium density matrix $\hat{\rho}$ by maximizing von Neumann entropy defined in (13.22) subject to the following constraints:

1. The trace of $\hat{\rho}$ is unity:

$$\text{Tr}(\hat{\rho}) = 1. \tag{13.34}$$

This constraint is always present.

2. Average values of certain set of observables $\hat{A}_k$ ($k = 1, 2, \ldots m$) is equal to some specified values:

$$\langle \hat{A}_k \rangle \equiv \text{Tr}(\hat{A}_k \hat{\rho}) = c_k. \tag{13.35}$$

Invoking the method of Lagrange multipliers, the desired $\hat{\rho}$ is obtained as the one which extremizes

$$F(\hat{\rho}) = \frac{S}{k_{\text{B}}} - \alpha_0 \text{Tr}(\hat{\rho}) - \sum_{k=1}^{m} \alpha_k \text{Tr}\left(\hat{A}_k \hat{\rho}\right)$$

$$= -\text{Tr}\left[\hat{\rho}\left(\ln(\hat{\rho}) + \alpha_0 + \sum_{k=1}^{m} \alpha_k \hat{A}_k\right)\right], \tag{13.36}$$

where $\{\alpha_k\}$ are Lagrange multipliers. The extremum of $F(\hat{\rho})$ is obtained by the condition $\delta F \equiv F(\hat{\rho} + \delta\hat{\rho}) - F(\hat{\rho}) = 0$. Now, if $\hat{X}$ is a matrix then $\delta\hat{X}^2 = (\delta\hat{X})\hat{X} + \hat{X}(\delta\hat{X}) \neq 2\hat{X}\delta\hat{X} \neq 2(\delta\hat{X})\hat{X}$. However, because of the cyclic property of trace, $\text{Tr}\left(\delta\hat{X}^2\right) = \text{Tr}\left((\delta\hat{X})\hat{X} + \hat{X}(\delta\hat{X})\right) = 2\text{Tr}\left(\hat{X}\delta\hat{X}\right) = 2\text{Tr}\left((\delta\hat{X})\hat{X}\right)$. In general

$$\delta\hat{X}^m = \sum_{k=1}^{m} \hat{X}^{m-k}(\delta\hat{X})\hat{X}^{k-1}, \tag{13.37}$$

so that

$$\text{Tr}\left(\delta\hat{X}^m\right) = m\text{Tr}\left(\hat{X}^{m-1}\delta\hat{X}\right) = m\text{Tr}\left(\delta\hat{X}\hat{X}^{m-1}\right). \tag{13.38}$$

Consequently, if $f(x)$ is expressible in terms of the powers of x then

$$\text{Tr}\left(\delta f\left(\hat{X}\right)\right) = \text{Tr}\left(\delta\hat{X}\frac{\text{d}f\left(\hat{X}\right)}{\text{d}\hat{X}}\right) = \text{Tr}\left(\frac{\text{d}f\left(\hat{X}\right)}{\text{d}\hat{X}}\delta\hat{X}\right). \tag{13.39}$$

On using the property above, the condition $\delta F(\hat{\rho}) = 0$, with $F(\hat{\rho})$ given by (13.36), leads to the equation

$$\text{Tr}\left[\delta\hat{\rho}\left(\ln(\hat{\rho}) + 1 + \alpha_0 + \sum_{k=1}^{m} \alpha_k \hat{A}_k\right)\right] = 0. \tag{13.40}$$

If the equation above is to hold for an arbitrary $\delta\hat{\rho}$, we must have

$$\ln(\hat{\rho}) + 1 + \alpha_0 + \sum_{k=1}^{m} \alpha_k \hat{A}_k = 0. \tag{13.41}$$

The solution of the equation above is evidently

$$\hat{\rho} = \frac{1}{Z} \exp\left(-\sum_{k=1}^{m} \alpha_k \hat{A}_k\right), \tag{13.42}$$

where $Z \equiv \exp(\alpha_0 + 1)$ is determined by the condition $\text{Tr}(\hat{\rho}) = 1$:

$$Z = \text{Tr}\left[\exp\left(-\sum_{k=1}^{m} \alpha_k \hat{A}_k\right)\right]. \tag{13.43}$$

The function $Z(\{\alpha_k\})$ is the partition function.

Next we derive expressions for the standard quantum distributions obtained before by the Gibbs–Shannon entropy approach.

13.4   Standard Distributions

The variables of interest from the point of view of thermodynamics application are energy and the number of particles. Derived below are equilibrium density matrices obtained from the general form (13.42) by specifying the said quantities deterministically or statistically in terms of their averages.

As in Sect. 5.3.2, we denote the Hamiltonian of the system for a fixed number N of particles by $\hat{H}_N$ and its eigenstates and eigenvalues respectively by $|E_{mN}\rangle$ and E_{mN}. The probability of finding the system in the state $|E_{mN}\rangle$, denoted by p_{mN}, is given in terms of the density matrix by

$$p_{mN} = \langle E_{mN}|\hat{\rho}|E_{mN}\rangle. \tag{13.44}$$

The theory of equilibrium statistical mechanics has been developed in earlier chapters in terms of p_{mN} by determining it for various ensembles using the principle of maximum entropy. We will show how those results follow from the general theory developed in terms of the density matrix.

13.4.1 *Microcanonical Ensemble*

Consider a system having fixed number N of particles and energy E specified to lie in the range $E_0 \leq E \leq E_0 + \Delta$. Recall that the ensemble describing such a system is called microcanonical. In this case, only constraint on the density matrix that maximizes entropy is on its trace. It is a special case of (13.42) with $\alpha_k = 0$ ($k \geq 1$) so that

$$\hat{\rho} = \frac{I}{Z}, \tag{13.45}$$

where $E_0 \leq E_{mN} \leq E_0 + \Delta$. If W is the number of states in the said interval, then the trace condition on $\hat{\rho}$ leads to

$$Z = \text{Tr}(I) = \sum_m \langle E_{mN} | I | E_{mN} \rangle = W. \tag{13.46}$$

Hence the density matrix describing microcanonical ensemble is

$$\hat{\rho} = \frac{I}{W}. \tag{13.47}$$

The probability of occupation of the level of energy E_{mN} is given, invoking (13.44), by

$$p_{mN} = \frac{1}{W}. \tag{13.48}$$

This is same as the expression (6.25) derived without use of the density matrix formalism.

13.4.2 *Canonical Ensemble*

When the number of particles in the system is fixed, but energy is specified in terms of average value then, apart from the trace condition, the constraint on the density matrix for maximizing entropy is

$$\langle \hat{H}_N \rangle \equiv \text{Tr}(\hat{H}_N \hat{\rho}) = U. \tag{13.49}$$

We know that the ensemble of such systems is called canonical ensemble. The equilibrium density matrix then is the special case of (13.42) with $\alpha_1 \rightarrow \beta$, $\hat{A}_1 \rightarrow \hat{H}_N$ and $\alpha_k = 0$ for $k > 1$:

$$\hat{\rho} = \frac{1}{Z_N} \exp(-\beta \hat{H}_N), \tag{13.50}$$

and the partition function Z_N is

$$Z_N = \text{Tr}\left(\exp(-\beta \hat{H}_N) \right). \tag{13.51}$$

The probability of occupation of level $|E_{mN}\rangle$ is given by

$$p_{mN} = \frac{1}{Z_N} \exp(-\beta E_{mN}). \tag{13.52}$$

This is same as the expression derived in (6.30) for canonical ensemble without the use of the density matrix formalism.

13.4.3   Grand Canonical Ensemble

For a system in which the number of particles as well as energy are given as averages then, apart from the trace condition, the constraints on the density matrix for maximizing the entropy are

$$\langle \hat{H} \rangle \equiv \text{Tr}(\hat{H} \hat{\rho}) = U, \qquad \langle \hat{N} \rangle \equiv \text{Tr}(\hat{N} \hat{\rho}) = \bar{N}, \tag{13.53}$$

where $\hat{N}$ is the number operator which is such that

$$\hat{N}|E_{mN}\rangle = N|E_{mN}\rangle. \tag{13.54}$$

Recall that the ensemble of such systems is called grand canonical ensemble. The equilibrium density matrix in the present case is the special case of (13.42) with $\alpha_1 \rightarrow \beta$, $\hat{A}_1 \rightarrow \hat{H}$; $\alpha_2 \rightarrow -\alpha$, $\hat{A}_2 \rightarrow \hat{N}$, and $\alpha_k = 0$ for $k > 2$:

$$\hat{\rho} = \frac{1}{Z_G} \exp(-\beta \hat{H} + \alpha \hat{N}), \tag{13.55}$$

and the partition function Z_G is

$$Z_G = \text{Tr}\left(\exp(-\beta \hat{H} + \alpha \hat{N}) \right). \tag{13.56}$$

Occupation probability of the state $|E_{mN}\rangle$, obtained using (13.44) is

$$p_m(N) = \frac{1}{Z_G} \exp(-\beta E_{mN} + \alpha N). \tag{13.57}$$

This is same as the corresponding expression derived in (6.59) without using the density matrix formalism. As stated in Sect. 6.3, change in notation from p_{mN} to $p_m(N)$ is to distinguish the probabilities for varying number of particles from those for fixed N.

The density matrix formalism thus leads to same standard distributions as were obtained using Gibbs–Shannon entropy. In Sect. 6.4 we used those distributions to establish relationship between the thermodynamic and statistical descriptions. The new aspect introduced by the density matrix description is the time evolution. In Sect. 13.6 we will show that the time evolution of entropy is consistent with the second law. For now we construct partition function for quantum harmonic oscillators which will be useful in the master equation formalism developed in Chap. 14.

13.5 Equilibrium Density Matrix of Harmonic Oscillators

A harmonic oscillator (h.o.) of frequency ω is described by the Hamiltonian (taking its ground state energy $\hbar\omega/2$ as the zero of energy)

$$\hat{H} = \hbar\omega\hat{a}^\dagger\hat{a}, \quad [\hat{a}, \hat{a}^\dagger] = 1. \tag{13.58}$$

Invoking (13.50), the equilibrium density matrix of the h.o. in contact with a heat reservoir at temperature T reads

$$\hat{\rho} = \frac{1}{Z}\exp(-\beta\hbar\omega\hat{a}^\dagger\hat{a}), \quad Z = \mathrm{Tr}\left[\exp(-\beta\hbar\omega\hat{a}^\dagger\hat{a})\right]. \tag{13.59}$$

Carrying trace in the basis of the number states (see Appendix G), the partition function Z reads

$$Z = \sum_{n=0}^{\infty}\langle n|\exp(-\beta\hbar\omega\hat{a}^\dagger\hat{a})|n\rangle = \sum_{n=0}^{\infty}\exp(-\beta\hbar\omega n)$$

$$= \frac{1}{1 - \exp(-\beta\hbar\omega)}. \tag{13.60}$$

The probability that the oscillator is in the state $|n\rangle$ is

$$p_n = \frac{1}{Z}\exp(-\beta\hbar\omega n). \tag{13.61}$$

Energy of the oscillator is

$$U = \frac{\hbar\omega}{\exp(\beta\hbar\omega) - 1}. \tag{13.62}$$

Exercises

Ex. 13. 1. Consider a system of N non-interacting harmonic oscillators of frequency ω at temperature T, each described by the Hamiltonian (13.58). Show that the probability that an oscillator is in the coherent state $|\alpha\rangle$ is

$$p(\alpha) = \frac{1}{Z} \exp\{-(1 - \exp(-\beta\hbar\omega))|\alpha|^2\}. \qquad (13.63)$$

Hint: Use (G.6).

13.6   Time Evolution of Entropy

Recall from (13.18) that, under the action of a Hamiltonian, the time evolution of the density matrix is governed by a unitary operator. Hence entropy at time t is given by

$$S(\hat{\rho}(t)) = k_B \text{Tr}\left\{\hat{\rho}(t)\ln(\hat{\rho}(t))\right\} = S(\hat{\rho}(0)), \qquad (13.64)$$

where last line is the consequence of (13.26) according to which entropy of two density matrices related by unitary transformation is same.

Thus entropy would remain unchanged as a function of time if the evolution is governed by a Hamiltonian. However, as we know, the molecular motion in macroscopic systems can be modeled as random. It may be described by assuming that the molecular evolution is governed by a Hamiltonian chosen randomly from a set of Hamiltonians $\{\hat{H}_j\}$ with probabilities $\{f_j\}$ (see [Balian]). The density matrix at time t may therefore be written as

$$\hat{\rho}(t) = \sum_i f_i \hat{U}_i(t)\hat{\rho}(0)\hat{U}_i^\dagger(t), \qquad \sum_{i=1}^{k} f_i = 1, \qquad (13.65)$$

where $\hat{U}_j(t)$ is as in (13.18) with $\hat{H} \to \hat{H}_j$. Hence the entropy at time t is

$$S\left(\hat{\rho}(t)\right) = S\left(\sum_i f_i \hat{U}_i(t)\hat{\rho}(0)\hat{U}_i^\dagger(t)\right). \qquad (13.66)$$

Recalling the inequality in (13.33) follows the relation

$$S\left(\sum_i f_i \hat{U}_j(t)\hat{\rho}(0)\hat{U}_j^\dagger(t)\right) \geq \sum_{i=1}^k f_i S\left(\hat{U}_j(t)\hat{\rho}(0)\hat{U}_j^\dagger(t)\right)$$

$$= \sum_{i=1}^k f_i S\left(\hat{\rho}(0)\right), \tag{13.67}$$

where the last line is due to (13.26) leading finally to the inequality

$$S(\hat{\rho}(t)) \geq S(\hat{\rho}(0)). \tag{13.68}$$

We thus see that entropy under random interaction cannot decrease. Though we have derived the result by assuming the Hamiltonian to be time independent in which case the time-evolution operator $\hat{U}(t)$ is given by (13.18), the derivation invokes only the unitarity of $\hat{U}$ which will hold even when $\hat{H}$ is time dependent though the form of $\hat{U}(t)$ will then be different from that in (13.18).

Reference

1. R.R. Puri, *Non-Relativistic Quantum Mechanics* (Cambridge University Press, 2017)

Chapter 14
Quantum Master Equation

The statistical mechanical equilibrium distribution is derived assuming maximum entropy principle. But the applicability of that principle must be ascertained by demonstrating that the asymptotic solution of the equation for the density matrix of a macroscopic system interacting with a reservoir is the one predicted by the maximum entropy principle. Starting from the equation of evolution of the density matrix of a system interacting with a thermal reservoir, called the master equation, in this chapter we show that it indeed approaches the canonical form in the asymptotic limit. The master equation is derived assuming weak interaction between the reservoirs and the system, an assumption which was used also to establish the relation between statistical mechanics and thermodynamics in Sect. 6.4.

14.1 Master Equation

Starting from the equation of evolution of the density matrix of the composite system, consisting of the system of interest, named A, interacting with a reservoir, named R, in this section we present the equation of evolution of the system of interest alone under various assumptions. See [1] for the details of the derivation.

Consider a system A coupled to a reservoir of harmonic oscillators such that the composite system is described by the Hamiltonian

$$\hat{H} = \hat{H}_A + \hat{H}_R + \hat{H}_{AR}. \tag{14.1}$$

Various terms in the expression above are:

1. $\hat{H}_A$ is the Hamiltonian of the system.
2. $\hat{H}_R$ is the Hamiltonian of the reservoir, assumed to consist of harmonic oscillators of frequencies $(\omega_1, \omega_2, \ldots) \equiv \{\omega_k\}$ described by the harmonic oscillator annihilation and creation operators $\{\hat{c}_k, \hat{c}_k^\dagger\}$ obeying the commutation relation

© The Author(s), under exclusive license to Springer Nature Switzerland AG 2024
R. R. Puri, *Modern Thermodynamics and Statistical Mechanics*, Undergraduate Lecture Notes in Physics, https://doi.org/10.1007/978-3-031-54310-4_14

$$[\hat{c}_k, \, \hat{c}_p^\dagger] = \delta_{kp}. \tag{14.2}$$

The $\hat{H}_R$ is given by

$$\hat{H}_R = \hbar \sum_k \omega_k \hat{c}_k^\dagger \hat{c}_k. \tag{14.3}$$

Modeling the reservoir as a collection of harmonic oscillators captures several real situations. For example, we know that the electromagnetic field is described as a collection of harmonic oscillators. It acts as a reservoir for atoms in free space.

3. The Hamiltonian $\hat{H}_{AR}$ describes the interaction of the system with the reservoir. It is assumed to be of the form

$$\hat{H}_{AR} = \hbar \sum_k \left(g_{ck} \hat{A}^\dagger \hat{c}_k + g_{ck}^* \hat{c}_k^\dagger \hat{A} \right), \tag{14.4}$$

where $\hat{A}$ is a system operator such that

$$[\hat{H}_A, \, \hat{A}] = -\hbar \omega_a \hat{A}. \tag{14.5}$$

We rewrite the equation above by introducing the notion of a "superoperator". Like an operator acts on a vector to transform it to another vector, a superoperator acts on an operator to transform it to another operator. We distinguish a superoperator from an operator by representing it by two carets on a letter. For example, the equation

$$\hat{\hat{O}} \hat{A} = \hat{B} \tag{14.6}$$

defines the superoperator $\hat{\hat{O}}$ which acting on some operator $\hat{A}$ transforms it to another operator $\hat{B}$. Let $\hat{\hat{\mathcal{H}}}_A$ be the superoperator defined by its action on an operator $\hat{P}$ by the equation

$$\hat{\hat{\mathcal{H}}}_A \hat{P} = (\hbar)^{-1} [\hat{H}_A, \, \hat{P}]. \tag{14.7}$$

In terms of $\hat{\hat{\mathcal{H}}}_A$, (14.5) assumes the form

$$\hat{\hat{\mathcal{H}}}_A \hat{A} = -\omega_a \hat{A}. \tag{14.8}$$

The evolution of the density matrix $\hat{\rho}^{(A+R)}(t)$ of the composite system is governed by the equation

$$\dot{\hat{\rho}}^{(A+R)}(t) = (i\hbar)^{-1}[\hat{H},\ \hat{\rho}^{(A+R)}(t)]. \qquad (14.9)$$

Our interest is in the dynamics of the system alone, described by the density matrix $\hat{\rho}^{(A)}(t)$. It is related with the density matrix $\hat{\rho}^{(A+R)}(t)$ of the composite system by the relation

$$\hat{\rho}^{(A)}(t) = \text{Tr}_R[\hat{\rho}^{(A+R)}(t)], \qquad (14.10)$$

with Tr_R standing for trace over the reservoir operators.

The equation for the time evolution of $\hat{\rho}^{(A)}(t)$ may in principle be derived by taking trace of the equation (14.9) for the density matrix of the composite system over the reservoir operators under the following initial conditions:

1. The oscillators constituting the reservoir are assumed to be in the state of thermal equilibrium at temperature T. Recalling that the density matrix of an oscillator in equilibrium with a reservoir at temperature T is given by (13.59), the density matrix of the reservoir is readily seen to be

$$\hat{\rho}^{(R)}(0) = \prod_k (1 - \exp(-\beta\hbar\omega_k)) \exp\left(-\beta\hbar\omega_k \hat{c}_k^\dagger \hat{c}_k\right). \qquad (14.11)$$

2. The system and the reservoir are assumed to be decoupled at $t = 0$:

$$\hat{\rho}^{(A+R)}(0) = \hat{\rho}^{(A)}(0) \otimes \hat{\rho}^{(R)}(0). \qquad (14.12)$$

However, except for some special systems, like the system consisting of one or more harmonic oscillators, it is not possible to derive an exact equation for $\hat{\rho}^{(A)}(t)$ starting from (14.9) for the composite system. Assuming weak system-reservoir interaction, it is derived under the so-called Born-Markov approximation outlined below:

1. Equation (14.9) is solved perturbatively to the second order in the system-reservoir coupling constant. This is called Born approximation.
2. It is assumed that the system and bath are uncorrelated at all times and that the state of the bath remains unchanged during the evolution, i.e. the density matrix of the combined system at time t is

$$\hat{\rho}^{(A+R)}(t) = \hat{\rho}^{(A)}(t) \otimes \hat{\rho}^{(R)}(0). \qquad (14.13)$$

3. It is assumed that the reservoir correlation time is much shorter than the time of observation. This usually requires the reservoir to have non-denumerably infinite number of degrees of freedom. In the present case it amounts to assuming that the reservoir oscillators' frequencies are spaced so closely that their distribution may

be described in terms of the density of frequencies $h(\omega)$ enabling one to replace sum over frequencies by integral:

$$\sum_k f(\omega_k) \rightarrow \int h(\omega) f(\omega) d\omega. \tag{14.14}$$

On the said time scale of observation the system loses the memory of its past. This is called Markov approximation.

The equation of evolution of $\hat{\rho}^{(A)}(t)$ under the conditions stated above turns out to be given by

$$\frac{d\hat{\rho}(t)}{dt} = \hat{\mathcal{W}}_A \hat{\rho}(t), \tag{14.15}$$

where, for notational convenience, we have removed the superscript A on $\hat{\rho}^{(A)}(t)$ and introduced the superoperator $\hat{\mathcal{W}}_A$ defined by

$$
\begin{aligned}
\hat{\mathcal{W}}_A \hat{\rho}(t) = {} & (i\hbar)^{-1} [\hat{H}_A, \hat{\rho}(t)] + i\Delta_1 [\hat{A}^\dagger \hat{A}, \hat{\rho}(t)] - i\Delta_2 [\hat{A}\hat{A}^\dagger, \hat{\rho}(t)] \\
& + \gamma(\bar{n} + 1) \left(2\hat{A}\hat{\rho}(t)\hat{A}^\dagger - \hat{A}^\dagger \hat{A}\hat{\rho}(t) - \hat{\rho}(t)\hat{A}^\dagger \hat{A} \right) \\
& + \gamma\bar{n} \left(2\hat{A}^\dagger \hat{\rho}(t)\hat{A} - \hat{A}\hat{A}^\dagger \hat{\rho}(t) - \hat{\rho}(t)\hat{A}\hat{A}^\dagger \right),
\end{aligned}
\tag{14.16}
$$

where

$$\bar{n} = \frac{1}{\exp(\hbar\omega_a \beta) - 1} \tag{14.17}$$

is the average number of reservoir oscillators at frequency ω_a defined in (14.5),

$$\gamma = \pi |g_{ck}|^2 h(\omega_k) \delta(\omega_k - \omega_a), \tag{14.18}$$

$$\Delta_1 = P \sum_k \frac{|g_{ck}|^2}{\omega_a - \omega_k} (\bar{n}(\omega_k) + 1), \quad \Delta_2 = P \sum_k \frac{|g_{ck}|^2}{\omega_a - \omega_k} \bar{n}(\omega_k), \tag{14.19}$$

where P denotes the principal part.

   Equation (14.15) is the so-called *master equation* for the density matrix of a system interacting with a thermal reservoir in the Born-Markov approximation. The meaning of various terms in it is:

1. First term on the right side of (14.16) describes evolution of the density matrix in the absence of its interaction with the reservoir.

2. Verify invoking the defining equation (14.5) that

$$\left[\hat{H}_A, \ \hat{A}^\dagger \hat{A}\right] = \left[\hat{H}_A, \ \hat{A}\hat{A}^\dagger\right] = 0. \tag{14.20}$$

The $\Delta_{1,2}$ terms therefore cause shift in the energy levels of the system due to its interaction with the reservoir. These terms are much smaller than the energies of non-interacting system and are often ignored.
3. The last two terms cause damping of the operator averages. They are responsible for driving the system toward the state of equilibrium.

Some properties of the equation (14.15) and its solution are:

1. On taking the trace of (14.15) it may be verified that

$$\frac{\mathrm{dTr}\{\hat{\rho}(t)\}}{\mathrm{d}t} = 0 \quad \Longrightarrow \quad \mathrm{Tr}\{\hat{\rho}(t)\} = \mathrm{Tr}\{\hat{\rho}(0)\} = 1. \tag{14.21}$$

This shows that (14.15) conserves trace of $\hat{\rho}$, as it must.
2. Let $\hat{\rho}_\lambda$ be an eigenoperator of $\hat{\mathcal{W}}_A$ corresponding to the eigenvalue λ in the sense that

$$\hat{\mathcal{W}}_A \hat{\rho}_\lambda = \lambda \hat{\rho}_\lambda. \tag{14.22}$$

If the eigenvalues are distinct, the solution of (14.15) can be written as

$$\hat{\rho}(t) = \sum_\lambda \hat{\rho}_\lambda \exp(\lambda t). \tag{14.23}$$

It can be shown that one of the eigenvalues is zero whereas the real part of all the others is negative. Hence, in the limit $t \to \infty$, $\hat{\rho}(t) \to \hat{\rho}_0$.

The density matrix, denoted by $\hat{\rho}_{ss} \equiv \hat{\rho}_0$, corresponding to the zero eigenvalue is called the *steady-state* density matrix. That is the state reached by the system as $t \to \infty$. It may be mentioned that similar conclusion is reached even when the eigenvalues are repeated.

In the following we determine the steady-state solution of (14.15).

14.2 Steady State Solution

As discussed above, the steady-state density matrix $\hat{\rho}_{ss}$ of (14.15) is the solution corresponding to the zero eigenvalue of $\hat{\mathcal{W}}_A \hat{\rho}_\lambda$:

$$\hat{\mathcal{W}}_A \hat{\rho}_{ss} = 0. \tag{14.24}$$

We show that the equation above is solved by

$$\hat{\rho}_{ss} = \frac{1}{Z}\exp(-\beta\hat{H}_A), \quad Z = \text{Tr}\{\exp(-\beta\hat{H}_A)\}. \tag{14.25}$$

To that end, recall (14.20) where it is shown that $\hat{A}^{\dagger}\hat{A}$ and $\hat{A}\hat{A}^{\dagger}$ commute with $\hat{H}_A$. Hence with $\hat{\rho} \to \hat{\rho}_{ss}$ in (14.16), first three terms therein give a vanishing contribution. To show that the remaining two terms therein, the damping terms, also reduce to zero when $\hat{\rho} \to \hat{\rho}_{ss}$, we need the identity

$$\hat{A}(\beta) \equiv \exp(-\beta\hat{H}_A)\hat{A}\exp(\beta\hat{H}_A) = \frac{\bar{n}+1}{\bar{n}}\hat{A}. \tag{14.26}$$

To prove this, differentiate (14.26) with respect to β by the chain rule of differentiation but without changing the order of $\hat{H}_A$ and $\hat{A}$ to get

$$\begin{aligned}
\frac{d\hat{A}(\beta)}{d\beta} &= -\exp(-\beta\hat{H}_A)\left[\hat{H}_A,\ \hat{A}\right]\exp(\beta\hat{H}_A) \\
&= \hbar\omega_a\exp(-\beta\hat{H}_A)\hat{A}\exp(\beta\hat{H}_A) = \hbar\omega_a\hat{A}(\beta),
\end{aligned} \tag{14.27}$$

where second equation is due to (14.5). This is solved by

$$\hat{A}(\beta) = \exp(\hbar\beta\omega)\hat{A}. \tag{14.28}$$

From the definition (14.17) of $\bar{n}$ we have

$$\exp(\hbar\beta\omega_a) = \frac{\bar{n}+1}{\bar{n}}. \tag{14.29}$$

Substitution of this in (14.28) gives the desired result (14.26). The hermitian conjugation of (14.28) with $\beta \to -\beta$ gives

$$\hat{A}^{\dagger}(-\beta) = \exp(-\hbar\beta\omega)\hat{A}^{\dagger} = \frac{\bar{n}}{\bar{n}+1}\hat{A}^{\dagger}. \tag{14.30}$$

This is same as

$$\exp(-\beta\hat{H}_A)\hat{A}^{\dagger}\exp(\beta\hat{H}_A) \equiv \hat{A}^{\dagger}(-\beta) = \frac{\bar{n}}{\bar{n}+1}\hat{A}^{\dagger}. \tag{14.31}$$

Now, with $\hat{\rho}(t) \to \hat{\rho}_{ss}$, rewrite the damping part in (14.16) as

$$(\bar{n} + 1)\left(2\hat{A}\hat{\rho}_{ss}\hat{A}^\dagger - \hat{A}^\dagger\hat{A}\hat{\rho}_{ss} - \hat{\rho}_{ss}\hat{A}^\dagger\hat{A}\right)$$

$$+\bar{n}\left(2\hat{A}^\dagger\hat{\rho}_{ss}\hat{A} - \hat{A}\hat{A}^\dagger\hat{\rho}_{ss} - \hat{\rho}_{ss}\hat{A}\hat{A}^\dagger\right)$$

$$=\left[(\bar{n} + 1)\hat{A}\hat{\rho}_{ss}\hat{A}^\dagger - \bar{n}\hat{A}\hat{A}^\dagger\hat{\rho}_{ss}\right] + \left[\bar{n}\hat{A}^\dagger\hat{\rho}_{ss}\hat{A} - (\bar{n} + 1)\hat{A}^\dagger\hat{A}\hat{\rho}_{ss}\right] + \text{h.c.},$$

$$(14.32)$$

where h.c. stands for hermitian conjugate. Consider first term on the right side of the equation above with $\hat{\rho}_{ss}$ given by (14.25):

$$(\bar{n} + 1)\hat{A}\hat{\rho}_{ss}\hat{A}^\dagger - \bar{n}\hat{A}\hat{A}^\dagger\hat{\rho}_{ss}$$

$$= \frac{1}{Z}\left[(\bar{n} + 1)\hat{A}\exp(-\beta\hat{H}_A)\hat{A}^\dagger - \bar{n}\hat{A}\hat{A}^\dagger\exp(-\beta\hat{H}_A)\right]$$

$$= \frac{1}{Z}\left[(\bar{n} + 1)\hat{A}\hat{A}^\dagger(-\beta) - \bar{n}\hat{A}\hat{A}^\dagger\right]\exp(-\beta\hat{H}_A) = 0, \qquad (14.33)$$

where last equation is due to (14.31). We can in this way show that each term in (14.32) vanishes.

We have thus demonstrated that the equilibrium state of a system interacting weakly with a thermal reservoir is, in accordance with statistical mechanics, described by the canonical density matrix having the same temperature as that of the reservoir.

We discuss next an exactly solvable model which is that of a harmonic oscillator interacting with the reservoir of harmonic oscillators.

14.3   Harmonic Oscillator Interacting with Reservoir of Harmonic Oscillators: Exact Solution

We consider a harmonic oscillator of frequency ω_a interacting with a reservoir of harmonic oscillators. The Hamiltonian of the system is given by (14.1) with

$$\hat{H}_A = \hbar\omega_a\hat{a}^\dagger\hat{a}, \qquad (14.34)$$

where $\hat{a}$, $\hat{a}^\dagger$ are the oscillator annihilation and creation operators obeying the commutation relation

$$[\hat{a}, \hat{a}^\dagger] = 1. \qquad (14.35)$$

The operator $\hat{A}$ in the system-reservoir interaction Hamiltonian (14.4) is assumed to be $\hat{a}$:

$$\hat{A} = \hat{a}. \qquad (14.36)$$

Clearly, with $\hat{H}_A$ given by (14.34) and $\hat{A}$ by (14.36), the relation (14.5) is obeyed. In Born-Markov approximation, the density matrix $\hat{\rho}(t)$ of the oscillator obeys the master equation (14.15). Ignoring the level-shift terms therein, in the present case it assumes the form

$$\frac{d\hat{\rho}(t)}{dt} = -i\omega_a[\hat{a}^\dagger\hat{a}, \ \hat{\rho}(t)]$$
$$+\gamma(\bar{n}+1)\left(2\hat{a}\hat{\rho}(t)\hat{a}^\dagger - \hat{a}^\dagger\hat{a}\hat{\rho}(t) - \hat{\rho}(t)\hat{a}^\dagger\hat{a}\right)$$
$$+\gamma\bar{n}\left(2\hat{a}^\dagger\hat{\rho}(t)\hat{a} - \hat{a}\hat{a}^\dagger\hat{\rho}(t) - \hat{\rho}(t)\hat{a}\hat{a}^\dagger\right), \tag{14.37}$$

and the corresponding steady-state solution (14.25) reads

$$\hat{\rho}_{ss} = \frac{1}{Z}\exp(-\beta\hbar\omega_a\hat{a}^\dagger\hat{a}), \quad Z = \frac{1}{1-\exp(-\beta\hbar\omega_a)}. \tag{14.38}$$

This is same as the solution (13.59) arrived at by the theory of equilibrium statistical mechanics.

The exact equation of evolution of the density matrix of the oscillator is also known. It will enable us to understand various approximations leading to its form (14.37).

The exact expression for the density matrix of the oscillator at time t in the coherent states representation is given by (see [2] for the derivation of the equation below and Appendix G for coherent states representation)

$$\rho(\alpha, \alpha^*, t) = \frac{1}{\pi\chi(t)}\int \exp\left[-|f(t)\alpha_o - \alpha|^2/\chi(t)\right]\rho(\alpha_0, \alpha_0^*, 0)d^2\alpha_0, \tag{14.39}$$

where $\rho(\alpha, \alpha^*, t) = \langle\alpha|\hat{\rho}(t)|\alpha\rangle$, is the density matrix of the oscillator at time t in the coherent state $|\alpha\rangle$ and $\rho(\alpha_0, \alpha_0^*, 0)$ that at $t = 0$ in the coherent state $|\alpha_0\rangle$. The functions $f(t)$ and $\chi(t)$ in (14.39) are:

$$f(t) = \frac{1}{2\pi}\int_C \frac{\exp(-izt)}{\eta(z)}dz, \quad \chi(t) = \sum_k |f_k|^2/\mu_k, \tag{14.40}$$

where

$$f_k(t) = \frac{g_k}{2\pi}\int_C \frac{\exp(-izt)}{(z-\omega_k)\eta(z)}, \quad \eta(z) = \omega_a - z + \sum_k \frac{|g_k|^2}{\omega_k - z}. \tag{14.41}$$

and

$$\mu_k = 1 - \exp(-\beta\hbar\omega_k). \tag{14.42}$$

The path C of integration in (14.41) is the line parallel to the real z-axis lying above the singularities of the integrand. The equation of evolution of $\rho(\alpha, \alpha^*, t)$ turns out to be given by

$$\frac{d\rho(\alpha, \alpha^*, t)}{dt} = -\left[\left\{\chi(A + A^*) - \frac{d\chi}{dt}\right\}\frac{\partial^2 \rho}{\partial\alpha\partial\alpha^*} + A\frac{\partial(\alpha\rho)}{\partial\alpha} + A^*\frac{\partial(\alpha^*\rho)}{\partial\alpha^*}\right],$$

$$(14.43)$$

where

$$A = \frac{1}{f(t)}\frac{df(t)}{dt}.$$

$$(14.44)$$

Equation (14.43) is in the form of the Fokker–Planck equation but with time-dependent coefficients. If the coefficients in (14.43) are independent of time, it would describe a Markov process, else it describes a non-Markov process. We will see that the coefficients in (14.43) become independent of time under the approximations similar to the Born-Markov approximation which led to the master equation (14.37).

The nature of time evolution of $\rho(\alpha, \alpha^*, t)$ is determined by that of the functions $f(t)$ and $\chi(t)$. The defining equation (14.40) of $f(t)$ shows that its behavior depends on the singularities of $1/\eta(z)$. The expression (14.41) of $\eta(z)$ shows that it will have isolated zeros if the number of oscillators in the reservoir is finite. Those zeros lie on the real axis. In that case $f(t)$ shall be a sum of the terms having time dependence of the form $\exp(-iz_i t)$ where z_i is a zero of $\eta(z)$. The $f_k(t)$ also will have similar behavior. It is then straightforward to see that under the operation of time reversal $t \to -t$, $f(-t) = f^*(t)$, $f_k(-t) = f_k^*(t)$ due to which $\chi(-t) = \chi(t)$. Also, under time reversal, $\alpha \to \alpha^*$. It follows that (14.43) governing the evolution of the density matrix is invariant under time reversal. In other words, the evolution is reversible. Hence, the reservoir consisting of finite number of degrees of freedom cannot give rise to irreversible behavior needed for approach to an equilibrium state.

On the other hand, in the thermodynamic limit $N \to \infty$, the spacing between the zeros of $\eta(z)$ becomes vanishingly small enabling one to describe their distribution by a density function $h(\omega)$ and to replace, as in (14.14), sum over k by the integral. The function $\eta(z)$ will then have branch cut along the real z-axis. In that case, under Born-Markov type of approximation, it has been shown that,

$$f(t) = \exp(-(\gamma + i\omega_a)t), \qquad \chi(t) = (\bar{n} + 1)(1 - \exp(-2\gamma t)). \quad (14.45)$$

Using the equations above, it can be seen that when $t \to \infty$, $\rho(\alpha, \alpha^*, t)$ in (14.39) approaches $\hat{\rho}_{ss}$ given by

$$\hat{\rho}_{ss} = \frac{1}{\bar{n} + 1}\exp\{-|\alpha|^2/(\bar{n} + 1)\}.$$

$$(14.46)$$

This is same as the canonical equilibrium density matrix in the coherent states representation derived in (13.63).

With $f(t)$, $\chi(t)$ given by (14.45), (14.43) reduces to

$$
\frac{d\rho(\alpha, \alpha^*, t)}{dt} = i\omega_a \left(\frac{\partial(\alpha\rho)}{\partial\alpha} - \frac{\partial(\alpha^*\rho)}{\partial\alpha^*} \right)
$$
$$
+ \gamma \left[2(\bar{n} + 1)\frac{\partial^2\rho}{\partial\alpha\partial\alpha^*} + \frac{\partial(\alpha\rho)}{\partial\alpha} + \frac{\partial(\alpha^*\rho)}{\partial\alpha^*} \right]. \qquad (14.47)
$$

Invoking (G.14), (G.15) it may be verified that, in the coherent states representation, the operator equation (14.37) assumes the form (14.47).

We see that the time-reversal breaks down and the evolution is irreversible if the reservoir has non-denumerably infinite number of degrees of freedom. The exact solution for the density matrix for an arbitrary number of oscillators interacting with a bath of harmonic oscillators is derived in [3]. It too predicts canonical form for the equilibrium density matrix of the system of oscillators under the assumptions similar to those under which the single oscillator described above does.

We have thus achieved the desired objective of demonstrating that, for certain form of weak system-reservoir interaction, the asymptotic solution of the equation for the density matrix of a macroscopic system is the one predicted by the maximum entropy principle.

Exercises

Ex. 14.1. Invoking (G.14), (G.15) show that (14.37) assumes the form (14.47) in the coherent states representation.

References

1. R.R. Puri, *Mathematical Methods of Quantum Optics* (Springer, 2001)
2. R.R. Puri, S.V. Lawande, Phys. Lett. **62A** 143 (1977)
3. R.R. Puri, S.V. Lawande, Phys. Lett. **62A** 143 (1977)

Appendix A
Some Relations Involving Partial Derivatives

In this appendix we derive some relations between the partial derivatives of functions under the transformations of the independent set of variables. The said task is facilitated by the use of the Jacobian notation.

Let $u(x, y)$, $v(x, y)$ be the functions of independent variables (x, y). The Jacobian of (u, v) with respect to (x, y) is defined by

$$\frac{\partial(u, v)}{\partial(x, y)} = \det \begin{pmatrix} \dfrac{\partial u}{\partial x} & \dfrac{\partial u}{\partial y} \\[2mm] \dfrac{\partial v}{\partial x} & \dfrac{\partial v}{\partial y} \end{pmatrix}, \tag{A.1}$$

where it is understood that y is constant in $\partial/\partial x$, and x is constant in $\partial/\partial y$. Some consequences of the definition (A.1) are:

1. It follows from (A.1) that

$$\frac{\partial(x, y)}{\partial(x, y)} = 1. \tag{A.2}$$

2. It is straightforward to verify that

$$\frac{\partial(u, v)}{\partial(x, y)} = -\frac{\partial(u, v)}{\partial(y, x)}, \qquad \frac{\partial(u, v)}{\partial(x, y)} = -\frac{\partial(v, u)}{\partial(x, y)}. \tag{A.3}$$

3. The partial derivative $\partial u/\partial x$ can be written in terms of the Jacobian as

$$\frac{\partial u}{\partial x} = \frac{\partial(u, y)}{\partial(x, y)}. \tag{A.4}$$

© The Editor(s) (if applicable) and The Author(s), under exclusive license to Springer Nature Switzerland AG 2024
R. R. Puri, *Modern Thermodynamics and Statistical Mechanics*, Undergraduate Lecture Notes in Physics, https://doi.org/10.1007/978-3-031-54310-4

4. If the set of independent variables (x, y) is transformed to the independent variables (w, z) then it can be shown that

$$\frac{\partial(u, v)}{\partial(x, y)} = \frac{\partial(u, v)}{\partial(w, z)} \frac{\partial(w, z)}{\partial(x, y)}. \tag{A.5}$$

This relation may be established by noting that, as functions of the transformed variables (w, z), the differentials of u and v read

$$du = \frac{\partial u}{\partial w} dw + \frac{\partial u}{\partial z} dz, \qquad dv = \frac{\partial v}{\partial w} dw + \frac{\partial v}{\partial z} dz, \tag{A.6}$$

where it is understood that z is constant in $\partial/\partial w$, and w is constant in $\partial/\partial z$. We can therefore rewrite $\partial(u, v)/\partial(x, y)$ as

$$\frac{\partial(u, v)}{\partial(x, y)} = \det \begin{pmatrix} \dfrac{\partial u}{\partial w}\dfrac{\partial w}{\partial x} + \dfrac{\partial u}{\partial z}\dfrac{\partial z}{\partial x} & \dfrac{\partial u}{\partial w}\dfrac{\partial w}{\partial y} + \dfrac{\partial u}{\partial z}\dfrac{\partial z}{\partial y} \\[2ex] \dfrac{\partial v}{\partial w}\dfrac{\partial w}{\partial x} + \dfrac{\partial v}{\partial z}\dfrac{\partial z}{\partial x} & \dfrac{\partial v}{\partial w}\dfrac{\partial w}{\partial y} + \dfrac{\partial v}{\partial z}\dfrac{\partial z}{\partial y} \end{pmatrix}$$

$$= \det \left[\begin{pmatrix} \dfrac{\partial u}{\partial w} & \dfrac{\partial u}{\partial z} \\[2ex] \dfrac{\partial v}{\partial w} & \dfrac{\partial v}{\partial z} \end{pmatrix} \begin{pmatrix} \dfrac{\partial w}{\partial x} & \dfrac{\partial w}{\partial y} \\[2ex] \dfrac{\partial z}{\partial x} & \dfrac{\partial z}{\partial y} \end{pmatrix} \right]$$

$$= \det \begin{pmatrix} \dfrac{\partial u}{\partial w} & \dfrac{\partial u}{\partial z} \\[2ex] \dfrac{\partial v}{\partial w} & \dfrac{\partial v}{\partial z} \end{pmatrix} \det \begin{pmatrix} \dfrac{\partial w}{\partial x} & \dfrac{\partial w}{\partial y} \\[2ex] \dfrac{\partial z}{\partial x} & \dfrac{\partial z}{\partial y} \end{pmatrix}$$

$$= \frac{\partial(u, v)}{\partial(w, z)} \frac{\partial(w, z)}{\partial(x, y)}. \tag{A.7}$$

5. Using (A.2) and (A.5) we can write

$$1 = \frac{\partial(x, y)}{\partial(x, y)} = \frac{\partial(x, y)}{\partial(z, y)} \frac{\partial(z, y)}{\partial(x, y)} = \left(\frac{\partial x}{\partial z}\right)_y \left(\frac{\partial z}{\partial x}\right)_y. \tag{A.8}$$

This is also called the chain rule.

6. We have

$$1 = \frac{\partial(x, y)}{\partial(x, y)} = \frac{\partial(x, y)}{\partial(z, y)} \frac{\partial(z, y)}{\partial(z, x)} \frac{\partial(z, x)}{\partial(x, y)} = -\frac{\partial(x, y)}{\partial(z, y)} \frac{\partial(y, z)}{\partial(x, z)} \frac{\partial(z, x)}{\partial(y, x)}$$

$$= -\left(\frac{\partial x}{\partial z}\right)_y \left(\frac{\partial y}{\partial x}\right)_z \left(\frac{\partial z}{\partial y}\right)_x, \tag{A.9}$$

where (A.3) has been invoked to write the third equation. Equation (A.9) is also called the cyclic rule.

7. Consider the variables (x, y, z, w) two of which are independent. We can write

$$\left(\frac{\partial x}{\partial y}\right)_w = \frac{\partial(x, w)}{\partial(y, w)} = \frac{\partial(x, w)}{\partial(z, w)} \frac{\partial(z, w)}{\partial(y, w)}$$

$$= \left(\frac{\partial x}{\partial z}\right)_w \left(\frac{\partial z}{\partial y}\right)_w. \tag{A.10}$$

Appendix B
Legendre Transform

Consider the function $y = f(x)$. Denote the slope of the tangent to it at the point x by $s(x)$:

$$s(x) = \frac{df(x)}{dx}. \tag{B.1}$$

Assume $s(x)$ to be a monotonic function of x. In many applications it is required to describe the behavior of $f(x)$ in terms of its slope. A straightforward way to construct such a description could be to invert (B.1) to express x in terms of s and rewrite $f(x)$ in terms of s to get the function $\tilde{f}(s)$ in terms of s, i.e.

$$\tilde{f}(s) = f(x(s)). \tag{B.2}$$

This procedure, however, leads to problems when one wants to invert the relation. To see that, consider the function $f(x)$ defined by

$$f(x) = \frac{x^2}{2}, \tag{B.3}$$

so that

$$s = \frac{df(x)}{dx} = x. \tag{B.4}$$

Replace x by s in (B.3) to obtain

$$\tilde{f}(s) \equiv f(x(s)) = \frac{s^2}{2}. \tag{B.5}$$

© The Editor(s) (if applicable) and The Author(s), under exclusive license to Springer
Nature Switzerland AG 2024
R. R. Puri, *Modern Thermodynamics and Statistical Mechanics*, Undergraduate Lecture
Notes in Physics, https://doi.org/10.1007/978-3-031-54310-4

Let us now construct the inverse of $\tilde{f}(s) = s^2/2$. We know that $s = df(x)/dx$ and $\tilde{f}(s) = f(x(s))$. Hence

$$\frac{df(x)}{dx} = \frac{df(x(s))}{ds} \frac{ds}{dx} \implies s = \frac{d\tilde{f}(s)}{ds} \frac{ds}{dx}. \tag{B.6}$$

Presently $\tilde{f}(s) = s^2/2$. Hence $s = s\,ds/dx$ which means $s = x + a$ where a is a constant. We therefore get the inverse transformation of $\tilde{f}(s)$ as $f(x) = \tilde{f}(s(x)) = (x+a)^2/2$ which is different from the function $f(x) = x^2/2$ which was transformed to $\tilde{f}(s)$. Indeed, if

$$f_1(x) = \frac{(x+a)^2}{2}, \quad \text{then} \quad s = \frac{df_1(x)}{dx} = x + a. \tag{B.7}$$

Hence $\tilde{f}_1(s) = s^2/2$. Thus both $f(x)$ and $f_1(x)$ lead to the same function $\tilde{f}(s) = s^2/2$ in terms of their slope. Hence, if we are given $\tilde{f}(s) = s^2/2$, we cannot say whether it corresponds to $f(x)$ or to $f_1(x)$.

We thus conclude that if $f(x)$ is transformed to $\tilde{f}(s)$ such that $\tilde{f}(s) = f(x(s))$ then the inverse of the transformation from the variable s to x need not give $f(x)$.

The reason for non-uniqueness may be traced to (B.6) which, for the purpose of constructing inverse transform, determines the relation between x and s as a solution of a differential equation (equation for ds/dx) relating s and x and hence its solution is determined up to an arbitrary constant which is fixed by specifying some condition on the relationship. Different constants define curves parallel to each other as a result of which, knowing only the slope of tangents to them, it is not possible to say to which curve they belong.

The said ambiguity is removed if we note that the intercept on a coordinate axis of a tangent to the curve $f(x)$ is different from that of a tangent of the same slope to the curve $f_1(x)$ parallel to it. We therefore define the intercepts of the tangents on the y-axis as the intended function of slope of $y = f(x)$. In order to find the said intercept, consider the curve $y = f(x)$ depicted in Fig. B.1. The tangent to it at the point $P \equiv (x, y) \equiv (x, f(x))$ has slope s. It intercepts the y-axis at Q. We know that the intercept of a line of slope m passing through the point (x_1, y_1) is $y_1 - mx_1$. In the present case $(x_1, y_1) \equiv (x, y) \equiv (x, f(x))$ and $m \equiv df(x)/dx = s$. Denoting the intercept OQ of the tangent on the y-axis by $G(s)$ we therefore have

$$G(s) = f(x(s)) - sx(s). \tag{B.8}$$

In the equation above, $x(s)$ stands for the expression of x in terms of s obtained by inverting the defining equation (B.1) of s. The function $G(s)$ is called the *Legendre transform* of $f(x)$.

Fig. B.1 Legendre transformation

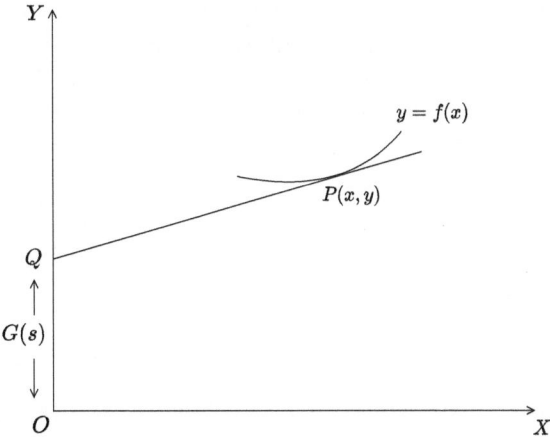

Equation (B.8) yields

$$dG(s) = df(x(s)) - sdx - x(s)ds = \frac{df(x)}{dx}dx - sdx - x(s)ds$$
$$= sdx - sdx - x(s)ds = -x(s)ds. \qquad (B.9)$$

Hence

$$\frac{dG(s)}{ds} = -x. \qquad (B.10)$$

This exhibits symmetry between $f(x)$ and its Legendre transform $G(s)$ in the sense that whereas the argument s of $G(s)$ is slope of $f(x)$, the argument x of $f(x)$ is negative of the slope of $G(s)$.

Rewrite (B.8) as

$$f(x) = G(s(x)) + xs(x). \qquad (B.11)$$

Noting that $-x$ is slope of $G(s)$, we see that the right side in the equation above is the Legendre transform of $G(s)$ which transforms it to the function of x which is same as the function of which $G(s)$ is the Legendre transform. We can say that $f(x)$ is the inverse of its Legendre transform $G(s)$.

To demonstrate that the Legendre transforms of two functions, which may have same functional form in terms of their slope, are different, we return to the function $f_1(x)$ defined in (B.7). It is straightforward to see that its Legendre transform is

$$G_1(s) = -\frac{s^2}{2} + sa. \qquad (B.12)$$

Thus, unlike $\tilde{f}_1(x(s))$, the Legendre transforms of $f_1(x)$ depends on a. The Legendre transform of $f(x) = x^2/2$ which corresponds to the $a = 0$ case of $f_1(x)$ is thus different from that of $f_1(x)$ corresponding to $a \neq 0$. Let us confirm that the inverse of $G_1(s)$ in (B.12) is indeed the function $f_1(x)$ in (B.7). To that end, using $s = x + a$ to express s in terms of x to obtain

$$G_1(s(x)) + s(x)x = \frac{(x+a)^2}{2} = f_1(x). \tag{B.13}$$

This confirms our expectation. See also [1].

B.1 Relations Between Second Derivatives of a Function and Its Legendre Transform: Single Variable Transform

Consider the function $f(x, y)$ of two variables. Let us construct the Legendre transform $G(s, y)$ of $f(x, y)$ with respect to the variable x with

$$s = \left(\frac{\partial f(x, y)}{\partial x} \right)_y, \tag{B.14}$$

$$G(s, y) = f(x(s), y) - sx, \tag{B.15}$$

$$dG(s, y) = -xds + \left(\frac{\partial f(x, y)}{\partial y} \right)_x dy, \tag{B.16}$$

and

$$x = -\left(\frac{\partial G(s, y)}{\partial s} \right)_y. \tag{B.17}$$

In certain applications it is desired to express second partial derivatives of $G(s, y)$ with respect to s, y in terms of those of $f(x, y)$ with respect to x, y. In the following we evaluate the said derivatives.

1. Evaluation of $(\partial^2 G(s, y)/\partial s^2)_y$. To that end, differentiate (B.17) with respect to s keeping y fixed to get

$$\left(\frac{\partial^2 G(s, y)}{\partial s^2}\right)_y = -\left(\frac{\partial x}{\partial s}\right)_y = -\left[\left(\frac{\partial s}{\partial x}\right)_y\right]^{-1}$$

$$= -\left[\left(\frac{\partial^2 f(x, y)}{\partial x^2}\right)_y\right]^{-1}, \tag{B.18}$$

where in writing the last equation we have invoked (B.14). This is the desired relationship. In particular, it shows that the sign of $(\partial^2 G(s, y)/\partial s^2)_y$ is opposite to that of $(\partial^2 f(x, y)/\partial x^2)_y$.

2. Evaluation of $(\partial^2 G(s, y)/\partial y^2)_s$. Equation (B.16) yields

$$\left(\frac{\partial G(s, y)}{\partial y}\right)_s = \left(\frac{\partial f(x, y)}{\partial y}\right)_x. \tag{B.19}$$

Differentiate this with respect to y keeping s constant to get

$$\left(\frac{\partial^2 G(s, y)}{\partial y^2}\right)_s = \left[\frac{\partial}{\partial y}\left(\frac{\partial f(x, y)}{\partial y}\right)_x\right]_s \equiv \frac{\partial((\partial f/\partial y)_x, s)}{(y, s)}. \tag{B.20}$$

Evaluate the right-hand side above using

$$\frac{\partial((\partial f/\partial y)_x, s)}{\partial(y, s)} = \frac{\partial((\partial f/\partial y)_x, s)}{\partial(y, x)}\frac{\partial(y, x)}{\partial(y, s)}, \tag{B.21}$$

and substitute the resulting expression in (B.20) to obtain

$$\left(\frac{\partial^2 G(s, y)}{\partial y^2}\right)_s = \left(\frac{\partial^2 f}{\partial x^2}\right)^{-1}\left(\frac{\partial^2 f}{\partial x^2}\frac{\partial^2 f}{\partial y^2} - \left(\frac{\partial^2 f}{\partial x \partial y}\right)^2\right). \tag{B.22}$$

3. Evaluation of $\partial^2 G(s, y)/\partial y \partial s$. Differentiate (B.19) with respect to s keeping y constant to obtain

$$\frac{\partial^2 G(s, y)}{\partial s \partial y} = \left[\frac{\partial}{\partial s}\left(\frac{\partial f(x, y)}{\partial y}\right)_x\right]_y = \frac{\partial((\partial f/\partial y)_x, y)}{\partial(s, y)}. \tag{B.23}$$

Evaluate the right-hand side of the equation above using

$$\frac{\partial((\partial f/\partial y)_x, y)}{\partial(s, y)} = \frac{\partial((\partial f/\partial y)_x, y)}{\partial(x, y)}\frac{\partial(x, y)}{\partial(s, y)}, \tag{B.24}$$

and substitute the resulting expression in (B.23) to get

$$\frac{\partial^2 G(s, y)}{\partial s \partial y} = \left(\frac{\partial^2 f(x, y)}{\partial x^2}\right)^{-1}\left(\frac{\partial^2 f(x, y)}{\partial x \partial y}\right). \tag{B.25}$$

4. On combining (B.18), (B.22), (B.25) to get

$$\frac{\partial^2 G(s, y)}{\partial s^2}\frac{\partial^2 G(s, y)}{\partial y^2} - \left(\frac{\partial^2 G(s, y)}{\partial s \partial y}\right)^2 = -\left(\frac{\partial^2 f}{\partial x^2}\right)^{-1}\left(\frac{\partial^2 f}{\partial y^2}\right). \quad \text{(B.26)}$$

We have thus at hand the relationship between the second derivatives of a function of two variables and those of its Legendre transforms with respect to one variable. We consider next the said relationships when both the variables are Legendre transformed.

B.2 Relations Between Second Derivatives of a Function and Its Legendre Transform: Two Variables Transformation

To that end we again consider the function $f(x, y)$ and define

$$s = \left(\frac{\partial f(x, y)}{\partial x}\right)_y, \quad t = \left(\frac{\partial f(x, y)}{\partial y}\right)_x. \quad \text{(B.27)}$$

The Legendre transform of $f(x, y)$ with respect to both the variables is defined by

$$G(s, t) = f(x, y) - xs - ty, \quad \text{(B.28)}$$

so that

$$dG(s, t) = -xds - ydt, \quad \text{(B.29)}$$

and

$$x = -\left(\frac{\partial G(s, t)}{\partial s}\right)_t, \quad y = -\left(\frac{\partial G(s, t)}{\partial t}\right)_s. \quad \text{(B.30)}$$

The relationships between the second derivatives of $G(s, t)$ and $f(x, y)$ with respect to their respective arguments can be derived as follows.

1. To evaluate of $\partial^2 G/\partial s^2$, differentiate first equation in (B.30) with respect to s keeping t fixed to get

$$\left(\frac{\partial^2 G}{\partial s^2}\right)_t = -\left(\frac{\partial x}{\partial s}\right)_t = -\left[\left(\frac{\partial s}{\partial x}\right)_t\right]^{-1}. \quad \text{(B.31)}$$

We can write

$$\left(\frac{\partial s}{\partial x}\right)_t = \frac{\partial(s,t)}{\partial(x,t)} = \frac{\partial(s,t)}{\partial(x,y)}\frac{\partial(x,y)}{\partial(x,t)}. \tag{B.32}$$

On evaluating the right-hand side above and substituting it in (B.31) we obtain

$$\left(\frac{\partial^2 G}{\partial s^2}\right)_t = -\left(\frac{\partial^2 f}{\partial y^2}\right)\left[\frac{\partial^2 f}{\partial x^2}\frac{\partial^2 f}{\partial y^2} - \left(\frac{\partial^2 f}{\partial x\partial y}\right)^2\right]^{-1}. \tag{B.33}$$

2. In similar way it can be shown that

$$\left(\frac{\partial^2 G}{\partial t^2}\right)_s = -\left(\frac{\partial^2 f}{\partial x^2}\right)\left[\frac{\partial^2 f}{\partial x^2}\frac{\partial^2 f}{\partial y^2} - \left(\frac{\partial^2 f}{\partial x\partial y}\right)^2\right]^{-1}. \tag{B.34}$$

3. Similarly the expression for $\partial^2 G/\partial s\partial t$ turns out to be given by

$$\frac{\partial^2 G}{\partial s\partial t} = \left(\frac{\partial^2 f}{\partial x\partial y}\right)\left[\frac{\partial^2 f}{\partial x^2}\frac{\partial^2 f}{\partial y^2} - \left(\frac{\partial^2 f}{\partial x\partial y}\right)^2\right]^{-1}. \tag{B.35}$$

4. On combining (B.33), (B.34), (B.35) follows the relation

$$\left(\frac{\partial^2 G}{\partial s^2}\right)_t\left(\frac{\partial^2 G}{\partial t^2}\right)_s - \left(\frac{\partial^2 G}{\partial s\partial t}\right)^2$$
$$= \left[\left(\frac{\partial^2 f}{\partial x^2}\right)_y\left(\frac{\partial^2 f}{\partial y^2}\right)_x - \left(\frac{\partial^2 f}{\partial x\partial y}\right)^2\right]^{-1}. \tag{B.36}$$

Reference

1. R.K.P. Zia, E.F. Redish, S.R. McKay, Am. J. Phys. **77**, 614 (2009)

Appendix C
Concave and Convex Functions

A function $f_1(x)$ which is such that its value at any point c lying between $[a, b]$ is higher than that on the line joining the said two points is called a *concave function*.

A function $f_2(x)$ which is such that its value at any point c lying between $[a, b]$ is lower than that on the line joining the said two points is called a *convex function*.

Referring to Fig. C.1, consider the point A at $x = a$, and the point B at $x = b$ on the curves $y = f_1(x)$ and $y = f_2(x)$. Let P be the point on the curves at $x = c$ lying between A and B so that the coordinates of A, B and P are $(a, f_i(a))$, $(b, f_i(b))$ and $(c, f_i(c))$, $(i = 1, 2)$. The point Q lies at $x = c$ on the line joining A and B, its coordinates being (c, y_c) where

$$y_c = \frac{c - a}{b - a}(f_i(b) - f_i(a)) + f_i(a). \tag{C.1}$$

The conditions of concavity and convexity then read

$$f_1(c) \geq \frac{c - a}{b - a}(f_1(b) - f_1(a)) + f_1(a), \tag{C.2}$$

and

$$f_2(c) \leq \frac{c - a}{b - a}(f_2(b) - f_2(a)) + f_2(a). \tag{C.3}$$

A useful way of writing the conditions above is to write c as

$$c = (1 - \theta)a + \theta b, \quad 0 \leq \theta \leq 1. \tag{C.4}$$

The (C.2) and (C.3) then assume the form

$$f_1((1 - \theta)a + \theta b) \geq (1 - \theta)f_1(a) + \theta f_1(b), \tag{C.5}$$

© The Editor(s) (if applicable) and The Author(s), under exclusive license to Springer Nature Switzerland AG 2024
R. R. Puri, *Modern Thermodynamics and Statistical Mechanics*, Undergraduate Lecture Notes in Physics, https://doi.org/10.1007/978-3-031-54310-4

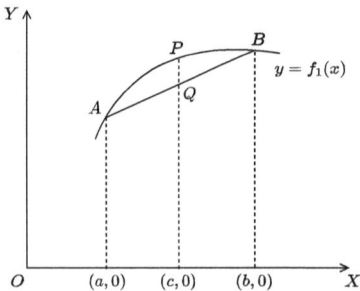 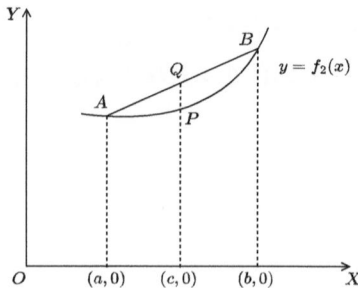

Fig. C.1 The figure on the left depicts a concave function $f_1(x)$ whereas that on the right depicts a convex function $f_2(x)$

and

$$f_2((1-\theta)a + \theta b) \le (1-\theta)f_2(a) + \theta f_2(b). \tag{C.6}$$

The question is, given a function, how to determine whether it is concave or convex in a given domain. The answer to it is provided by the criterion based on the sign of the second derivative of $f_i(x)$. To that end consider the concave function $f_1(x)$. Let $a = x, b = x + \delta x$ in (C.5) to obtain

$$f_1(x + \theta \delta x) \ge f_1(x) + \theta(f_1(x + \delta x) - f_1(x)). \tag{C.7}$$

Taylor expand $f_1(x + \theta \delta x)$ and $f_1(x + \delta x) - f_1(x)$ to get

$$(\delta x)^2 \theta (1 - \theta) \frac{d^2 f_1(x)}{dx^2} \le 0. \tag{C.8}$$

Since $0 \le \theta \le 1$ it follows that, if $f_1(x)$ is concave then

$$\frac{d^2 f_1(x)}{dx^2} \le 0. \tag{C.9}$$

It can similarly be shown that if $f_2(x)$ is convex then

$$\frac{d^2 f_2(x)}{dx^2} \ge 0. \tag{C.10}$$

Though we do not prove it, the converse of the results proved above also hold.

Appendix D
Some Combinatorics Formulas

In this appendix we address the problem of finding the number of different ways of distributing L identical objects, called balls, in N distinguishable boxes subject to the condition that not more than P balls are in one box. This means the maximum number of balls that can be accommodated in N boxes is when each is filled to its capacity of holding P balls i.e. we must ave

$$L \leq NP. \tag{D.1}$$

Clearly, there is only one way of distribution when $L = NP$.

Let the boxes be numbered $1, 2, \ldots, N$ and let p_i denote the number of balls in the ith box such that

$$\sum_{i=1}^{N} p_i = L, \quad 0 \leq p_i \leq P. \tag{D.2}$$

The set $(p_1, p_2, \ldots, p_N)$ denotes an ordered set in which there are p_i balls in the ith box. Since the boxes are distinguishable, the set obtained by interchange of p_i and p_k ($p_i \neq p_k$) is to be counted as a different set. The number of different sets of the values of the p_i's satisfying the condition above, called configurations, are different ways of distributing L balls in N boxes such that there are at most P balls in one box. We determine the number of configurations, denoted by $D(N, P, L)$, as follows.

Since exchange of balls between the boxes having the same number of balls does not lead to a new set, we identify the boxes having the same number of balls as one group including the group having zero number. The number of distinct configurations is obtained as the number of different ways of choosing the boxes for each group. To compute the said number, let n_k be the number of boxes having same number k of balls where, since maximum possible number of balls in a box is P, k can take values ($k = 0, 1, 2, \ldots, P$) such that

© The Editor(s) (if applicable) and The Author(s), under exclusive license to Springer Nature Switzerland AG 2024
R. R. Puri, *Modern Thermodynamics and Statistical Mechanics*, Undergraduate Lecture Notes in Physics, https://doi.org/10.1007/978-3-031-54310-4

$$\sum_{k=0}^{P} n_k = N, \quad \sum_{k=0}^{P} k n_k = L. \tag{D.3}$$

The desired number of configurations is obtained by counting first the number of ways in which we can choose the set of numbers $(n_0, n_1, \ldots, n_P) \equiv \{n_k\}_P$. Since the boxes are distinguishable, the number of ways of choosing a particular set $(n_0, n_1, \ldots, n_P)$ may be obtained by choosing first n_0 boxes from N boxes, followed by choosing n_1 from the remaining $N - n_0$ and so on. The said number, denoted by $D(N, P, \{n_k\}_P)$ is given by

$$
\begin{aligned}
D(N, P, \{n_k\}_P) \\
= {}^{N}C_{n_0} \, {}^{N-n_0}C_{n_1} \, {}^{N-n_0-n_1}C_{n_2} \cdots {}^{N-n_0-n_1-\cdots-n_{P-2}}C_{n_{P-1}} \\
= \frac{N!}{n_0! n_1! n_2! \cdots n_P!}, \quad \sum_{k=0}^{P} n_k = N.
\end{aligned} \tag{D.4}
$$

Total number of balls contained in the set $\{n_k\}_P$ is evidently

$$M(\{n_k\}_P) = \sum_{k=0}^{P} k n_k. \tag{D.5}$$

Different sets $\{n_k\}_P$ of the values of the n_k's, all satisfying the summation condition in (D.4), would yield different values of $M(\{n_k\}_P)$. Hence the value of $M(\{n_k\}_P)$ above need not be the number L of the balls desired to be distributed.

The desired number $D(N, P, L)$ is given by summing $D(N, P, \{n_k\}_P)$ over those values of the n_k's which satisfy, in addition to the summation condition in (D.4), also the condition (D.5) with $M(\{n_k\}_P) = L$:

$$D(N, P, L) = \sum_{\{n_k\}_P} D(N, P, \{n_k\}_P), \quad \sum_{k=0}^{P} n_k = N, \quad \sum_{k=0}^{P} k n_k = L. \tag{D.6}$$

The maximum possible value of L is $L_{\max} = NP$. Before outlining the procedure for evaluating (D.6), we find the sum of the number of distributions $D(N, P)$ for all possible values of L up to its maximum possible value NP. It is given by summing $D(N, P, \{n_k\}_P)$ over all values of the n_k's satisfying the first summation condition, but not the second condition in (D.6):

$$D_{\mathrm{T}}(N, P) \equiv \sum_{\{n_k\}_P} D(N, P, \{n_k\}_P) = \sum_{\{n_k\}_P} \frac{N!}{n_0! n_1! n_2! \cdots n_P!}, \quad \sum_{k=0}^{P} n_k = N.$$

$$\tag{D.7}$$

To evaluate the expression above, invoke the multinomial summation formula

$$(x_0 + x_1 + x_2 + \cdots + x_P)^N$$

$$= \sum_{\{n_k\}_P} \frac{N!}{n_0! n_1! n_2! \cdots n_P!} x_0^{n_0} x_1^{n_1} \cdots x_P^{n_P}, \qquad \sum_{k=0}^{P} n_k = N. \qquad (D.8)$$

The choice $x_k = 1$ for all k reduces the right-hand side of (D.8) to that in (D.7) yielding

$$D_T(N, P) = (P + 1)^N. \qquad (D.9)$$

This is, as stated before, the sum of the number of ways of distributing M balls ($M = 0, 1, 2, \ldots, NP$) in N boxes such that each box can have at most P balls in it.

Our interest, however, is in the number of ways of distributing L balls. That number is obtained by restricting the summation over those sets of values of $n'_k s$ for which (D.5) holds with $M = L$. That end can be achieved by noting that if in (D.8) we let $x_k = x^k$ then

$$\left(1 + x + x^2 + \cdots + x^P\right)^N = \sum_{\{n_k\}_P} \frac{N!}{n_0! n_1! n_2! \cdots n_P!} x^{n_1 + 2n_2 + \cdots P n_P}$$

$$\equiv \sum_{\{n_k\}_P} D(N, P, \{n_k\}_P) x^{M(\{n_k\}_P)}. \qquad (D.10)$$

Clearly, the coefficient of $x^{M(\{n_k\}_P)}$ for $M(\{n_k\}_P) = L$ is the desired number $D(N, P, L)$:

$$D(N, P, L) - \text{the coefficient of } x^L \text{ in } f(x), \qquad (D.11)$$

where $f(x)$ is the function on the left side of (D.10), called the *generating function*,

$$f(x) = \left(1 + x + x^2 + \cdots + x^P\right)^N. \qquad (D.12)$$

The coefficient of x^L in $f(x)$ is

$$\text{coefficient of } x^L \text{ in } f(x) = \frac{1}{L!} \frac{d^L f(x)}{dx^L} \bigg|_{x=0}. \qquad (D.13)$$

Hence

$$D(N, P, L) = \frac{1}{L!} \frac{d^L f(x)}{dx^L} \bigg|_{x=0}. \qquad (D.14)$$

Some properties of $D(N, P, L)$ are:

1. If the number L of balls to be distributed is less than the maximum number P that can be accommodated in a box then, since maximum available number of balls is L, P in $f(x)$ can be replaced by L:

$$f(x) = \left(1 + x + \cdots + x^L\right)^N. \tag{D.15}$$

In particular, if $L = 1$,

$$f(x) = (1 + x)^N. \tag{D.16}$$

Invoking (D.14), it is straightforward to see that

$$D(N, P, 1) = N. \tag{D.17}$$

This is consistent with the fact that if one ball is to be placed in one of the N boxes, there are N ways of choosing the box.

2. When $L = NP$, the only term in $f(x)$ that contributes in the expression (D.14) for $D(N, P, L)$ is the highest power term x^{NP} leading to

$$D(N, P, NP) = 1. \tag{D.18}$$

This is consistent with the fact that there is only one way to place NP balls in N boxes, namely, to place maximum of P balls in each box.

3. Since the maximum number of balls that can be placed in the said boxes is NP, we must have

$$D(N, P, L) = 0, \qquad L \geq NP + 1. \tag{D.19}$$

This can be seen to be consistent with the expression (D.14) of $D(N, P, L)$ as the maximum power of x in $f(x)$ in (D.12) is NP.

4. Total number of distributions summed over all values of L is given by

$$D_{\mathrm{T}}(N, P) = \sum_{L=0}^{NP} D(N, P, L) = \sum_{L=0}^{\infty} D(N, P, L), \tag{D.20}$$

where the last equation is due to (D.19). On using the expression (D.14) for $D(N, P, L)$, (D.20) assumes the form

$$D_{\mathrm{T}}(N, P) = \left[\exp(\mathrm{d}/\mathrm{d}x)\right] f(x)\Big|_{x=0} = f(x+1)\Big|_{x=0} = (P+1)^N. \tag{D.21}$$

This is same as the result derived in (D.9) by another method.

For present we evaluate $D(N, P, L)$ in (D.14).

Evaluation of $D(N, P, L)$

Let $L \leq P$. As shown in (D.14), $D(N, P, L)$ is the Lth derivative of $f(x)$ at $x = 0$. Since the Lth derivative of x^{L+k} ($k \geq 1$) at $x = 0$ is zero, the x^{L+k} ($k \geq 1$) terms will not contribute to the Lth derivative of $f(x)$ at $x = 0$. Due the assumption that $P \geq L$, we can therefore replace the polynomial of degree P in the expression (D.12) of $f(x)$ by an infinite series to rewrite it as

$$f(x) = (1 - x)^{-N}, \tag{D.22}$$

which on substitution in (D.14) yields

$$D(N, P, L \leq P) = \frac{(N + L - 1)!}{L!(N - 1)!}. \tag{D.23}$$

This is the number of ways of distributing L indistinguishable balls in N distinguishable boxes in each of which P balls can be placed when $L \leq P$.

The expression (D.23) may alternatively be arrived at as follows: Represent L balls by points on a line and draw $N - 1$ vertical lines at arbitrary positions between the points partitioning L points in N groups (see Fig. D.1). The space between the $(l - 1)$th and the lth partitions is the lth box and the number n_l of points in that space is the number of balls in the lth box. We thus have $L + N - 1$ objects which we call positions from which we choose $N - 1$ objects or positions to place $N - 1$ partitions. The number of ways of distributing L balls in N boxes may then be viewed as the number of ways of picking $N - 1$ positions for the partitions from $N + L - 1$ positions. That number, given by $^{N+L-1}C_{N-1}$, is same as the expression in (D.23).

Since we do not need it, we do not evaluate $D(N, P, L)$ for $P < L$.

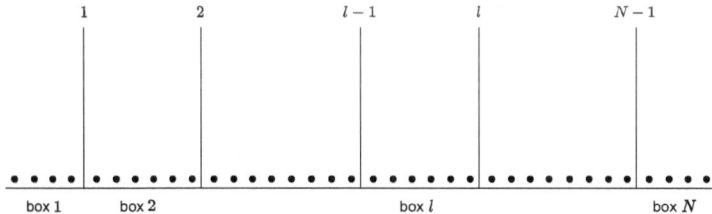

Fig. D.1 Figure to evaluate $D(N, L, P)$, $P \geq L$

Exercises

Ex. D. 1 Show that the number of ways of distributing L identical balls in N distinguishable boxes which can accommodate any number of balls such that there is at least one ball in a box is

$$D_1(N, L) = \frac{(L-1)!}{(N-1)!(L-N)!}.$$ (D.24)

Hint: One straightforward way of arriving at the formula above is to place one ball in each box so that the problem reduces to finding number of ways of distributing $L - N$ identical balls in N boxes which is given by (D.23) with L therein replaced by $L - N$. Another way is to put $n_0 = 0$ in (D.8) so that

$$(x_1 + x_2 + \cdots + x_P)^N$$

$$= \sum_{\{n_k\}_P} \frac{N!}{n_0! n_1! n_2! \cdots n_P!} x_1^{n_1} \cdots x_P^{n_P}, \qquad \sum_{k=1}^{P} n_k = N.$$ (D.25)

Let $x_k = x^k$ so that the equation above reduces to

$$\left(x + x^2 + \cdots + x^P\right)^N$$

$$= \sum_{\{n_k\}_P} \frac{N!}{n_0! n_1! n_2! \cdots n_P!} x^{n_1 + 2n_2 + \cdots P n_P}, \qquad \sum_{k=1}^{P} n_k = N.$$ (D.26)

The desired result is obtained by using (D.14) with $f(x)$ therein given by the left-hand side of (D.26) with $P \to \infty$:

$$f(x) = x^N (1-x)^{-N}.$$ (D.27)

Ex. D. 2 Show that the number of ways of distributing L identical balls in N distinguishable boxes which can accommodate any number of balls when there is given number n_0 of boxes having no ball is

$$C(N, L; n_0) = \frac{(L-1)!}{(L - N + n_0)!(N - n_0 - 1)!} \, {}^{N}C_{n_0}.$$ (D.28)

Hint: Since there are n_0 boxes having no ball, the problem reduces to finding number of ways of distributing L balls in remaining $N - n_0$ boxes such that there is at least one ball in each of them. This number is given by (D.24) with $N \to N - n_0$. The desired result (D.28) follows by noting that n_0 boxes can be chosen in ${}^{N}C_{n_0}$ ways.

Appendix E
Cubic Equation

Consider the cubic equation

$$x^3 + a_2 x^2 + a_1 x + a_0 = 0, \qquad (\text{E.1})$$

where a_i's are real. Define

$$p = a_1 - \frac{1}{3}a_2^2, \qquad q = a_0 - \frac{1}{3}a_1 a_2 + \frac{2}{27}a_2^3, \qquad (\text{E.2})$$

and

$$A = -\frac{q}{2} + \sqrt{\Delta}, \qquad B = -\frac{q}{2} - \sqrt{\Delta}, \qquad (\text{E.3})$$

where

$$\Delta = \frac{q^2}{4} + \frac{p^3}{27}. \qquad (\text{E.4})$$

The roots of (E.1) may then be shown to be given by

$$
\begin{aligned}
x_1 &= A^{1/3} + B^{1/3} - \frac{1}{3}a_2, \\
x_2 &= \frac{A^{1/3} + B^{1/3}}{2} + i\sqrt{3}\frac{A^{1/3} - B^{1/3}}{2} - \frac{1}{3}a_2 \\
x_3 &= \frac{A^{1/3} + B^{1/3}}{2} - i\sqrt{3}\frac{A^{1/3} - B^{1/3}}{2} - \frac{1}{3}a_2.
\end{aligned}
\qquad (\text{E.5})
$$

© The Editor(s) (if applicable) and The Author(s), under exclusive license to Springer
Nature Switzerland AG 2024
R. R. Puri, *Modern Thermodynamics and Statistical Mechanics*, Undergraduate Lecture
Notes in Physics, https://doi.org/10.1007/978-3-031-54310-4

The nature of the roots depends on the sign of Δ as follows:

1. If $\Delta = 0$ then (E.3) shows that A, B are real such that $A = B$. It then follows from (E.5) that in that case the roots are real and $x_2 = x_3$ i.e. two of the roots are the same.
2. If $\Delta < 0$ then (E.3) shows that A and B are complex and are such that $B = A^*$. In that case $A^{1/3} + B^{1/3}$ is real and $A^{1/3} - B^{1/3}$ is imaginary. Hence (E.5) shows that all the three roots are real.
3. If $\Delta > 0$ then (E.3) shows that A and B are real. In that case $A^{1/3} \pm B^{1/3}$ are real. Hence (E.5) shows that y_1 is real and y_2, y_3 are complex such that $y_3 = y_2^*$.

Appendix F
Thermodynamic Properties of Blackbody Radiation

The Blackbody radiation is characterized by the *Stefan–Boltzmann law* according to which the electromagnetic field energy inside a blackbody of volume V kept at temperature T is given by

$$U = \alpha V T^4, \quad \alpha = 4\sigma/c, \tag{F.1}$$

where σ is Stefan–Boltzmann constant. Starting from (F.1), we derive expressions for various thermodynamic quantities:

1. The specific heat at constant volume is

$$C_V = 4\alpha V T^3. \tag{F.2}$$

2. Invert (F.1) to express T in terms of U:

$$T = \left(\frac{U}{\alpha V}\right)^{1/4} \tag{F.3}$$

and integrate $(\partial S/\partial U)_{V,N} = 1/T$ under the condition $S = 0$ when $U = 0$ to obtain

$$S = \frac{4}{3}\alpha^{1/4}U^{3/4}V^{1/4}. \tag{F.4}$$

3. Using $(\partial S/\partial V)_{U,N} = P/T$, this leads to the following expression for pressure:

$$P = \frac{U}{3V}. \tag{F.5}$$

4. Similarly, the relation $(\partial S/\partial N)_{U,V} = \mu/T$ implies $\mu = 0$.

© The Editor(s) (if applicable) and The Author(s), under exclusive license to Springer   387
Nature Switzerland AG 2024
R. R. Puri, *Modern Thermodynamics and Statistical Mechanics*, Undergraduate Lecture
Notes in Physics, https://doi.org/10.1007/978-3-031-54310-4

Appendix G
Harmonic Oscillator Number and Coherent States

The set of operators $\hat{a}$, $\hat{a}^\dagger$, obeying the commutation relations

$$[\hat{a}, \ \hat{a}^\dagger] = 1, \tag{G.1}$$

along with $\hat{N} = \hat{a}^\dagger \hat{a}$ are said to be harmonic oscillator (h.o.) operators. Of interest to us are two sets of basis states in the Hilbert space of states on which these operators act: (a) the number states and (b) coherent states.

Number States

The number states $|n\rangle$ are the eigenstates of $\hat{a}^\dagger \hat{a}$:

$$\hat{a}^\dagger \hat{a} |n\rangle, \quad n = 0, 1, 2, \ldots \tag{G.2}$$

The number states constitute complete orthonormal set:

$$\sum_{n=0}^{\infty} |n\rangle\langle n| = I, \quad \langle m|n\rangle = \delta_{mn}. \tag{G.3}$$

The action of $\hat{a}$, $\hat{a}^\dagger$ on $|n\rangle$ is given by

$$\hat{a}|n\rangle = \sqrt{n}|n-1\rangle, \quad \hat{a}^\dagger|n\rangle = \sqrt{n+1}|n+1\rangle. \tag{G.4}$$

Since acting on a number state, $\hat{a}$ generates the state of lower number and $\hat{a}^\dagger$ generates that of higher number, $\hat{a}$ is called the h.o. annihilation and $\hat{a}^\dagger$ the creation operator respectively.

© The Editor(s) (if applicable) and The Author(s), under exclusive license to Springer 389
Nature Switzerland AG 2024
R. R. Puri, *Modern Thermodynamics and Statistical Mechanics*, Undergraduate Lecture Notes in Physics, https://doi.org/10.1007/978-3-031-54310-4

Coherent States

A coherent state $|\alpha\rangle$ is the eigenstates of the annihilation operator:

$$\hat{a}|\alpha\rangle = \alpha|\alpha\rangle, \tag{G.5}$$

where α is a complex number. The right normalizable eigenstates of $\hat{a}^{\dagger}$ do not exist. The expression of $|\alpha\rangle$ in terms of the number states is given by

$$|\alpha\rangle = \exp(-|\alpha|^2/2) \sum_{m=0}^{\infty} \frac{\alpha^m}{\sqrt{m!}}|m\rangle. \tag{G.6}$$

Some properties of the coherent states and the representation of operators in terms of them are:

1. The coherent states obey the completeness relation

$$\frac{1}{\pi} \int |\alpha\rangle\langle\alpha|d^2\alpha = I. \tag{G.7}$$

2. The coherent states are not orthonormal. For, invoking (G.6) it is readily seen that

$$\langle\beta|\alpha\rangle = \exp(-(|\alpha|^2 + |\beta|^2)/2 + \beta^*\alpha). \tag{G.8}$$

3. Using (G.6) it can be seen that

$$\hat{a}^{\dagger}|\alpha\rangle = \exp(-|\alpha|^2/2) \sum_{m=0}^{\infty} \frac{\sqrt{m+1}\alpha^m}{\sqrt{m!}}|m+1\rangle$$

$$= \exp(-|\alpha|^2/2) \sum_{m=0}^{\infty} \frac{m\alpha^{m-1}}{\sqrt{m!}}|m\rangle. \tag{G.9}$$

This may be rewritten as

$$\hat{a}^{\dagger}|\alpha\rangle = \exp(-|\alpha|^2/2)\frac{\partial}{\partial\alpha}\exp(|\alpha|^2/2)|\alpha\rangle. \tag{G.10}$$

4. If $\hat{\rho}$ is a function of the h.o. operators then

$$\rho(\alpha, \alpha^*) = \langle\alpha|\hat{\rho}|\alpha\rangle \tag{G.11}$$

is the coherent states representation of $\hat{\rho}$.

5. It is straightforward to see that the coherent states representations of $\hat{\rho}\hat{a}$ and $\hat{a}^\dagger\hat{\rho}$ in terms of the coherent states are

$$\langle\alpha|\hat{\rho}\hat{a}|\alpha\rangle = \alpha\rho(\alpha,\alpha^*), \qquad \langle\alpha|\hat{a}^\dagger\hat{\rho}|\alpha\rangle = \alpha^*\rho(\alpha,\alpha^*). \qquad (G.12)$$

6. To derive the coherent states representations of $\hat{\rho}\hat{a}^\dagger$, recall (G.10) to write

$$\langle\alpha|\hat{\rho}\hat{a}^\dagger|\alpha\rangle = \exp(-|\alpha|^2/2)\langle\alpha|\hat{\rho}|\frac{\partial}{\partial\alpha}\exp(|\alpha|^2/2)|\alpha\rangle$$

$$= \exp(-|\alpha|^2)\{\exp(|\alpha|^2/2)\}\langle\alpha|\hat{\rho}|\frac{\partial}{\partial\alpha}\exp(|\alpha|^2/2)|\alpha\rangle$$

$$= \exp(-|\alpha|^2)\frac{\partial}{\partial\alpha}\exp(|\alpha|^2)\langle\alpha|\hat{\rho}|\alpha\rangle. \qquad (G.13)$$

The last equation is due to the fact that since $\exp(|\alpha|^2/2)\langle\alpha|$ is a function only of α^*, it is like a constant for derivative with respect to α. Hence

$$\langle\alpha|\hat{\rho}\hat{a}^\dagger|\alpha\rangle = \left(\alpha^* + \frac{\partial}{\partial\alpha}\right)\rho(\alpha,\alpha^*). \qquad (G.14)$$

7. Similarly it can be shown that

$$\langle\alpha|\hat{a}\hat{\rho}|\alpha\rangle = \left(\alpha + \frac{\partial}{\partial\alpha^*}\right)\rho(\alpha,\alpha^*). \qquad (G.15)$$

Appendix H
Some Mathematical Formulas

-

$$\sum_{n=0}^{\infty} n^k \exp(-nx) = (-)^k \frac{d^k}{dx^k} \sum_{n=0}^{\infty} \exp(-nx)$$

$$= (-)^k \frac{d^k}{dx^k} (1 - \exp(-x))^{-1}. \tag{H.1}$$

- The Gamma function $\Gamma(z)$ is defined by

$$\Gamma(z+1) = z\Gamma(z), \qquad \Gamma(1) = 1. \tag{H.2}$$

$$\Gamma(z) = \int_0^{\infty} x^{z-1} \exp(-x)dx, \qquad \text{Re}(z) > 0. \tag{H.3}$$

If m is a positive integer or zero then

$$\Gamma(m+1) = m!, \qquad \frac{1}{\Gamma(-m)} \to 0. \tag{H.4}$$

$$\Gamma\left(\frac{1}{2}\right) = \sqrt{\pi}. \tag{H.5}$$

- The approximate value of $N!$ when $N \gg 1$ is given by

$$\ln(N!) = N\ln(N) - N. \tag{H.6}$$

This is Stirling's approximation in its simplest form.

© The Editor(s) (if applicable) and The Author(s), under exclusive license to Springer
Nature Switzerland AG 2024
R. R. Puri, *Modern Thermodynamics and Statistical Mechanics*, Undergraduate Lecture
Notes in Physics, https://doi.org/10.1007/978-3-031-54310-4

- If $\mathrm{Re}(\alpha) > 0$,

$$\int_{-\infty}^{\infty} \exp(-\alpha x^2 + \beta x)\mathrm{d}x = \sqrt{\frac{\pi}{\alpha}} \exp(\beta^2/4\alpha), \tag{H.7}$$

$$\int_{0}^{\infty} x^m \exp(-\alpha x^2)\mathrm{d}x = \frac{1}{2\sqrt{\alpha^{m+1}}} \Gamma\left(\frac{m}{2} + \frac{1}{2}\right). \tag{H.8}$$

- The Riemann zeta function, denoted by $\zeta(p)$, is defined by

$$\zeta(p) = \sum_{k=1}^{\infty} \frac{1}{k^p}, \qquad \mathrm{Re}(p) > 1. \tag{H.9}$$

Some values of $\zeta(2p)$ ($p = 1, 2, 3$) are:

$$\zeta(2) = \frac{\pi^2}{6}, \quad \zeta(4) = \frac{\pi^4}{90}, \quad \zeta(6) = \frac{\pi^6}{945}. \tag{H.10}$$

Some values of $\zeta(2p+1)$ ($p = 1, 2, 3$) are:

$$\zeta(3) = 1.202, \quad \zeta(5) = 1.036, \quad \zeta(7) = 1.008. \tag{H.11}$$

-

$$\int_{0}^{\infty} \frac{x^p\,\mathrm{d}x}{\exp(x) - 1} = \Gamma(p+1)\zeta(p+1). \tag{H.12}$$

- $p > 1$

$$\int_{0}^{\infty} \frac{x^p\,\exp(x)\mathrm{d}x}{(\exp(x) - 1)^2} = \Gamma(p+1)\zeta(p). \tag{H.13}$$

- $p > 1$

$$\int_{0}^{\infty} \frac{x^{p-1}\mathrm{d}x}{\exp(x) + 1} = \left(1 - 2^{-p+1}\right)\Gamma(p)\zeta(p). \tag{H.14}$$

Bibliography

1. Balian Roger, *From Microphysics to Macrophysics*, Volume I (Springer-Verlag, 2007).
2. Balian Roger, *From Microphysics to Macrophysics*, Volume II (Springer-Verlag, 2007).
3. Callen Herbert B. *Thermodynamics and an Introduction to Thermostatics* (Wiley, 1985).
4. Huang Kerson *Statistical Mechanics* (John Wiley & Sons, 1987).
5. Kardar Mehran *Statistical Physics of Particles* (Cambridge University Press, 2007).
6. Landau L.D., and Lifshitz E.M. *Statistical Physics* Part 1 (Elsevier, 2014).
7. Pathria R.K. *Statistical Mechanics* (Buuterworth-Heinamann, 1996).
8. H.E. Stanley *Introduction to Phase Transitions and Critical Phenomena* (Oxford University Press, 1971).

© The Editor(s) (if applicable) and The Author(s), under exclusive license to Springer
Nature Switzerland AG 2024
R. R. Puri, *Modern Thermodynamics and Statistical Mechanics*, Undergraduate Lecture
Notes in Physics, https://doi.org/10.1007/978-3-031-54310-4

Index

© The Editor(s) (if applicable) and The Author(s), under exclusive license to Springer
Nature Switzerland AG 2024
R. R. Puri, *Modern Thermodynamics and Statistical Mechanics*, Undergraduate Lecture
Notes in Physics, https://doi.org/10.1007/978-3-031-54310-4